IET ENERGY ENGINEERING SERIES 242

AI for Power Electronics and Renewable Energy Systems

Other volumes in this series:

Volume 1 **Power Circuit Breaker Theory and Design** C.H. Flurscheim (Editor)
Volume 4 **Industrial Microwave Heating** A.C. Metaxas and R.J. Meredith
Volume 7 **Insulators for High Voltages** J.S.T. Looms
Volume 8 **Variable Frequency AC Motor Drive Systems** D. Finney
Volume 10 **SF_6 Switchgear** H.M. Ryan and G.R. Jones
Volume 11 **Conduction and Induction Heating** E.J. Davies
Volume 13 **Statistical Techniques for High Voltage Engineering** W. Hauschild and W. Mosch
Volume 14 **Uninterruptible Power Supplies** J. Platts and J.D. St Aubyn (Editors)
Volume 15 **Digital Protection for Power Systems** A.T. Johns and S.K. Salman
Volume 16 **Electricity Economics and Planning** T.W. Berrie
Volume 18 **Vacuum Switchgear** A. Greenwood
Volume 19 **Electrical Safety: A guide to causes and prevention of hazards** J. Maxwell Adams
Volume 21 **Electricity Distribution Network Design, 2nd Edition** E. Lakervi and E.J. Holmes
Volume 22 **Artificial Intelligence Techniques in Power Systems** K. Warwick, A.O. Ekwue and R. Aggarwal (Editors)
Volume 24 **Power System Commissioning and Maintenance Practice** K. Harker
Volume 25 **Engineers' Handbook of Industrial Microwave Heating** R.J. Meredith
Volume 26 **Small Electric Motors** H. Moczala *et al.*
Volume 27 **AC-DC Power System Analysis** J. Arrillaga and B.C. Smith
Volume 29 **High Voltage Direct Current Transmission, 2nd Edition** J. Arrillaga
Volume 30 **Flexible AC Transmission Systems (FACTS)** Y.-H. Song (Editor)
Volume 31 **Embedded Generation** N. Jenkins *et al.*
Volume 32 **High Voltage Engineering and Testing, 2nd Edition** H.M. Ryan (Editor)
Volume 33 **Overvoltage Protection of Low-Voltage Systems, Revised Edition** P. Hasse
Volume 36 **Voltage Quality in Electrical Power Systems** J. Schlabbach *et al.*
Volume 37 **Electrical Steels for Rotating Machines** P. Beckley
Volume 38 **The Electric Car: Development and future of battery, hybrid and fuel-cell cars** M. Westbrook
Volume 39 **Power Systems Electromagnetic Transients Simulation** J. Arrillaga and N. Watson
Volume 40 **Advances in High Voltage Engineering** M. Haddad and D. Warne
Volume 41 **Electrical Operation of Electrostatic Precipitators** K. Parker
Volume 43 **Thermal Power Plant Simulation and Control** D. Flynn
Volume 44 **Economic Evaluation of Projects in the Electricity Supply Industry** H. Khatib
Volume 45 **Propulsion Systems for Hybrid Vehicles** J. Miller
Volume 46 **Distribution Switchgear** S. Stewart
Volume 47 **Protection of Electricity Distribution Networks, 2nd Edition** J. Gers and E. Holmes
Volume 48 **Wood Pole Overhead Lines** B. Wareing
Volume 49 **Electric Fuses, 3rd Edition** A. Wright and G. Newbery
Volume 50 **Wind Power Integration: Connection and system operational aspects** B. Fox *et al.*
Volume 51 **Short Circuit Currents** J. Schlabbach
Volume 52 **Nuclear Power** J. Wood
Volume 53 **Condition Assessment of High Voltage Insulation in Power System Equipment** R.E. James and Q. Su
Volume 55 **Local Energy: Distributed generation of heat and power** J. Wood
Volume 56 **Condition Monitoring of Rotating Electrical Machines** P. Tavner, L. Ran, J. Penman and H. Sedding
Volume 57 **The Control Techniques Drives and Controls Handbook, 2nd Edition** B. Drury
Volume 58 **Lightning Protection** V. Cooray (Editor)
Volume 59 **Ultracapacitor Applications** J.M. Miller
Volume 62 **Lightning Electromagnetics** V. Cooray

Volume 63 **Energy Storage for Power Systems, 2nd Edition** A. Ter-Gazarian
Volume 65 **Protection of Electricity Distribution Networks, 3rd Edition** J. Gers
Volume 66 **High Voltage Engineering Testing, 3rd Edition** H. Ryan (Editor)
Volume 67 **Multicore Simulation of Power System Transients** F.M. Uriate
Volume 68 **Distribution System Analysis and Automation** J. Gers
Volume 69 **The Lightening Flash, 2nd Edition** V. Cooray (Editor)
Volume 70 **Economic Evaluation of Projects in the Electricity Supply Industry, 3rd Edition** H. Khatib
Volume 72 **Control Circuits in Power Electronics: Practical issues in design and implementation** M. Castilla (Editor)
Volume 73 **Wide Area Monitoring, Protection and Control Systems: The enabler for smarter grids** A. Vaccaro and A. Zobaa (Editors)
Volume 74 **Power Electronic Converters and Systems: Frontiers and applications** A.M. Trzynadlowski (Editor)
Volume 75 **Power Distribution Automation** B. Das (Editor)
Volume 76 **Power System Stability: Modelling, analysis and control** A.A. Sallam and B. Om P. Malik
Volume 78 **Numerical Analysis of Power System Transients and Dynamics** A. Ametani (Editor)
Volume 79 **Vehicle-to-Grid: Linking electric vehicles to the smart grid** J. Lu and J. Hossain (Editors)
Volume 81 **Cyber-Physical-Social Systems and Constructs in Electric Power Engineering** S. Suryanarayanan, R. Roche and T.M. Hansen (Editors)
Volume 82 **Periodic Control of Power Electronic Converters** F. Blaabjerg, K. Zhou, D. Wang and Y. Yang
Volume 86 **Advances in Power System Modelling, Control and Stability Analysis** F. Milano (Editor)
Volume 87 **Cogeneration: Technologies, optimisation and implementation** C.A. Frangopoulos (Editor)
Volume 88 **Smarter Energy: From smart metering to the smart grid** H. Sun, N. Hatziargyriou, H.V. Poor, L. Carpanini and M.A. Sánchez Fornié (Editors)
Volume 89 **Hydrogen Production, Separation and Purification for Energy** A. Basile, F. Dalena, J. Tong and T.N. Veziroğlu (Editors)
Volume 90 **Clean Energy Microgrids** S. Obara and J. Morel (Editors)
Volume 91 **Fuzzy Logic Control in Energy Systems with Design Applications in MATLAB®/Simulink®** İ.H. Altaş
Volume 92 **Power Quality in Future Electrical Power Systems** A.F. Zobaa and S.H.E.A. Aleem (Editors)
Volume 93 **Cogeneration and District Energy Systems: Modelling, analysis and optimization** M.A. Rosen and S. Koohi-Fayegh
Volume 94 **Introduction to the Smart Grid: Concepts, technologies and evolution** S.K. Salman
Volume 95 **Communication, Control and Security Challenges for the Smart Grid** S.M. Muyeen and S. Rahman (Editors)
Volume 96 **Industrial Power Systems With Distributed and Embedded Generation** R Belu
Volume 97 **Synchronized Phasor Measurements for Smart Grids** M.J.B. Reddy and D.K. Mohanta (Editors)
Volume 98 **Large Scale Grid Integration of Renewable Energy Sources** A. Moreno-Munoz (Editor)
Volume 100 **Modeling and Dynamic Behaviour of Hydropower Plants** N. Kishor and J. Fraile-Ardanuy (Editors)
Volume 101 **Methane and Hydrogen for Energy Storage** R. Carriveau and D.S-K. Ting
Volume 104 **Power Transformer Condition Monitoring and Diagnosis** A. Abu-Siada (Editor)
Volume 106 **Surface Passivation of Industrial Crystalline Silicon Solar Cells** J. John (Editor)
Volume 107 **Bifacial Photovoltaics: Technology, applications and economics** J. Libal and R. Kopecek (Editors)

Volume 108 **Fault Diagnosis of Induction Motors** J. Faiz, V. Ghorbanian and G. Joksimović
Volume 109 **Cooling of Rotating Electrical Machines: Fundamentals, modelling, testing and design** D. Staton, E. Chong, S. Pickering and A. Boglietti
Volume 110 **High Voltage Power Network Construction** K. Harker
Volume 111 **Energy Storage at Different Voltage Levels: Technology, integration, and market aspects** A.F. Zobaa, P.F. Ribeiro, S.H.A. Aleem and S.N. Afifi (Editors)
Volume 112 **Wireless Power Transfer: Theory, technology and application** N. Shinohara
Volume 114 **Lightning-Induced Effects in Electrical and Telecommunication Systems** Y. Baba and V.A. Rakov
Volume 115 **DC Distribution Systems and Microgrids** T. Dragičević, F. Blaabjerg and P. Wheeler
Volume 116 **Modelling and Simulation of HVDC Transmission** M. Han (Editor)
Volume 117 **Structural Control and Fault Detection of Wind Turbine Systems** H.R. Karimi
Volume 118 **Modelling and Simulation of Complex Power Systems** A. Monti and A. Benigni
Volume 119 **Thermal Power Plant Control and Instrumentation: The control of boilers and HRSGs, 2nd Edition** D. Lindsley, J. Grist and D. Parker
Volume 120 **Fault Diagnosis for Robust Inverter Power Drives** A. Ginart (Editor)
Volume 121 **Monitoring and Control Using Synchrophasors in Power Systems with Renewables** I. Kamwa and C. Lu (Editors)
Volume 123 **Power Systems Electromagnetic Transients Simulation, 2nd Edition** N. Watson and J. Arrillaga
Volume 124 **Power Market Transformation** B. Murray
Volume 125 **Wind Energy Modeling and Simulation, Volume 1: Atmosphere and plant** P. Veers (Editor)
Volume 126 **Diagnosis and Fault Tolerance of Electrical Machines, Power Electronics and Drives** A.J.M. Cardoso
Volume 127 **Lightning Electromagnetics, Volume 1: Return stroke modelling and electromagnetic radiation, 2nd Edition** V. Cooray, F. Rachidi, M. Rubinstein (Editors)
Volume 127 **Lightning Electromagnetics, Volume 2: Electrical processes and effects, 2nd Edition** V. Cooray, F. Rachidi, M. Rubinstein (Editors)
Volume 128 **Characterization of Wide Bandgap Power Semiconductor Devices** F. Wang, Z. Zhang and E.A. Jones
Volume 129 **Renewable Energy From the Oceans: From wave, tidal and gradient systems to offshore wind and solar** D. Coiro and T. Sant (Editors)
Volume 130 **Wind and Solar Based Energy Systems for Communities** R. Carriveau and D. S.-K. Ting (Editors)
Volume 131 **Metaheuristic Optimization in Power Engineering** J. Radosavljević
Volume 132 **Power Line Communication Systems for Smart Grids** I.R.S Casella and A. Anpalagan
Volume 134 **Hydrogen Passivation and Laser Doping for Silicon Solar Cells** B. Hallam and C. Chan (Editors)
Volume 139 **Variability, Scalability and Stability of Microgrids** S.M. Muyeen, S.M. Islam and F. Blaabjerg (Editors)
Volume 142A **Wind Turbine System Design, Volume 1: Nacelles, drive trains and verification** J. Wenske (Editor)
Volume 142B **Wind Turbine System Design, Volume 2: Electrical systems, grid integration, control and monitoring** J. Wenske (Editor)
Volume 143 **Medium Voltage DC System Architectures** B. Grainger and R.D. Doncker (Editors)
Volume 145 **Condition Monitoring of Rotating Electrical Machines** P. Tavner, L. Ran and C. Crabtree
Volume 146 **Energy Storage for Power Systems, 3rd Edition** A.G. Ter-Gazarian
Volume 147 **Distribution Systems Analysis and Automation, 2nd Edition** J. Gers
Volume 151 **SiC Power Module Design: Performance, robustness and reliability** A. Castellazzi and A. Irace (Editors)

Volume 152 **Power Electronic Devices: Applications, failure mechanisms and reliability** F. Iannuzzo (Editor)
Volume 153 **Signal Processing for Fault Detection and Diagnosis in Electric Machines and Systems** M. Benbouzid (Editor)
Volume 155 **Energy Generation and Efficiency Technologies for Green Residential Buildings** D. Ting and R. Carriveau (Editors)
Volume 156 **Lithium-ion Batteries Enabled by Silicon Anodes** C. Ban and K. Xu (Editors)
Volume 157 **Electrical Steels, 2 Volumes** A. Moses, K. Jenkins, Philip Anderson and H. Stanbury
Volume 158 **Advanced Dielectric Materials for Electrostatic Capacitors** Q Li (Editor)
Volume 159 **Transforming the Grid Towards Fully Renewable Energy** O. Probst, S. Castellanos and R. Palacios (Editors)
Volume 160 **Microgrids for Rural Areas: Research and case studies** R.K. Chauhan, K. Chauhan and S.N. Singh (Editors)
Volume 161 **Artificial Intelligence for Smarter Power Systems: Fuzzy logic and neural networks** M.G. Simoes
Volume 165 **Digital Protection for Power Systems, 2nd Edition** Salman K Salman
Volume 166 **Advanced Characterization of Thin Film Solar Cells** N. Haegel and M. Al-Jassim (Editors)
Volume 167 **Power Grids With Renewable Energy: Storage, integration and digitalization** A.A. Sallam and B. Om P. Malik
Volume 169 **Small Wind and Hydrokinetic Turbines** P. Clausen, J. Whale and D. Wood (Editors)
Volume 170 **Reliability of Power Electronics Converters for Solar Photovoltaic Applications** F. Blaabjerg, A.l Haque, H. Wang, Z. Abdin Jaffery and Y. Yang (Editors)
Volume 171 **Utility-scale Wind Turbines and Wind Farms** A. Vasel-Be-Hagh and D.S.-K. Ting
Volume 172 **Lightning Interaction With Power Systems, 2 Volumes** A. Piantini (Editor)
Volume 174 **Silicon Solar Cell Metallization and Module Technology** T. Dullweber (Editor)
Volume 175 **n-Type Crystalline Silicon Photovoltaics: Technology, applications and economics** D. Munoz and R. Kopecek (Editors)
Volume 178 **Integrated Motor Drives** Xu Deng and B. Mecrow (Editors)
Volume 180 **Protection of Electricity Distribution Networks, 4th Edition** J. Gers and E. Holmes
Volume 182 **Surge Protection for Low Voltage Systems** A. Rousseau (Editor)
Volume 184 **Compressed Air Energy Storage: Types, systems and applications** D. Ting and J. Stagner
Volume 186 **Synchronous Reluctance Machines: Analysis, optimization and applications** N. Bianchi, C. Babetto and G. Bacco
Volume 190 **Synchrophasor Technology: Real-time operation of power networks** N. Kishor and S.R. Mohanty (Editors)
Volume 191 **Electric Fuses: Fundamentals and new applications, 4th Edition** N. Nurse, A. Wright and P.G. Newbery
Volume 193 **Overhead Electric Power Lines: Theory and practice** S. Chattopadhyay and A. Das
Volume 194 **Offshore Wind Power: Reliability, availability and maintenance, 2nd Edition** P. Tavner
Volume 196 **Cyber Security for Microgrids** S. Sahoo, F. Blaabjerg and T. Dragicevic
Volume 198 **Battery Management Systems and Inductive Balancing** A. Van den Bossche and A. Farzan Moghaddam
Volume 199 **Model Predictive Control for Microgrids: From power electronic converters to energy management** J. Hu, J.M. Guerrero and S. Islam
Volume 201 **Design, Control and Monitoring of Tidal Stream Turbine Systems** M. Benbouzid (Editor)
Volume 202 **Hydrogen From Seawater Splitting: Technology and outlook** A. Ray and M. Patel (Editors)

Volume 204 **Electromagnetic Transients in Large HV Cable Networks: Modeling and calculations** Ametani, Xue, Ohno and Khalilnezhad
Volume 208 **Nanogrids and Picogrids and Their Integration with Electric Vehicles** S. Chattopadhyay
Volume 210 **Superconducting Magnetic Energy Storage in Power Grids** M.H. Ali
Volume 211 **Blockchain Technology for Smart Grids: Implementation, management and security** H.L. Gururaj, K.V. Ravi, F. Flammini, H. Lin, B. Goutham, K.B.R. Sunil and C. Sivapragash
Volume 212 **Battery State Estimation: Methods and models** S. Wang
Volume 215 **Industrial Demand Response: Methods, best practices, case studies, and applications** H.H. Alhelou, A. Moreno-Muñoz and P. Siano (Editors)
Volume 213 **Wide Area Monitoring of Interconnected Power Systems, 2nd Edition** A.R. Messina
Volume 214 **Distributed Energy Storage in Urban Smart Grids** P.F. Ribeiro and R.S. Salles (Editors)
Volume 217 **Advances in Power System Modelling, Control and Stability Analysis, 2nd Edition** F. Milano (Editor)
Volume 222 **Large Scale Grid Integration of Renewable Energy Sources, 2nd Edition: Solutions and technologies** A. Moreno-Munoz
Volume 225 **Fusion–Fission Hybrid Nuclear Reactors: For enhanced nuclear fuel utilization and radioactive waste reduction** W.M. Stacey
Volume 227 **Processing and Manufacturing of Electrodes for Lithium-Ion Batteries** J. Li and C. Jin (Editors)
Volume 228 **Digital Technologies for Solar Photovoltaic Systems: From general to rural and remote installations** S. Motahhir (Editor)
Volume 235 **Clean Energy Past and Future** P. Tavner
Volume 236 **Green HV Switching Technologies for Modern Power Networks,** K. Niayesh (Editor)
Volume 237 **High Voltage Power Transformers: State of the art and technological innovations** M. Sforna, S. Zunino and F. Ferrari
Volume 238 **AI for Status Monitoring of Utility Scale Batteries** S. Wang, K. Liu, Y. Wang, D. Stroe, C. Fernandez and J.M. Guerrero
Volume 239 **Intelligent Control of Medium and High Power Converters** M. Bendaoud, Y. Maleh and S. Padmanaban (Editors)
Volume 241A **Power Electronic Converters and Systems, 2nd Edition, Volume 1: Converters and machine drives** M. Simões and T. Busarello (Editors)
Volume 241B **Power Electronic Converters and Systems, 2nd Edition, Volume 2: Applications** M. Simões and T. Busarello (Editors)
Volume 245 **Agrivoltaics: Technical, ecological, commercial and legal aspects** C. Klyk and S. Schindele (Editors)
Volume 247 **Power System Strength: Evaluation methods, best practice, case studies, and applications** H. Alhelou, N. Hosseinzadeh and B. Bahrani (Editors)
Volume 250 **Interactions of Wind Turbines with Aviation Radio and Radar Systems** Alan Collinson
Volume 256 **Wireless Power Transfer Technologies: Theory and technologies, 2nd Edition** N. Shinohara (Editor)
Volume 905 **Power System Protection, 4 Volumes**

AI for Power Electronics and Renewable Energy Systems

Weihao Hu, Guozhou Zhang, Zhenyuan Zhang, Sayed Abulanwar and Frede Blaabjerg

The Institution of Engineering and Technology

Published by The Institution of Engineering and Technology, London, United Kingdom

The Institution of Engineering and Technology is registered as a Charity in England & Wales (no. 211014) and Scotland (no. SC038698).

First published 2024

The Institution of Engineering and Technology
Futures Place
Kings Way, Stevenage
Hertfordshire SG1 2UA, United Kingdom

www.theiet.org

British Library Cataloguing in Publication Data
A catalogue record for this product is available from the British Library

ISBN 978-1-83953-774-5 (hardback)
ISBN 978-1-83953-775-2 (PDF)

Typeset in India by MPS Limited

Cover Image: Marian/Moment via Getty Images

Contents

About the authors

Weihao Hu is a full professor at the School of Mechanical and Electrical Engineering, University of Electronic Science and Technology of China. He has participated in several national and international research projects and authored more than 200 publications in his technical field, which has been cited over 8,000 times. His research interests include intelligent energy systems and renewable power generation. Dr Hu is an IET fellow and a senior member of IEEE. Dr Hu is an associate editor for *IET Renewable Power Generation* and *Journal of Modern Power Systems and Clean Energy*. He served as the conference chair, conference co-chair, and technical program chair for several IEEE conferences and he is currently serving as the chair for IEEE Chengdu Section PELS Chapter.

Guozhou Zhang is a lecturer at the College of Engineering and Technology, Southwest University, China. He received his PhD degree from the University of Electronic Science and Technology of China, Chengdu, China, in 2022. He has published more than 30 peer-reviewed journal papers. His research interests include energy management, reinforcement learning, wind farm control, distributed energy systems, power stability, and energy storage control.

Zhenyuan Zhang is a professor at the School of Mechanical and Electrical Engineering, University of Electronic Science and Technology of China, Chengdu, China. Since 2010, he has been the project associate for the IEEE/NFPA Arc Flash Research Project. A senior member of IEEE, his research interests include power system analysis and power system optimization, but he has also been involved in smart grids, renewable energy, electrical safety analysis, arc flash analysis, and power market research works.

Sayed Abulanwar is an associate professor at the Electrical Engineering Department, Faculty of Engineering, Mansoura University, Mansoura, Egypt. He is also with the Faculty of Engineering, Horus University-Egypt. Dr Abulanwar is a senior member of IEEE. He received the BSc and MSc degrees in electrical engineering from Mansoura University, Mansoura, Egypt, in 2005 and 2010, respectively, and the PhD degree from the Energy Technology Department, Aalborg University, Denmark, in 2016. He is an associate editor for *IET Renewable Power Generation* and *Alexandria Engineering Journal*, and guest editor for *IET Renewable Power Generation* for Special Issue: Applications of Artificial Intelligence in Renewable Energy Systems. He is also a guest editor of the Special

Issue: Planning and Operation of Hybrid Renewable Energy Systems, Volume II, *Frontiers in Energy Research*. He has participated in several national and international research projects. His research interests include hybrid AC/DC microgrids, intelligent energy systems, protection of AC and DC systems, wind energy conversion systems and grid-connected converters.

Frede Blaabjerg is a full professor at the Faculty of Engineering and Science of Aalborg University, Denmark. Prior positions include ABB-Scandia. His current research interests include power electronics and its applications such as in wind turbines, PV systems, reliability, harmonics, and adjustable speed drives. He has authored more than 800 journal papers on power electronics and its applications, he is co-author of six monographs and editor of 15 books. He has received 40 IEEE paper awards and numerous distinguished international research awards. He has been the President of the IEEE Power Electronics Society.

Chapter 1
Introduction to AI in power system

Weihao Hu[1], Guozhou Zhang[2], Zhenyuan Zhang[1], Sayed Abulanwar[3,4] and Frede Blaabjerg[5]

1.1 Overview of power system

In recent decades, with the rapid development of the economy, electrical power demand has significantly increased, and the per capita electricity consumption has been much greater than in any previous cultures throughout history. To this end, massive fossil fuels (crude oil, natural gas, and coal) are utilized in gas- and coal-fired power plants for electrical power generation, which leads to uncontrolled carbon emissions and the greenhouse effect. Recent studies show that if the global temperature continues to rise with the following current trends and no other strategies are adopted, it will cause losses of up to 10% of the global economy by the middle of this century [1]. To address it, more and more governments are paying attention to sustainable development and combating climate change. In this context, due to the clean, low carbon emission, and recycled nature characteristic of renewable energy sources, it is considered a foremost promising alternative to traditional fossil fuels in the transition toward a low-carbon economy. Specifically, renewable energy sources like wind, solar, and hydropower have been widely utilized for electrical power generation, as shown in Figures 1.1 and 1.2.

However, renewable energy sources are variable, intermittent, and uncertain, as the proportion of renewable energy in electrical power generation continues to rise. This may increase the uncertainty of the power system and make it difficult to dispatch optimal generator settings to meet daily load variations and schedule maintenance. On the other hand, the uncertainty of renewable energy generation will increase the nonlinear characteristics of the power system. In addition, different from traditional thermal-dominated systems, to maximize and optimize renewable energy sources such as solar and wind, various types of power electronics (AC–DC, DC–AC) are applied in

[1]School of Mechanical and Electrical Engineering, University of Electronic Science and Technology of China, China
[2]College of Engineering and Technology, Southwest University, China
[3]Electrical Engineering Department, Faculty of Engineering, Mansoura University, Egypt
[4]Faculty of Engineering, Horus University-Egypt, Egypt
[5]Department of Energy Technology, Aalborg University, Denmark

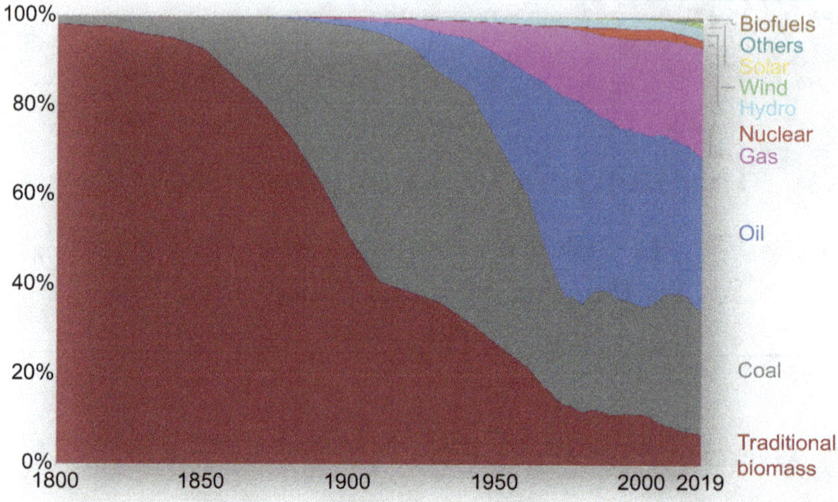

Figure 1.1 Global direct primary energy consumption (relative) in the past two centuries [2]

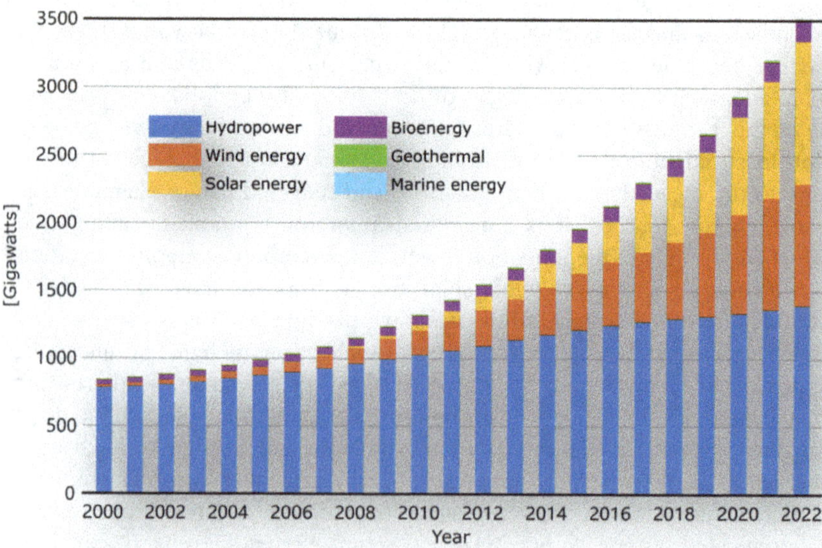

Figure 1.2 Global accumulative capacity of renewable energy from 2000 to 2021 [2]

wind turbines and photovoltaic plants for power conversion. Such widespread application of power electronics may strengthen this phenomenon and make the traditional stability control strategies (voltage control, small signal stability control) ineffective, indicating a big challenge to the stability control of renewable energy systems.

Furthermore, the power electronics devices of renewable energy generators are generally made up of semiconductor switches and energy storage elements, which are much different from traditional electrical equipment. To this end, novel optimization and modulation strategies should be adopted for power electronics devices to achieve more efficient and reliable power electronic systems.

To address these challenges, different advanced control methods are applied to enhance the efficiency, stability, availability, and reliability of such complex systems. Most of all, with the arrival of the era of Big Data and the leap in computer performance, artificial intelligence (AI) technology has developed rapidly and shows great advantages in data processing, nonlinear fitting, and decision-making, which makes it attract more attention and become a promising alternative solution to address some issues in renewable energy systems. Some studies explore the possibilities to improve the accessibility and efficiency of the renewable energy system with the help of AI technologies, such as electric machine fault diagnosis, power electronic reliability enhancement, dual active bridge modulation, voltage control, small signal stability control, and energy management. Simulation results show that it can better deal with the uncertainty of the renewable energy system in comparison with traditional methods and achieve a satisfactory performance. Therefore, it can be observed that AI-enabled strategies are regarded as an important tool to help operate power electronics and renewable energy systems to ensure optimal and secure operation.

1.2 Renewable energy system

As in the power system mentioned above, considering the environment-friendly and pecuniary characteristics of renewable energy resources, they can be applied to replace conventional fossil fuels with clean energy generation. Figure 1.3 shows the comparison of classical power systems and renewable energy systems. Among them, hydropower is the leading source of renewable energy. Over the past decade, with the maturity of wind power and solar power generation technology, the cost of electricity from these renewable energies has continued to decrease. In this context, wind turbines and photovoltaics (PV) attract more attention. From a historical perspective, the first utility-scale wind turbine (WT) was installed in the 1980s, and the rated power of this WT was only 50 kW. After that, WT manufacturers

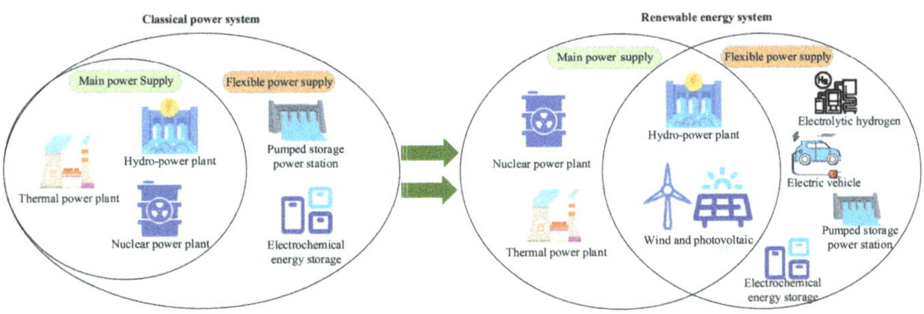

Figure 1.3 Comparison of classical power system and renewable energy system

continuously improved the technology and pushed the maximum output power of WT. As of 2022, the rated power of WT has reached up to 15 MW and is available in the market. Furthermore, WT with a rated power of 15–20 MW is also being developed and is expected to be available for deployment soon. Hence, the installed capacity of wind turbines is increasing year by year. In China, the rated installed wind power growth exceeds 10%. Similarly, PV manufacturers have been striving to increase the conversion efficiency of PV panels, and at the same time, minimize production costs, resulting in a rapid improvement in PV installed capacity worldwide. According to the International Renewable Energy Agency (IRENA) survey report [3], in 2022, the share of renewable energy generation will be 29%. Among them, the proportion of hydropower installed capacity remained the highest, reaching 39.69%, but decreased by 2.84% from the previous year. The proportion of solar energy was 30.01%, significantly increasing by 3.29% compared to the previous year. Wind energy ranks third, accounting for 25.62%. In summary, the hydro-wind-PV accounts for over 90% of renewable energy power generation.

However, renewable energy sources such as wind and solar energy have some shortcomings. For example, the output power of a PV system depends on solar radiation. The same is true in that the output power of wind turbines varies with the captured wind speed. Therefore, the output of wind turbines and PV-integrated renewable energy systems is easily affected by natural conditions. This is different from conventional power system generation capacity and reduces the reliability, controllability, and availability of the system. To solve these issues, energy storage devices, such as batteries, are used to improve the reliability of renewable energy systems for all operating conditions. Specifically, the battery can absorb surplus electricity in the system and discharge it when the power system has a shortage of electricity to ensure a basic balancing of power and energy. Moreover, electric vehicles are considered mobile batteries, which can be utilized to achieve power energy management through a demand response strategy. In this way, the random characteristics of wind and solar energy can be dealt with very well. In fact, some recent studies further coordinate it with some other forms of energy systems (water systems, heating systems, and vessel systems) to form a renewable energy-powered hybrid energy system, which contributes to achieving better economic and environmental performance as well as good energy utilization. In this context, the structure and functionality of the system will become more complex. Achieving and maintaining energy management and stability within such a system, ensuring it remains within defined parameters for angle, voltage, and frequency during both normal operation and after events like faults or failures, presents significant challenges.

More importantly, different from conventional hydroelectric and thermal plants, power electronic converters are utilized in WT and PV. These power electronic converters will play an important role in the energy conversion and control system, as shown in Figure 1.4. To achieve efficient and secure access to the power grid, the power electronic converters in WT and PV should be carefully optimized and designed to achieve high efficiency and reliability under several normal and faulty operating conditions [2]. As the proportion of WT and PV continues to increase, the system will become a power electronics-enabled power system.

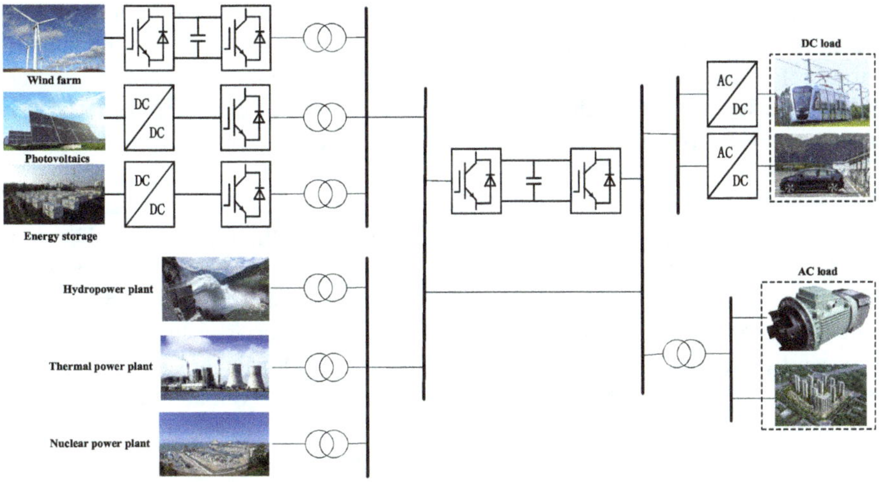

Figure 1.4 Power electronic converters driven renewable energy system

However, integrating and operating so many power electronic converters in a power electronic-dominated power system is a big challenge. Traditional optimization and design methods are difficult to apply to such complex systems and can often not achieve satisfactory performance.

1.3 Development of artificial intelligence in power systems

AI is one of the most popular technologies currently, and it is considered an important driving force of the fourth industrial revolution. In many research studies, AI has become the focus of attention on most topics to solve engineering problems, especially power systems. For AI, the artificial neural networks (ANNs) were constructed based on the human nervous system, and that is the foundational part. Specifically, the ANN model was first proposed in 1943 by Warren McCulloch and Walter Pitts [4], who arranged it in layers, and each layer includes multiple neurons. The motivation for this is to imitate the structure of the human brain to obtain the ability to learn, reason, and perceive to achieve specific targets through trial and error.

Machine learning is an integral part of AI, and it has become a key technology to solve real tasks. Reviewing the history of machine learning, it was first introduced in 1959. Arthur Samuel defined it as: “it gives computers the ability to learn without being explicitly programmed” [5]. In general, machine learning involves encoding rules via computers and computational techniques, and it allows us to extract powerful knowledge from the processed data to make decisions or predictions. In this way, it becomes possible to apply machine learning to learn the action or reaction from big data to achieve better adaptability, efficiency, and scalability. Recent studies have utilized such methods to address the intermittency and uncertainty in renewable energy systems, such as power load forecasting [6], energy management [7], angle stability control [8], and so on. Among the machine learning family, three significant types are most famous: supervised machine learning, unsupervised machine learning, and reinforcement learning (RL). The main machine-learning algorithms and their relationships are shown in Figure 1.5.

Supervised machine learning: First, the expected output was pre-labeled for some input data to form a labeled dataset [9]. This process is important, and the labeled data contains all the features needed to generate the final output. Then, this set would be utilized to train in a supervised manner to approximate the relationship between the input and the output, such as decision tree networks, linear regression, and nearest neighbor. In real-life tasks, supervised learning has been successfully utilized for spam email detection, fault diagnosis, and stability assessment. More specifically, the application of supervised machine learning can be divided into two types: classification and regression. Among them, the classification algorithm is suitable for conditions where the output result is categorical. Some crucial classification algorithms include: random forest, decision trees, and logistic regression. Different from this, the regression algorithms are suitable to be

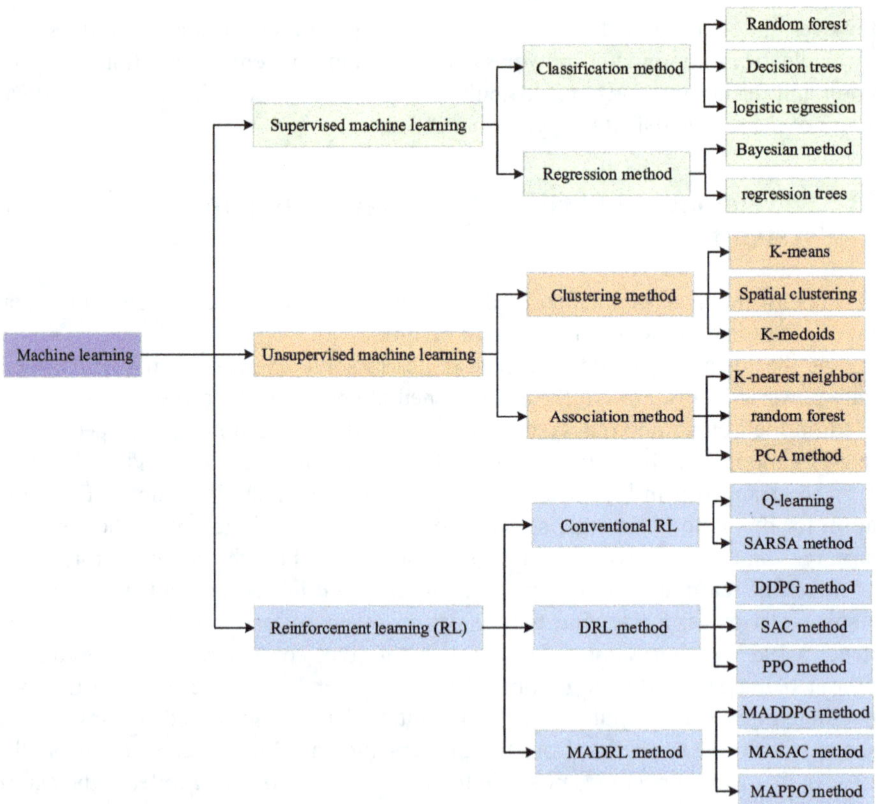

Figure 1.5 Main machine learning algorithms and their relationships

utilized to map the relation between input variables and output variables to predict continuous output variables. Some common regression algorithms include: Bayesian linear regression and regression trees.

Unsupervised machine learning: Different from supervised machine learning, this kind of algorithm does not rely on certain related responses but rather trains the models based on the unlabeled data, which means proceeding with the data in an unsupervised manner, and the obtained results are determined only by the data [10]. In fact, this is similar to the process of humans learning new knowledge. In real-life tasks, unsupervised learning has always been applied to search for important insights and hidden patterns in the dataset. In general, unsupervised machine learning can also be divided into two categories: clustering and association. Among them, the clustering algorithm would analyze the feature similarities of the data and divide the data into different clusters depending on them. Specifically, the data with higher feature similarities will be assigned to the same cluster, and the data with lower or no feature similarities will be assigned to a new cluster. Some crucial algorithms include k-means clustering, density-based spatial clustering of applications with noise, and the K-medoids method. The association method is always utilized to search for the correlation of data to achieve dimensionality reduction. Some crucial algorithms include principal component analysis (PCA), k-nearest neighbors, and random forest.

Reinforcement learning (RL): RL is a class of methods inspired by behavioral psychology, and it can extract optimal operational knowledge from historical data through trial and error even if the global optimum is unknown. Specifically, an RL-based agent would first interact with the environment to observe the state. Then, based on the obtained state, the agent would provide the corresponding action, and the immediate feedback could be calculated. Finally, the feedback would guide optimizing the parameter settings of agents to reach the desired output [11]. The represented RL algorithm includes Q-learning, state-action-reward-state-action, and deep Q-learning. However, conventional RL is only applicable for scenarios that have a low-dimensional and discrete action domain. When it is used to solve a continuous problem, the action domain needs to be discretized, and the control action is modeled as several discretized decisions. This may cause the following problems: a rough discretization of the action domain is prone to weakening the control effect; a detailed discretization would lead to dimensional disaster problems. To address it, Google DeepMind combined the RL algorithm and deep neural network to form a novel algorithm, deep reinforcement learning (DRL), indicating its applicability to scenarios with continuous action domains. Examples of these approaches are deep deterministic policy gradient (DDPG), soft actor critic (SAC), and proximal policy optimization (PPO). However, the mentioned DRL methods belong to single-agent DRL algorithms in a centralized control manner. When it is applied to larger-scale systems, all the system states must be sent to the central DRL agent. Then, based on the obtained states, the DRL agent will provide the corresponding action and send it to the system. In this context, it easily encounters communication delays and single-point failure issues. To address it, a centralized training and decentralized executing architecture is proposed to form a multi-agent DRL (MADRL) algorithm, such as multi-agent

DDPG, multi-agent SAC, and multi-agent PPO. Recent studies have applied these multi-agent DRL algorithms to power systems and achieved great success.

In Chapter 2, the application of AI in electric machine fault diagnosis is analyzed. Chapter 3 discusses AI in power electronic reliability. Chapter 4 summarizes the AI in active bridge modulation. AI in active distributed network voltage control is concluded in Chapter 5. AI in the energy management of hybrid energy systems and microgrids is discussed in Chapters 6 and 7, respectively. Chapter 8 reviews the AI in renewable energy systems' small signal stability control. The conclusion and outlook of this book are given in Chapter 9.

References

[1] A. K. Tiwari, S. Nasreen, and M. A. Anwar, "Impact of equity market development on renewable energy consumption: Do the role of FDI, trade openness and economic growth matter in Asian economies," *Journal of Cleaner Production*, vol. 334, p. 130244, 2022.

[2] B. Frede, Y. Yang, K. A. Kim, and J. Rodriguez. "Power electronics technology for large-scale renewable energy generation," *Proceeding of IEEE*, vol. 111, no. 4, pp. 335–355, 2023.

[3] International Energy Agency: Renewable capacity statistics 2023. https://www.irena.org/Publications/2023/Mar/Renewable-capacity-statistics-2023.

[4] W. McCulloch, and W. Pitts. "A logical calculus of the ideas immanent in nervous activity." *Bulletin of Mathematical Biophysics*, vol. 5, pp. 115–133, 1943.

[5] G. Wiederhold, and J. McCarthy, "Arthur Samuel: Pioneer in machine learning," *IBM Journal of Research and Development*, vol. 36, no. 3, pp. 329–331, 1992.

[6] J. Li, D. Deng, J. Zhao, *et al.*, "A novel hybrid short-term load forecasting method of smart grid using MLR and LSTM neural network," *IEEE Transactions on Industrial Informatics*, vol. 17, no. 4, pp. 2443–2452, 2021.

[7] G. Zhang, W. Hu, D. Cao, *et al.*, "A multi-agent deep reinforcement learning approach enabled distributed energy management schedule for the coordinate control of multi-energy hub with gas, electricity, and freshwater," *Energy Conversion and Management*, vol. 255, p. 115340, 2022.

[8] G. Zhang, W. Hu, D. Cao, *et al.*, "Deep reinforcement learning-based approach for proportional resonance power system stabilizer to prevent ultra-low-frequency oscillations," *IEEE Transactions on Smart Grid*, vol. 11, no. 6, pp. 5260–5272, 2020.

[9] S. Ma and C. Ji, "Performance and efficiency: recent advances in supervised learning," *Proceedings of the IEEE*, vol. 87, no. 9, pp. 1519–1535, 1999.

[10] M. Usama, J. Qadir, A. Raza, *et al.*, "Unsupervised machine learning for networking: techniques, applications and research challenges," *IEEE Access*, vol. 7, pp. 65579–65615, 2019.

[11] D. Cao, W. Hu, J. Zhao, *et al.*, "Reinforcement learning and its applications in modern power and energy systems: A review," *Journal of Modern Power Systems and Clean Energy*, vol. 8, no. 6, pp. 1029–1042, 2020.

Chapter 2

Artificial intelligence for electric machine fault diagnosis

Weihao Hu[1], Guozhou Zhang[2], Zhenyuan Zhang[1], Sayed Abulanwar[3,4] and Frede Blaabjerg[5]

2.1 Introduction

Electric machines (EMs) play an important role in industrial production, and their reliability is directly related to enterprise safety and economic benefits. With the continuous development of science and technology, the application fields of electric machines have been widely expanded, such as electric vehicles, wind power generation, rail transit, etc. State monitoring and fault diagnosis can provide a reliable guarantee for the normal operation of industrial systems.

Common EM faults can be divided into two main categories: electrical faults and mechanical faults [1]. According to research data from the American Electric Power Research Institute (EPRI) [2] on EM faults, approximately 53% of motor faults are due to mechanical faults, such as bearing faults, unbalance faults, etc.; and approximately 47% of motor faults are electrical faults, of which about 10% are due to the rotor, such as rotor casting defects, rotor broken bars, and other factors caused by the air gap magnetic field imbalance. In addition, about 37% of the faults are due to turn-to-turn and phase-to-phase short circuits and other faults caused by damage to the insulation of the stator windings.

Traditional motor fault diagnosis methods mainly rely on the accuracy of modeling or the experience of diagnostic experts to manually extract fault features for judgment. These methods not only have low efficiency but also make it difficult to detect potential fault hazards. Therefore, the development of automated and intelligent motor fault diagnosis technology has become an inevitable trend. According to the development of motor fault diagnosis technology, diagnostic

[1]School of Mechanical and Electrical Engineering, University of Electronic Science and Technology of China, China
[2]College of Engineering and Technology, Southwest University, China
[3]Electrical Engineering Department, Faculty of Engineering, Mansoura University, Egypt
[4]Faculty of Engineering, Horus University-Egypt, Egypt
[5]Department of Energy Technology, Aalborg University, Denmark

methods can be mainly divided into two categories: traditional methods [3,4] and artificial intelligence (AI) based methods [5,6]. Traditional methods mainly include Variational Mode Decomposition [7], Wavelet Transform [8], Empirical Mode Decomposition [9], Hilbert Transform [4], etc. And AI-based methods mainly include Deep Belief Networks [5,10], Deep Auto-Encoder [11], Convolutional Neural Networks [12,13], Long Short-Term Memory [6], etc. In general, a large number of experts and scholars have conducted a lot of exploration and research on motor fault diagnosis strategies based on traditional methods and artificial intelligence methods, and have achieved a series of good results. However, there are still some problems that need to be solved: (1) The generalization problem of motor fault diagnosis models under different operating conditions. When the motor operating conditions change, the performance of most fault diagnosis models will significantly decrease or even fail. The working conditions of a machine are usually variable, where conventional fault diagnosis models with weak generalization ability can only be used for fault detection under constant working conditions. (2) The overfitting problem caused by limited fault data samples is still a problem that needs to be solved. Although transfer learning or Generative Adversarial Networks can solve these problems to a certain extent. However, the role of the transfer learning models is to provide an initial parameter for fault diagnosis models, which focus on the pre-training tasks. Generative adversarial networks make it difficult to ensure the sample quality of the generated "fake data," and the training of network models is also relatively difficult. (3) Due to the different working environments of motors, the sampled operating data may be mixed with different levels of noise, which can also cause a decrease in the performance of intelligent fault diagnosis models. Moreover, traditional AI-based fault detection approaches need a large amount of data for model learning. However, in a real-world system, it is very difficult and expensive to obtain massive labeled fault data.

To this end, a novel intelligent deep-learning-based multi-fault detection method for EMs under varying working conditions is proposed in Section 2.2. This method involves two steps: first, a 2D convolution network without a pooling layer is proposed to extract features from raw EM data from the operation. In addition, a Long Short-Term Memory (LSTM) network is applied to extract the fault features for comparison. Second, a capsule network (Caps-Net) based on a dynamic routing algorithm is used as a classifier to realize intelligent multi-fault detection and improve the generalization performance of the proposed model. The proposed method applies to the raw physical signals of EMs, which improves the overall efficiency of fault detection. In Section 2.3, a novel meta-learning-enabled method for the detection of faults in the rolling bearings of electric machines under varying working conditions with limited data is proposed. The fault diagnosis under various working conditions is cast as a few-shot classification problem, which is solved using a model-agnostic meta-learning-based model. Specifically, a meta-learner is first trained using a series of interrelated fault-diagnosis tasks under various working conditions. During this stage, a gradient-by-gradient rule is utilized for parameter optimization to achieve an effective representation of these tasks. Then, the parameters of the meta-learner are refined on a new task. This technique can achieve fast adaptation to new tasks by utilizing only a few shot samples. The proposed method can obtain high fault-detection accuracy under various working conditions when only a limited amount of data is available. In Section 2.4, a novel deep Gaussian process (GP) kernel transfer-based few-shot learning method (RNGPT) is proposed for the fault detection of electric machines. Here, a deep residual network (ResNet) is used to extract the features of the raw data. Then, the encoded latent feature vector is fed into the GP with kernel transfer ability to make motor fault detection and classification possible.

2.2 Capsule network for electric machine fault diagnosis

In this section, the method involves two steps: first, a 2D convolution network without a pooling layer is proposed to extract features from raw EM data. In addition, a long short-term memory (LSTM) network is applied to extract the fault features for comparison. Second, a capsule network (Caps-Net) based on a dynamic routing algorithm is used as a classifier to realize intelligent multi-fault detection and improve the generalization performance of the proposed model. The proposed method applies to the raw physical signals of EMs, which improves the overall efficiency of fault detection.

2.2.1 Fundamental theoretical basis of capsule network

2.2.1.1 CNN and LSTM

Convolutional computation models have been widely applied to big data for feature learning, and CNNs are now widely used in various fields, such as video processing and speech recognition [14], etc. The architecture of the traditional CNN is shown in Figure 2.1; there are three main layers: the convolution, pooling, and full connection layers. Through the convolution and pooling operations of the input data, the feature maps can be obtained by the full connection layer for further processing.

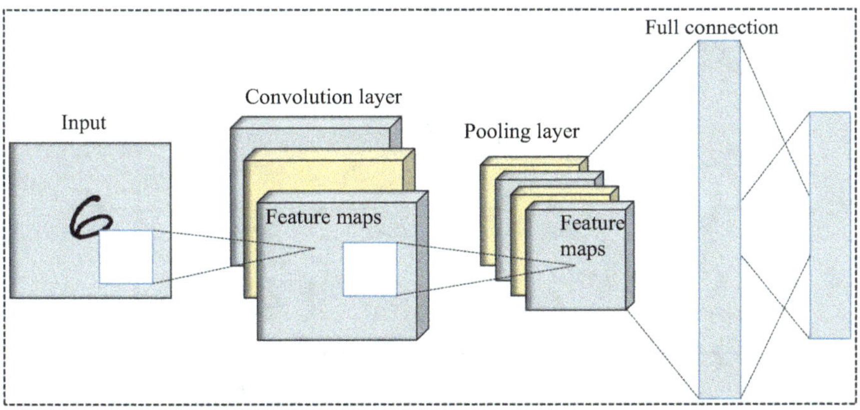

Figure 2.1 Architecture of traditional convolution neural network (CNN) [15]

For comparison, long short-term memory (LSTM) is applied to extract the features of raw time series data of an EM. The LSTM is an improved recurrent neural network (RNN) with better performance [16]. The powerful learning ability of LSTM for time series data mitigates the limitations of traditional handcrafted methods sufficiently. The inner architecture of an LSTM is shown in Figure 2.2. The three gating units (Input, Forget, Output) and cell are the most important parts of the LSTM. Compared with a general neural network, LSTMs are capable of learning long-term dependencies.

The LSTM cell state, C_t, can be described as follows:

$$C_t = m_t + n_t \tag{2.1}$$

where m_t and n_t are the intermediate states of the LSTM:

$$m_t = i_t \cdot a_t, \tag{2.2}$$

$$m_t = i_t \cdot a_t, \tag{2.3}$$

where i_t and f_t are the input and forget control gates of the LSTM, respectively; their values are between 0 and 1; a_t is the input state of the LSTM, and C_{t-1} represents the last cell state:

$$i_t = \sigma\left(W_i \begin{bmatrix} h_{t-1} \\ X_t \end{bmatrix} + b_i\right), \tag{2.4}$$

$$f_t = \sigma\left(W_f \begin{bmatrix} h_{t-1} \\ X_t \end{bmatrix} + b_f\right), \tag{2.5}$$

$$a_t = \tanh\left(W_{t1} \begin{bmatrix} h_{t-1} \\ X_t \end{bmatrix} + b_{t1}\right), \tag{2.6}$$

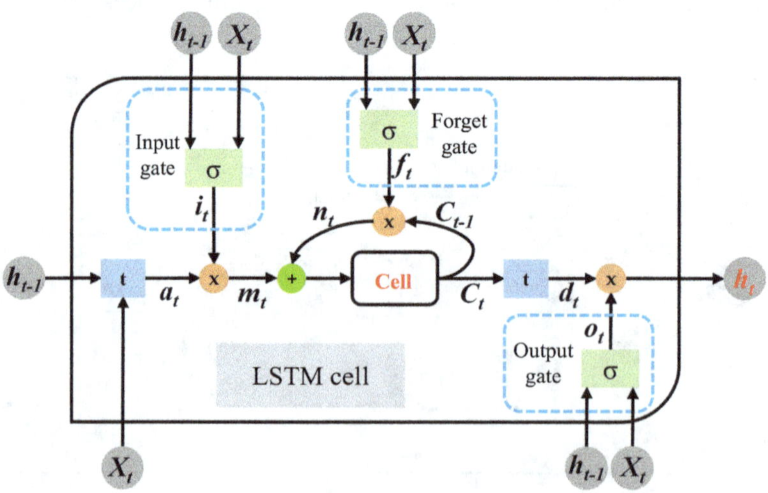

Figure 2.2 Inner architecture of Long Short-Term Memory (LSTM) cell [15]

where W_i, W_f, and W_{t1} are the weight matrices; b_i, b_f, and b_{t1} are the biases of the neural network; σ is the sigmoid activation function, and tanh is the tanh activation function. The role of f_t is to forget the information in the LSTM cell selectively, and the function of i_t is to record new information into the LSTM cell selectively.

Moreover, the remaining intermediate state d_t can be represented as follows:

$$d_t = \tanh(C_t) \tag{2.7}$$

Finally, the output of the LSTM, h_t, is as follows:

$$h_t = o_t \cdot d_t, \tag{2.8}$$

where o_t is the output gate of the LSTM (value between 0 and 1). The function of o_t is to determine the output information:

$$o_t = \sigma\left(W_o \begin{bmatrix} h_{t-1} \\ X_t \end{bmatrix} + b_o\right) \tag{2.9}$$

In this study, the feature output of the LSTM is used as the feature input of the fault classifier.

2.2.1.2 Caps-Net

The traditional CNN includes an important pooling layer. However, the max-pooling operation allows neurons in one layer to consider only the most active information in a local pool of the next layer, which causes CNNs to lose important information in the feature transformation. Caps-Net is used to overcome the deficiency of the pooling layer, which replaces the scalar-output features of CNNs with vector output capsules and the max-pooling operation with a dynamic routing algorithm. The architecture of a Caps-Net is shown in Figure 2.3.

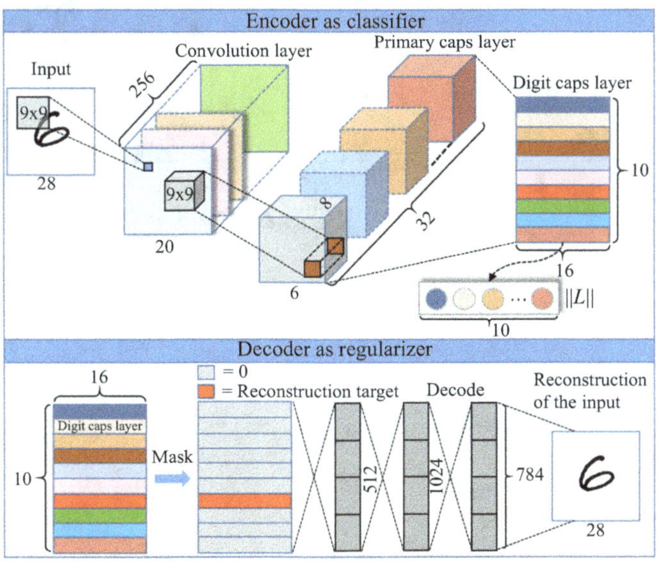

Figure 2.3 Architecture of original Caps-Net [15]

The original Caps-Net consists of two parts: a classifier and a decoder. The architecture of a capsule classifier has three parts: convolution, primary capsule, and digit capsule layers. Unlike traditional intelligent classifiers, Caps-Net uses the length of the vector in the digit capsule layer to represent the probability of the existence of the entity and the orientation of the vector to present the properties of the entity [17]. In addition, the function of the decoder is to reconstruct the input data. Finally, the total loss of the overall Caps-Net includes two parts: the weighted sum of the classification and reconstruction losses.

2.2.2 Proposed framework

To establish a strong generalization method for EM fault diagnosis, an intelligent approach (CCaps-Net) and a comparison method (LSTM–Caps-Net) are introduced in this section. The architecture of the proposed EM fault detection model is shown in Figure 2.4. The proposed method can be classified into two parts: the raw data feature-learning layer based on convolution operation without pooling layer, or LSTM, and the capsule feature classifier layer based on the proposed Caps-Net.

2.2.2.1 Learning of raw data features

In this chapter case, a 9×9 convolution kernel is applied to extract fault features. In the pooling layer of traditional CNNs, features obtained by the convolution operation are replaced by their average or maximum, which may lead to the loss of

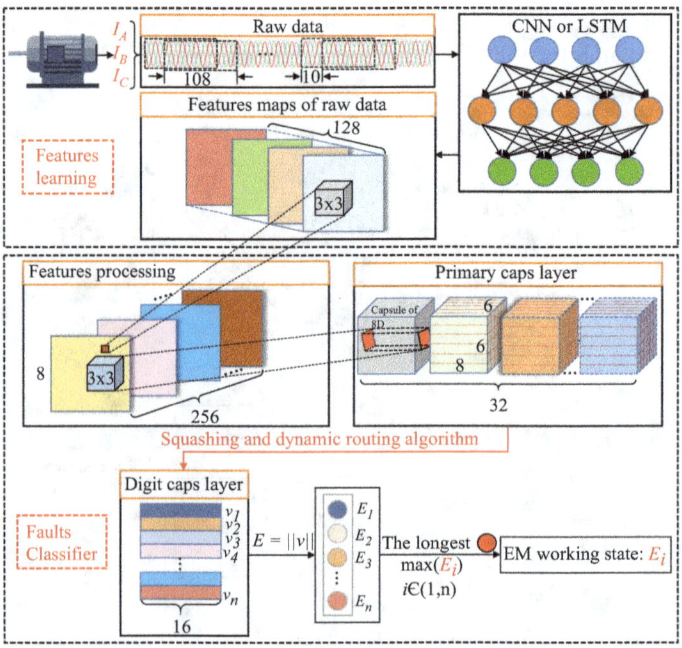

Figure 2.4 Architecture of EM multi-fault detection model [15]

important information. Consequently, the pooling layer of the CNN is removed to ensure the integrity of features. The convolution kernel for EM data feature learning can be mathematically expressed as follows:

$$f_c = act(W_{EM}M_t + b_{EM}), \tag{2.10}$$

where f_c is used to store the output features of the convolution kernel, which can be applied to further analyses; *act* represents the activation function; W_{EM} and b_{EM} are the weight matrix and bias, respectively; and M_t is the input EM signal. The feature output, h_t, of LSTM, is equal to that in Equation (2.8), whereas the input of the LSTM network is the motor operation data M_t (current I_A, I_B, I_C in Figure 2.4).

The output feature shapes of the convolution kernel and LSTM are reshaped to $s\times18\times18$, where the variable parameter s is the batch size of the raw data. In general, the output of the last time step of the LSTM is used for subsequent processing. By contrast, the proposed model uses the output features of all time steps. The features of CNN and LSTM learning are directly used as the input of the proposed fault classifier model without any additional processing.

2.2.2.2 Fault classifier

An improved Caps-Net classifier, which is more suitable for EM fault detection, is proposed. In fault diagnosis, reconstructing the input data is generally not required. When the complexity of the input data increases, the neural network has difficulties converging owing to the reconstruction branch, which increases the computational cost [18]. Consequently, to enhance the performance of the fault detection model, a more efficient classifier based on the original capsule classifier is proposed.

As shown in Figure 2.4, two 2D CNNs (two 3×3 convolution kernels with a stride of (1) are applied after using the feature learning model to map the EM features to higher dimensions. After the convolution with two CNNs, the raw data features are classified into [$6\times6\times32$] primary capsules, and each capsule is an 8D vector. Moreover, each capsule in the [6×6] grid is weight-sharing. The calculation of the 8D vector e_i can be expressed as follows:

$$g_1 = \tanh\left(W_{g1}f_c + b_{g1}\right), \tag{2.11}$$

or

$$g_1 = \tanh\left(W_{g1}h_t + b_{g1}\right), \tag{2.12}$$

$$e_i = \tanh\left(W_{ei}g_1 + b_{ei}\right). \tag{2.13}$$

As shown in Figure 2.4, the primary caps layer generates a series of 16D feature vectors in the digit caps layer through "squashing" and "dynamic routing" operations. The length of these vectors represents the probability of the existence of the corresponding working states. The non-linear "squashing" function ensures that the length of these vectors of the digit caps layer is compressed to a value between zero and one. Moreover, the squashing function squash (v_j) can be described as follows:

$$v_j = \left[||l_j||^2/\left(1 + ||l_j||^2\right)\right]l_j/||l_j||, \tag{2.14}$$

where v_j is the output vector of the capsule j ($j \in (1, n)$), n is the number of working states of the EM, and l_j the total input vector. From the primary to digit caps layer, the total input l_j is a weighted sum of all prediction vectors $\hat{e}_{j|i}$, and the prediction vector is calculated by multiplying the output 8D vector e_i of a capsule in the primary caps layer by a weight matrix W_{ij}. Therefore, l_j and $\hat{e}_{j|i}$ can be calculated as follows:

$$l_j = \sum_i a_{ij} \hat{e}_{j|i}, \tag{2.15}$$

$$\hat{e}_{j|i} = W_{ij} e_i, \tag{2.16}$$

where a_{ij} ($i \in (1, 6\times6\times32)$) are coupling coefficients that are determined by the operation process softmax(b_{ij}):

$$a_{ij} = \exp(b_{ij}) / \sum_k \exp(b_{ik}), \tag{2.17}$$

where b_{ij} are intermediate variables with initial values of zero, which can be updated discriminatively simultaneously such as the weights W_{ij}. Updating b_{ij} works as follows:

$$b_{ij}^r = b_{ij}^{r-1} + \hat{e}_{j|i} \cdot v_j^r, \tag{2.18}$$

where r denotes the number of iterations of the dynamic routing algorithm. The entire previously presented v_j calculation process is called a "dynamic routing algorithm." The procedures of dynamic routing are summarized in Algorithm 1.

Algorithm 1: Dynamic routing algorithm

1: **Procedure** routing ($\hat{e}_{j|i}$, r, primary caps, digit caps)
2: **for** all capsules i in the primary caps layer and capsules j in the digit caps layer: set b_{ij} to 0, T to 3.
3: **for** $r = 1 : T$ **do**
4: for all capsules i in the primary caps layer: set a_i to softmax(b_i);
5: for all capsules j in digit caps layer: set l_j to $\sum_i a_{ij} \hat{e}_{j|i}$;
6: for all capsules j in digit caps layer: set v_j to squash(v_j);
7: for all capsules i in the primary caps layer and capsules j in the digit caps layer: set b_{ij} to $b_{ij} + \hat{e}_{j|i} \cdot v_j$.
return v_j

After the operation of the dynamic routing algorithm between the primary and digit caps layers, the proposed model calculates the lengths of the vectors v_j to present the probability that different EM working states exist, and the state corresponding to the longest vector is chosen.

To increase the distances of the interclass and simultaneously reduce intra-class variations of the proposed model, a margin loss function is introduced to optimize the proposed model. The model is trained by minimizing the margin loss function (2.19).

$$L_k = Y_n \max(0, m^+ - ||v_n||)^2 + \lambda(1 - Y_n)\max(0, ||v_n|| - m^-)^2, \tag{2.19}$$

where Y_n represents the index value of the data sample labels; different values represent different working states of EMs. In this study, m^+, m^-, and λ are set to 0.9, 0.1, and 0.25, respectively [17], and the Adam optimizer with a learning rate of 0.001 is applied to optimize the model and parameter update. Algorithm 2 presents the training procedure of the proposed model.

Algorithm 2: Training procedures of proposed model for EM

Input: Three-phase current I_A, I_B, I_C; learning rate lr; batch size b; number of iterations for training t; number of iterations in routing algorithm T;
Output: fault detection accuracy; states of motor predicted by model; real states of a motor;
Initialize: initialize the weights and biases of the CNN and LSTM $W_{EM}, W_i, W_f, W_o, W_{t1}, b_{EM}, b_i, b_f, b_o$; initialize the classifier parameters $W_{g1}, W_{ei}, b_{g1}, b_{ei}$ and routing algorithm parameters b_{ij}, W_{ij};
  for iterations in range t **do**
      # obtain motor working state feature:
    $f_c = act(W_{EM}M_t + b_{EM})$ or $h_t = o_t \cdot d_t$;
    # feature processing
      $g_1 = \tanh\left(W_{g1}f_c + b_{g1}\right)$
      or $g_1 = \tanh\left(W_{g1}h_t + b_{g1}\right)$;
    # primary caps layer: obtain 8D vectors e_i
      $e_i = \tanh\left(W_{ei}g_1 + b_{ei}\right)$;
      # digit caps layer: obtain fault feature vectors v_j
  for $r = 1 : T$ **do**
        routing ($\hat{e}_{j|i}$, r, Primary caps, Digit caps)
    return v_j;
    # calculate the length of vector v_j
    $S_j = ||v_j||$;
    # determine the longest vector and output the index j, which corresponds
  to different motor states
    $J =$ tf.argmax(S_j, axis $= 1$);
  end for
  output the fault detection accuracy and EM working states.
The model is trained by minimizing the margin loss function and optimized by the Adam optimizer.

2.2.3 Analysis of results

A case study and analysis of the results are presented in this section. The numerical simulation and model training are completed on a 64-bit PC with an Intel Core i9-7900X CPU of 3.8 GHz, 32 GB RAM, and an RTX2080 Ti GPU with 11 GB VRAM. The proposed model is implemented in the Pycharm software platform with the GPU version TensorFlow.

The proposed method is compared with LSTM–Caps-Net and other state-of-the-art methods: CNN [19], SVM [20], LSTM-ATT [21], and ResNet [22]. More specifically, a two-layer CNN with a max-pooling operation model is applied for comparison. The kernel function and parameters *c* and gamma of the SVM are RBF, 0.9, and 20, respectively. Only the SVM without a combination with other feature extraction methods is applied in this study. The LSTM–ATT model is a combination of the attention mechanism model and a three-layer LSTM. A ResNet with ten residual units is also applied for the performance comparison, each residual unit contains four layers of neurons. It evaluates the performance of the proposed and other state-of-the-art methods based on the average accuracy of the fault identification. The accuracy can be calculated as follows:

$$A = \sum \frac{n_{\text{dec}}}{n_{\text{batch}}} \times 100\%, \tag{2.20}$$

where n_{dec} is the number of EM states correctly detected by the model, n_{batch} represents the number of batches of data input into the neural network in each test round. The hyperparameter tuning of the models is based on the Grid Search method [23]. The learning rate is set to be in the range of $0.1–10^{-8}$ and reduced ten times at each tuning time, which is a general range of neural network parameter tuning. In addition, we set the number of neural layers in the range of 1–5 and increase by 1 layer at each tuning time. Similarly, other hyperparameters are also set to a general range. After all the hyperparameters are set to a specific range, these parameters are tuned by Grid Search [23].

2.2.3.1 EM working states and training dataset description

The proposed method is applied for a multi-fault diagnosis of a three-phase asynchronous induction motor under varying working conditions. The data of the motor under different working conditions are simulated by commercial software, *ANSYS Electronics Suite*, which can uniquely simulate the electromagnetic performance of the circuit and system design and evaluate the temperature, vibration, and other key mechanical effects to help users design innovative electrical and electronic products faster and more economically [24]. The detailed motor parameters are shown in Table 2.1.

Three different types of loads are applied to the motor: constant power (CP), constant torque (CT), and linear torque (LT) loads. In addition, three varying load torques are applied. The loads are set to 68, 70, and 72 Nm, respectively. Therefore, there are six

Table 2.1 Motor parameters used for condition monitoring

No.	Parameter	Setting
1	Rated power	11 kW
2	Rated speed	1,458 rpm
3	Rated load torque	75 Nm
4	Rated voltage	380 V
5	Rated frequency	50 Hz
6	Efficiency and power factor	0.85 and 0.86

Table 2.2 Description of seven different working states of EM

Label	Health state	Description
0	Normal	Normal working state
1	ISC	Phase A with an inter-turn short circuit of two turns
2	BRB	Four broken rotor bars in the induction motor, two neighboring broken bars, and another two bars in the opposite direction
3	LM	When the motor works normally for 0.4 s, the load is suddenly missing
4	OP	When the motor works normally for 0.4 s, phase A experiences a phase failure
5	RDE	Rotor dynamic eccentricity: the rotation track of the rotor is a circle, and the center of the circle is a point
6	RSE	Rotor static eccentricity; the center of the rotation track of the rotor is a circle

different kinds of motor operation datasets for fault detection model training and testing. In addition, seven induction motor health states are considered, including one normal state and six fault states: inter-turn short circuit (ISC), broken rotor bar (BRB), load missing (LM), open-phase (OP), and rotor dynamic and static eccentricity (RDE and RSE, respectively) states. The health states of EM are simulated by changing the external circuit structure of the EM model in *ANSYS Electronics Suite* software [24]. The descriptions of the seven EM health states are listed in Table 2.2.

The motor current is one of the most commonly used and easily measured values. Different motor fault types can be detected by analyzing the features of the three-phase current (I_A, I_B, I_C) of the motor in different fault states with the neural network model. The sampling frequency of the EM working signal is 1 kHz. There are six EM working conditions in total, and each working condition has seven health states. The length of each time-series data point is 4,000. In total, there are 168,000 samples. It can be observed in Figure 2.4 that a sub-sampling window containing 108 raw data points is collected to be one sample, and there are 10 data point intervals between each sub-sampling window. A dataset is a combination of all these samples with different EM health states. In this study, when the training and test data of the model are in the same condition, 80% of the motor operation dataset is selected for training, 10% for validation under the training phase, and the remaining 10% for testing. When the training and test data come from different working conditions, 80% of the dataset is selected for model training and 20% for validation during the training phase. Under this condition, 10% of the samples generated under other loading conditions are randomly selected as the test set. Validation is similar to the test, but different from the model final test. The purpose of validation is to observe the performance of the fault diagnosis model in real-time during training. In each training round, 450 batches of EM data are sent to the fault detection model. The shape of each batch of EM raw signal input sent to the fault detection model is 108×3.

2.2.3.2 Analysis of multi-fault detection result

First, the performance of different fault detection methods under different load types is compared. The results are shown in Table 2.3. The average test time of

Table 2.3 Fault classification accuracy of the proposed and other state-of-the-art methods under different load type conditions at 72 Nm load torque. CP: Constant power; CT: Constant torque; LT: Linear torque

Methods	Fault detection accuracy under different load types				Average test time (s)
	CP	CT	LT	Average	
CCaps-Net	98.33±0.11%	98.56±0.07%	98.28±0.09%	98.39±0.09%	0.08
LSTM–Caps-Net	94.76±0.12%	94.22%±0.21%	94.07%±0.15%	94.35±0.16%	0.62
LSTM–ATT	82.45±1.12%	85.16%±0.86	83.59±0.89%	83.73±0.96%	0.48
SVM	79.17±1.33%	81.73±1.14%	81.26±1.08%	80.72±1.18%	0.12
CNN	92.88±0.43%	91.56±0.39%	92.66±0.47%	92.37±0.43%	0.06
ResNet	95.13±0.22%	95.34±0.18%	94.85±0.21%	95.11±0.20%	0.33

each batch of CNN data is 0.06 s, which takes the least time. The proposed method follows closely, which has a 0.08 s average test time and is much less than the LSTM–Caps-Net and LSTM–ATT models. Despite the average test time of each batch of CNN data being 0.02 s shorter than the proposed method, the fault detection performance of CCaps-Net is much better than that of the CNN method.

The proposed method has a better fault detection performance, with an average accuracy of 98.39±0.09%. Moreover, LSTM–Caps-Net has a high average accuracy of 94.35±0.16%. The performance of the CNN and ResNet methods is relatively high but also weaker than that of the proposed method; the average fault detection accuracy of the CNN is 92.37±0.43% and the ResNet is 95.11±0.20% because the output of these methods is scalar, whereas the output of the capsule network is a vector with direction. Therefore, the capsule network can carry more data features. In addition, because some important data features are lost owing to the existence of a pooling layer in the traditional CNN, its accuracy of fault detection is lower than that of the proposed method. When the number of fault types increases, the dimension of the data label increases. Consequently, the classification ability of the SVM that has not been combined with other feature extraction models is weakened. This is the reason why the average accuracy of the SVM is only 80.72±1.18%. Compared with that of the proposed model, the performance of the LSTM–ATT model is only 83.73±0.96% under the three different load type conditions. This is because the feature extraction ability of the LSTM is weaker than that of the proposed method. Furthermore, the attention mechanism model causes over-fitting, which decreases the EM fault detection accuracy of the LSTM–ATT model and the activation function of the LSTM cell affects the fault detection accuracy of the LSTM-based model. For example, under CP conditions with a 72 Nm load, for the LSTM–Caps-Net model, when the activation function has been changed to ReLU (Figure 2.5 shows the illustrations of ReLU and the default LSTM activation function Tanh), the average fault detection accuracy of the model decreases from 94.76% to 81.22%.

Figure 2.6 shows the accuracy of the LSTM–Caps-Net with different activation functions at the last 20 episodes. It can be observed from the figure that when the activation function is changed to ReLU, the fault detection accuracy decreases a

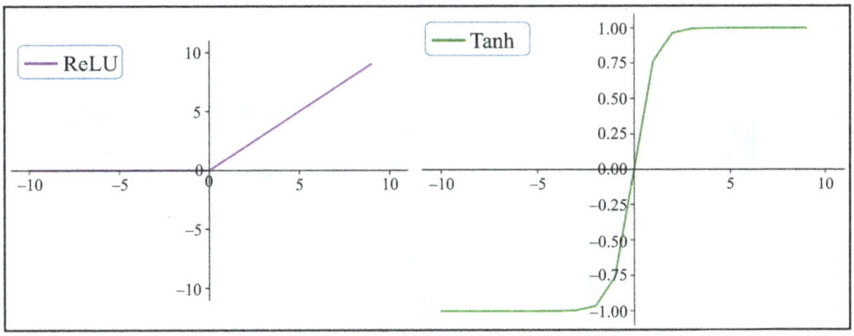

Figure 2.5 Illustrations of ReLU and Tanh functions [15]

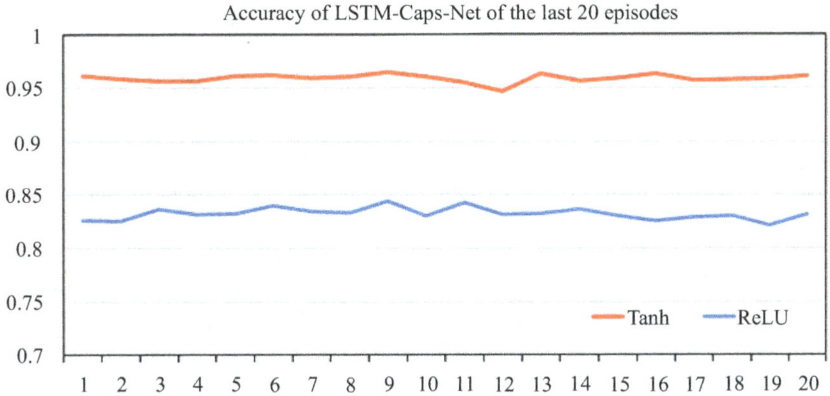

Figure 2.6 Impact of LSTM cell activation function on performance of LSTM–Caps-Net fault detection accuracy under CP condition with 72 Nm load [15]

lot. This is because the gradient explosion occurred in the neural network, which leads to a decrease in the fault detection accuracy of the model. Moreover, the input data of the model is a three-phase sinusoidal current, and there are many negative values in the sampling data, which will cause some neurons to be 0 when using ReLU. Tanh can solve this problem, which is why its fault detection accuracy is higher than the LSTM-based model using the ReLU function.

To explore the efficiency of the proposed method, a five-fold cross-validation is carried out. Take the dataset under constant power conditions with a 72 Nm load as an example, the proposed method is compared with the ResNet model, and the results are shown in Figure 2.7. It shows that the fault detection performance of the proposed method is higher than that of the ResNet.

The results of different methods under varying load types are shown in Table 2.4, where "CP→CT" indicates that the fault detection model is trained under the CP load dataset and tested under the CT load dataset. Due to limited space, this

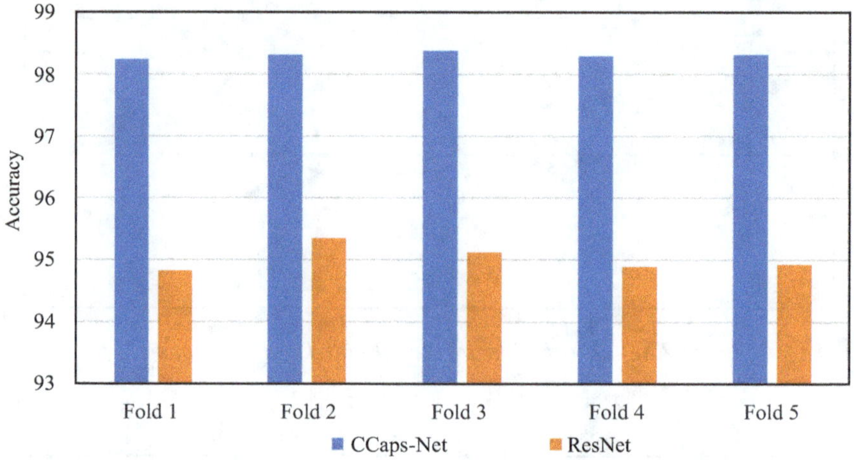

Figure 2.7 Fault detection accuracy of five-fold cross-validation [15]

chapter only presents the results under different load type conditions of 72 Nm load torque. The fault detection average accuracy of the proposed model is 94.72±0.12% and the accuracy of LSTM–Caps-Net is 91.78±0.39%, respectively. In addition, the fault classification accuracy of the other four state-of-the-art models is 81.22±0.94%, 79.59±1.21%, 91.33±1.30%, and 92.91±0.69%, respectively, which is lower than the proposed method. It can be seen from Table 2.4 that when the load type changes, the accuracy of the other four models is lower than that of the CCaps-Net model. This might be because when the load type of EM changes, the features of the EM data also change. These state-of-the-art models have weak generalization, which leads to a decrease in fault detection accuracy. CNN without pooling operations retains all features of EM data, which are further processed and stored in the capsule vectors of Caps-Net. Therefore, the CCaps-Net model has a stronger generalization ability. However, due to the limited learning ability of LSTM on a large number of data, the accuracy of the LSTM-based method is still lower than that of CCaps-Net despite the existence of Caps-Net.

The confusion matrices of CCaps-Net under different load type conditions are shown in Figure 2.8. In the confusion matrix (a), the misjudgment of the OP state as an ISC state causes the main error of the "CT→LT" scenario. As shown in the distribution of the confusion matrices (b), (c), (d), and (f), the fault detection error of the CCaps-Net model originates mainly from the fact that the ISC and RDE states are misjudged as OP and RSE states in "CP→LT," "CP→LT," "CT→LT," and "LT→CT" scenarios. In matrices (a), (c), (d), and (e), the misjudgment of the LM and ISC states as normal states is the error source of the CCaps-Net model.

The fault classification accuracy of different methods under varying working conditions is shown in Table 2.5. The results demonstrate that the proposed approach has the best performance; its average fault detection accuracy reaches 98.63%. Compared with Table 2.3, except for CCaps-Net, the accuracy of the other

Table 2.4 Accuracy of different methods under varying load type conditions with 72 Nm load torque. CP: Constant power; CT: Constant torque; LT: Linear torque

Methods	Fault detection accuracy under different load conditions					
	CP→CT	CP→LT	CT→CP	CT→LT	LT→CP	LT→CT	Average
CCaps-Net	94.52±0.07%	95.29±0.14%	94.74±0.09%	94.67±0.11%	94. 48±0.21%	94.44±0.07%	94.72±0.12%
LSTM–Caps-Net	88.61±0.33%	89.73±0.42%	92.44±0.34%	92.77±0.51%	94. 63±0.28%	92.66±0.46%	91.78±0.39%
LSTM–ATT	80.79±1.21%	81.81±0.98%	81.28±0.88%	82.49±0.74%	79.29±1.01%	81.64±0.79%	81.22±0.94%
SVM	77.91±1.55%	77.02±1.23%	80.75±1.08%	81.57±1.02%	80.55±0.98%	79.75±1.37%	79.59±1.21%
CNN	93.33±1.02%	93.77±1.13%	90.22±1.53%	90.44±1.42%	89.33±1.62%	90.89±1.08%	91.33±1.30%
ResNet	92.79±0.42%	94.94±0.84%	93.11±0.63%	94.08±0.58%	91.82±0.79%	90.71±0.86%	92.91±0.69%

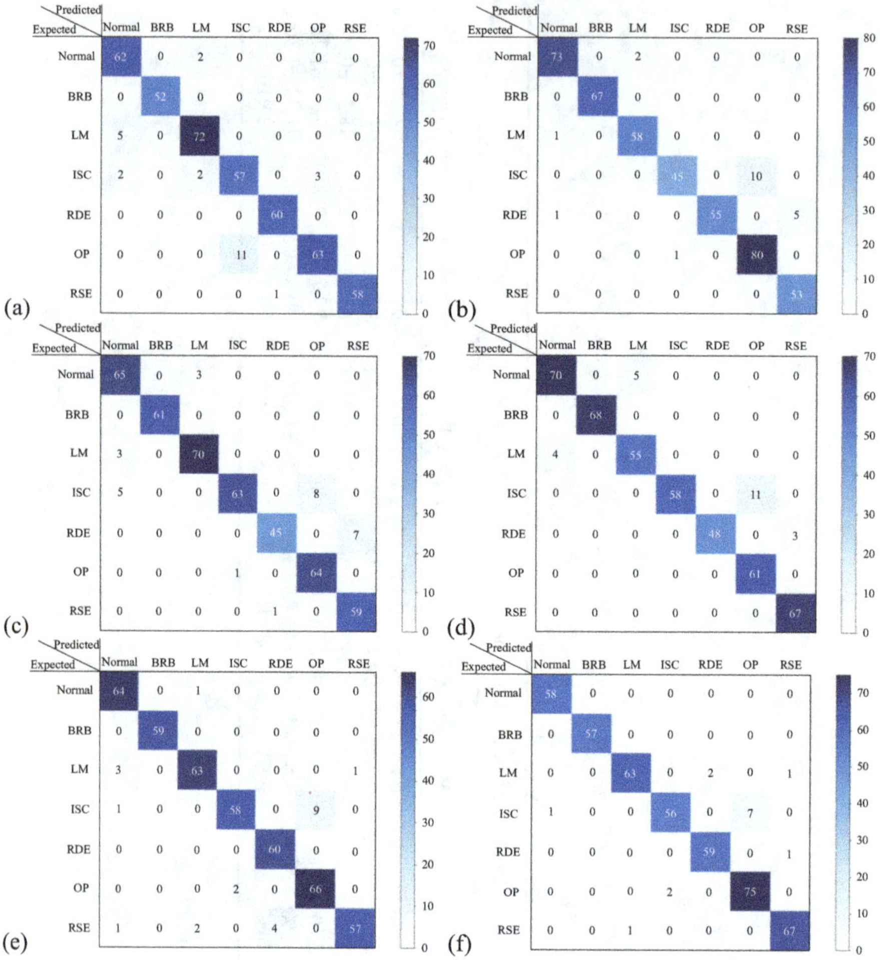

Figure 2.8 Confusion matrices of CCaps-Net fault detection results under varying load type conditions: (a) "CP→CT," (b) "CP→LT," (c) "CT→CP," (d) "CT→LT," (e) "LT→CP," and (f) "LT→CT" [15]

four fault diagnosis models has decreased. This might be because the features of the motor data are more complex under different load size conditions.

The results mentioned above demonstrate that, compared with other state-of-the-art methods, the proposed method has a higher fault detection accuracy and stronger generalization ability. Thus, it can be used for EM multi-fault detection under varying working conditions. In addition, in this study, data features are directly extracted without complex handcrafted mathematical operations, which increases the efficiency of fault diagnosis and reduces the impact of human interventions on data processing. In practical applications, the proposed method can detect various types of faults in EMs by pure current signal acquisition without changing the original

Table 2.5 Fault classification accuracy of different methods under different working conditions with constant power load

Methods	**Fault detection accuracy under different working conditions**			
	68 Nm	**70 Nm**	**72 Nm**	**Average**
CCaps-Net	98.88±0.05%	98.67±0.12%	98.33±0.07%	98.63±0.08%
LSTM–Caps-Net	94.18±0.23%	90.89±0.37%	94.76±0.34%	93.28±0.31%
LSTM–ATT	80.62±0.76%	80.66±0.83%	82.45±0.69%	81.24±0.76%
SVM	80.85±0.98%	76.80±1.65%	79.17±1.23%	78.94±1.29%
CNN	90.11±0.34%	90.48±0.55%	92.88±0.29%	91.16±0.39%
ResNet	93.49±0.18%	95.97±0.09%	94.52±0.15%	94.66±0.14%

controller structure or using external measuring equipment (such as an oscilloscope). This reduces measurement and time costs effectively.

2.3 Few-shot-learning-enabled method for rolling bearing fault diagnosis

Effective detection of faults in rolling bearings with a limited amount of data is essential for the safe operation of electric machines. This study proposes a novel meta-learning-enabled method for the detection of faults in the rolling bearings of electric machines under varying working conditions with limited data. The fault diagnosis under various working conditions is called a few-shot classification problem, which is solved using a model-agnostic meta-learning-based (MAML) model [25].

2.3.1 Basic method and proposed model

2.3.1.1 Basic theory of MAML

The purpose of a traditional deep learning (DL) model is to develop a mathematical prediction model. Unlike traditional DL, the goal of meta-learning is to train a model that can learn a mathematical model faster and better from various prior tasks and can quickly adapt to a new task using only a few samples and training iterations. Inspired by meta-learning for supervised learning, a MAML-enabled method for electric machine fault detection is proposed. In supervised learning, the meta-learning method can be regarded as a few-shot technique.

An illustration of few-shot learning is shown in Figure 2.9, where it can be observed that the training and testing of the few-shot learning model are composed of various tasks T_i. Each task has its own corresponding dataset (D_{train}, D_{test}). In these tasks, the training set (support set) $D_{train} = \{(x_i, y_i)\}$ and testing set (query set) $D_{test} = \{(x_{test}, y_{test})\}$ contain a series of input–output pairs. Assuming that the input of classifier f_θ is the feature vector x, the output is the probability $P_\theta(y|x)$ of class y, where θ is the parameter of the classifier.

The MAML method initializes the model parameters for various tasks using the gradient descent-based rule in the meta-training phase, and it can achieve

Figure 2.9 Illustration of few-shot learning [26]

promising results by fine-tuning over a few steps in the meta-test phase. The key to MAML is to obtain the initial parameters under the new task, that is, meta-training, which is completed in two steps: the basic-learner learning phase and the meta-learner learning phase. The basic-learner mainly learns the attributes of a specific task. In each task of this study, each basic learner is a multiclass classifier. In the basic-learner-learning phase, each batch of data contains Q tasks. When the model adapts to a new task, T_i $(1 \leq i \leq Q)$, the parameter θ of each task becomes $\theta_i^{'}$, which is updated using the gradient decent method:

$$\theta_i^{'} = \theta - \alpha \nabla_\theta L_{T_{i_{support}}} f(\theta), \tag{2.21}$$

where α is the learning rate of the basic learner and $\nabla_\theta L_{T_{i_{support}}} f(\theta)$ is the loss gradient of the support set of task T_i. After obtaining Q parameters $\theta_i^{'}(1 \leq i \leq Q)$, the loss value of the query set can be evaluated as follows:

$$f_{loss\_}T_{i_{query}} = L_{T_{i_{query}}} f(\theta_i^{'}), \tag{2.22}$$

In this paper, the loss function of the basic learner is cross-entropy, which is common in classification tasks. The parameters $\theta_i^{'}$ of each basic learner are trained by optimizing the performance of each batch of classification tasks at each training epoch, which can be calculated as follows:

$$f_{loss} = \min_{\theta} \sum_{i=1}^{Q} L_{T_{i_{query}}} f(\theta_i^{'}), \tag{2.23}$$

The meta-learner mainly learns the commonality among different tasks, which provides the basic learner with the initial model parameter θ. Then, θ is calculated based on the basic learner parameters, which can be described as

$$\theta = g(\theta_1^{'}, \theta_2^{'}, \cdots, \theta_i^{'}), \tag{2.24}$$

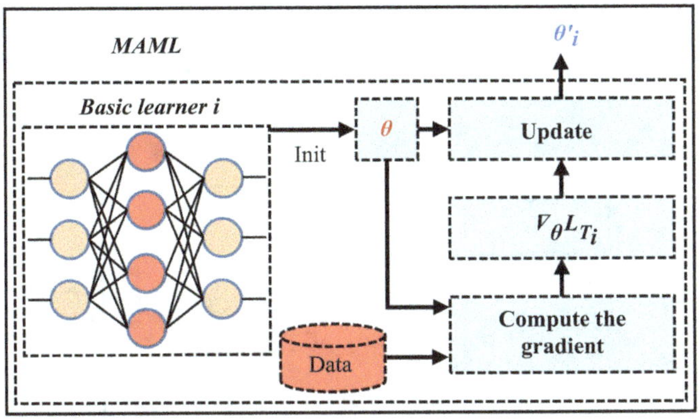

Figure 2.10 Illustration of MAML model [26]

where $g(\cdot)$is the mapping function for the parameter θ_i' of different tasks. The parameter θ is updated as follows:

$$\theta = \theta - \beta \nabla_\theta \sum_{i=1}^{Q} L_{T_{iquery}} f(\theta_i'), \quad (2.25)$$

where β is the learning rate of the meta-learner. The MAML model is illustrated in Figure 2.10.

Inspired by the aforementioned discussion, the fault detection requirements under any working condition can be regarded as separate meta-learning tasks. The trained model is broadly applicable to fault detection tasks under a wide range of working conditions, rather than a single individual detection task. Moreover, when an unseen fault detection task appears, the model only needs to use a limited amount of data for one or a small number of fine-tunings using the gradient-based learning rule, which causes the model to achieve maximum performance. The architecture of the proposed model is discussed in the following section.

2.3.1.2 Framework and training process of the proposed model

The overall architecture of the proposed MAML-based model is illustrated in Figure 2.11. The basic-learner of the proposed model is applied to data feature extraction and electric machine state classification under various working conditions. The meta-learner is applied to the model parameters for fast adaptation when the working conditions are changed.

Feature extraction and state classification

A 4-layer CNN is used for the feature extraction of an electric machine vibration signal. A simple convolution structure makes the training of the model more efficient. As shown in Figure 2.11, in each fault-detection task, the vibration series data are fed into the neural network and mapped into a nonlinear space. The mapping process can be described as follows:

$$M(x) = ReLU(m(x, W) + b), \quad (2.26)$$

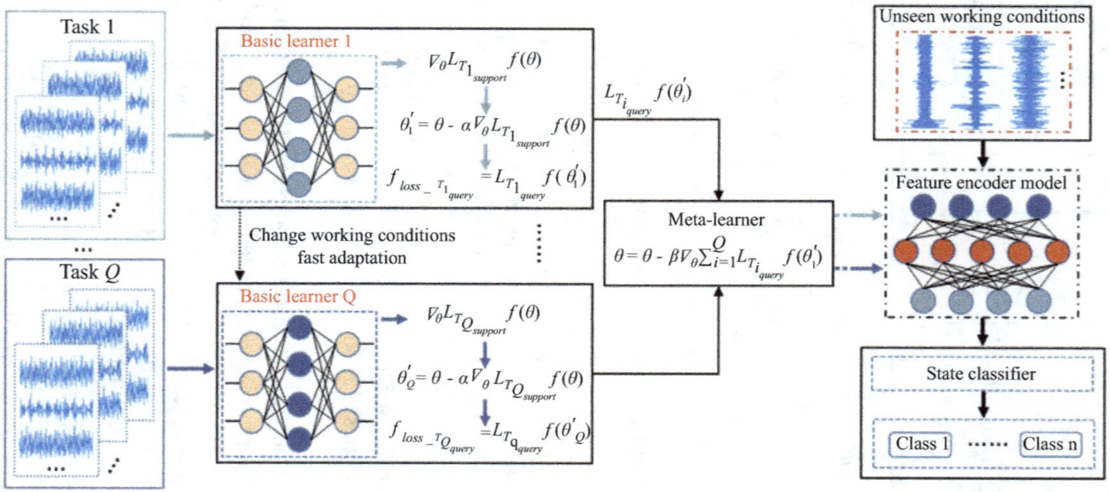

Figure 2.11 Overall architecture of the proposed electric machine fault diagnosis model [26]

where $m(x, W)$ is the feature mapping of the vibration signal x and neural network weight parameter W, and b is the bias vector. In this study, the parameters W and b are represented by θ. $ReLU(\cdot)$ is a rectified linear unit activation function that follows the convolution operation. The architecture of the 4-layer CNN is shown in Figure 2.12, and detailed descriptions of the parameter setup are shown in Table 2.6, where the "NF" represents the number of faults.

After extracting the inner features of the input data, the extracted features are fed into a classifier to identify the electric machine fault types. The classifier is trained by minimizing the cross-entropy loss function, which can be described as:

$$L_{T_i} = \sum_{(x_i, y_i) \sim T_i} y_i \log M(x) + (1 - y_i) \log(1 - M(x)), \tag{2.27}$$

where (x_i, y_i) is an input–output pair of task T_i. In an N-way K-shot classification task, there are K input–output pairs for each sample. Therefore, there are in total NK samples for each task.

Fast adaptation

It is also shown in Figure 2.11 that when another batch of tasks is sampled and fed into the model, the model parameters need to quickly adapt to the new batch of tasks with a small number of fine-tuning updates. Except for the first batch of tasks, the initial model parameter θ of the current training phase is obtained from the training of the previous batch of tasks. The mathematical description of how θ is

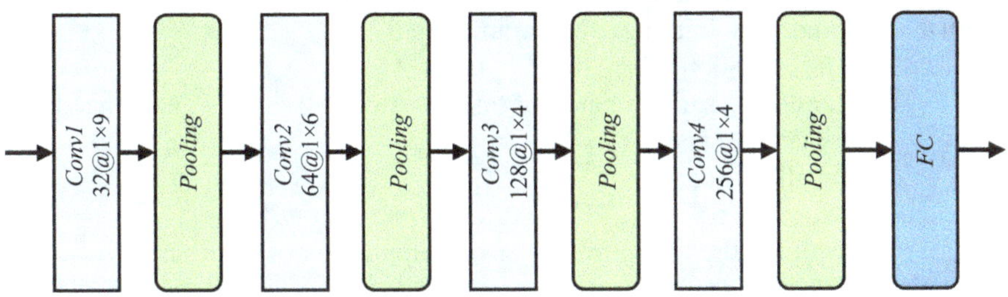

Figure 2.12 Illustration of the 4-layer CNN [26]

Table 2.6 Detailed parameters of the CNN architecture

Layer	**Type**	**Number of channels @ Kernel size**	**Padding**
Conv1	Convolution1	32@1×9	SAME
Pooling	Maxpooling1	32@1×4	VALID
Conv2	Convolution2	64@1×6	SAME
Pooling	Maxpooling2	64@1×4	VALID
Conv3	Convolution3	128@1×4	SAME
Pooling	Maxpooling3	128@1×2	VALID
Conv4	Convolution4	256@1×4	SAME
Pooling	Maxpooling4	256@1×2	VALID
FC	Fully-connected	NF	–

calculated is shown in (2.25). Each batch of tasks contains Q tasks, and the parameter fast adaptation between different tasks can be expressed as given in (2.21).

The meta-training and meta-testing processes of the proposed fault-diagnosis strategy are introduced in Algorithm 3. The model parameters of the fault detection framework are represented as $\theta = \{W_1, b_1, W_2, b_2, W_3, b_3, W_4, b_4\}$, where W_i and b_i $(i = 1, 2, 3, 4)$ are the weight and bias parameters, respectively, of a 4-layer CNN shown in Figure 2.12.

The meta-training process primarily completes the knowledge accumulation of prior tasks. The fast learning phase is shown in Processes 3 to 9 in Algorithm 3, which primarily learns the properties of each batch of tasks. Process 10 is the meta-learning phase, in which the learned attributes of the prior tasks are used to initialize the model parameters of the next batch of tasks. In the meta-testing phase, the trained model initialization parameter is obtained from various batches of prior task training. The trained parameters are applied to unseen tasks, and the trained model can obtain fast convergence with a small number of parameter updates, or without any update.

Algorithm 3: Meta-training and meta-testing procedures

Meta-training

Input: Electric machine vibration data under various working conditions
Output: Electric machine working states
Setting: Parameters of the proposed model: n-way, m-shot, learning rate of basic learner and meta-learner: α, β

Initialize: Randomly initialize the parameters θ

1: **while** not done **do:**
2: Randomly sample batch of tasks T_i from all tasks T, each batch contains Q tasks
3: **for** all T_i **do:**
4: Feature extraction: $M(x) = ReLU(m(x, W) + b)$
5: Evaluate the loss gradient of the support set of each task T_i: $\nabla_\theta L_{T_{isupport}} f(\theta)$
6: Update the parameters of different tasks: $\theta_i' = \theta - \alpha \nabla_\theta L_{T_{isupport}} f(\theta)$
7: Evaluate the loss of each task: $f_{loss\_T_{iquery}} = L_{T_{iquery}} f(\theta_i')$
8: Record the cumulative sum of the loss values of all Q tasks: $\sum_{i=1}^{Q} L_{T_{iquery}} f(\theta_i')$
9: **end for**
10: Update the parameters of meta-learner: $\theta = \theta - \beta \nabla_\theta \sum_{i=1}^{Q} L_{T_{iquery}} f(\theta_i')$
11: **end while**

Meta-testing

Required: Trained parameters θ, new task T_*

12: Randomly sample dataset D_{train} from test task T_*
13: Evaluate loss gradient: $\nabla_\theta Lf(\theta, D_{train})$
14: Evaluate adapted parameters: $\theta_{test}' = \theta - \alpha \nabla_\theta Lf(\theta, D_{train})$
15: **end**

2.3.2 Experiments and analysis of results

Rolling bearings are one of the most important components of electric machines, and they are widely used in many types of mechatronics equipment. Because of the long-term rotating wear and high temperature that the bearings are subjected to during the operation process, bearing failure is a common fault. This study focuses on fault detection in rolling bearings in electric machines under varying working conditions with a limited number of samples. To verify the performance and effectiveness of the proposed strategy, a series of case studies are carried out using two popular datasets: the Case Western Reserve University (CWRU) Bearing Dataset [27] and the Paderborn University Rolling Bearing Dataset [28].

Moreover, several state-of-the-art methods have been applied for comparison, including the capsule network (Cap-Net) [15,18], a transfer-learning-based pre-trained model (Pre-trained) [29], prototypical networks (PN) [30], CNN with wide first layer (WDCNN) [31], and a WDCNN-based few-shot learning model (WDCNN-few-shot) [32]. The Cap-Net, Pre-trained, and WDCNN are conventional DL-based methods, PN and WDCNN-Few-shot are meta-learning-based methods. The model training is completed on a 64-bit computer with one Intel(R) Core(TM) i7-9800X CPU at 3.8 GHz and one RTX2080 Ti GPU with 11 GB VRAM, that also has 32.0 GB RAM. The code for the proposed model is written in the PyCharm platform with the GPU version of Pytorch.

2.3.2.1 Case 1: CWRU bearing datasets

Dataset description

The vibration signal at the drive end of an electric motor is collected with a sampling frequency of 12 kHz under various bearing health states. The test stand is shown in Figure 2.13 and contains three parts: a 1.5-kW (2-hp) reliance electric motor, a torque transducer, and a dynamometer. Various faulty bearings are installed in the test stand, and the vibration signal is collected.

A detailed description of the selected rolling CWRU bearing dataset is presented in Table 2.7, which shows that there are four different types of electric machine health states, including one normal state and three fault states. The

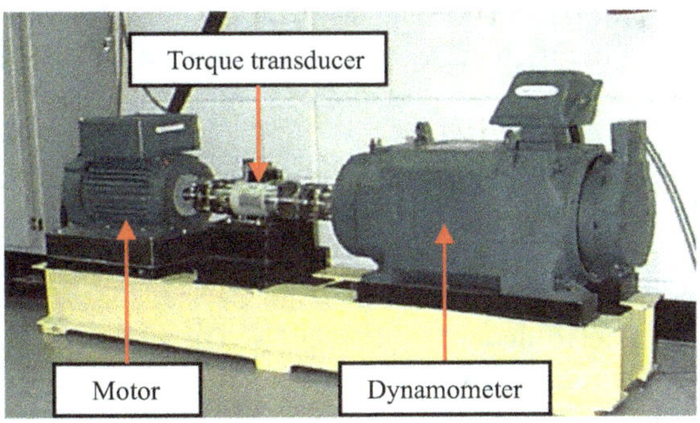

Figure 2.13 Illustration of the test stand from CWRU [26]

Table 2.7 Detailed description of the selected dataset

Load (hp)	Health state	Descriptions
0 to 3	Normal	Normal working state
	IR	Faults ranging from 0.007 inches in diameter to 0.021 inches in diameter at the inner raceway
	RE	Faults ranging from 0.007 inches in diameter to 0.021 inches in diameter at the rolling element
	OR	Faults ranging from 0.007 inches in diameter to 0.021 inches in diameter at the outer raceway

Table 2.8 Relationship between motor load and motor speed

Load (hp)	Motor speed (rpm)
0	1,797
1	1,772
2	1,750
3	1,730

diameters of the faults range from 0.007 inches to 0.021 inches. In addition, four different load sizes are considered, and the rotational speed of the motor is related to its load.

The detailed relationship between the motor load and motor speed is shown in Table 2.8. Therefore, there are 10 different types of health states under each load size and a total of 40 (10×4) sets of samples for model training and testing.

Effectiveness of the proposed model under varying working conditions with a limited number of samples

As mentioned in this section, fault detection under various working conditions can be regarded as different tasks, and each task is a multiclass classification problem. Each dataset obtained in different health states is divided into 100 subsamples by a sub-sampling window. Detailed descriptions of each task are presented in Table 2.9.

In summary, 1,000 subsamples of each task are used for model learning. T_0, T_1, T_2, and T_3 indicate that the dataset of each task is obtained from the motor with a load of 0, 1, 2, and 3 hp, respectively. In this case, the training of the meta-learning-based models is completed under the datasets T_0, T_1, and T_3, and the model testing is completed under the dataset T_2. The proposed strategy is trained for 400 epochs to obtain a promising fault-detection performance. To enable the classifier of the learner to diagnose the health state of the electric machine more accurately, the learning rate of the learner is set to 0.01. Moreover, the meta-learner

Table 2.9 *Detailed descriptions of each task (T) using CWRU dataset*

Task	Load (hp)	Number of health states	Sample size	Number of samples
T_0	0	10	1,500	100*10
T_1	1	10	1,500	100*10
T_2	2	10	1,500	100*10
T_3	3	10	1,500	100*10

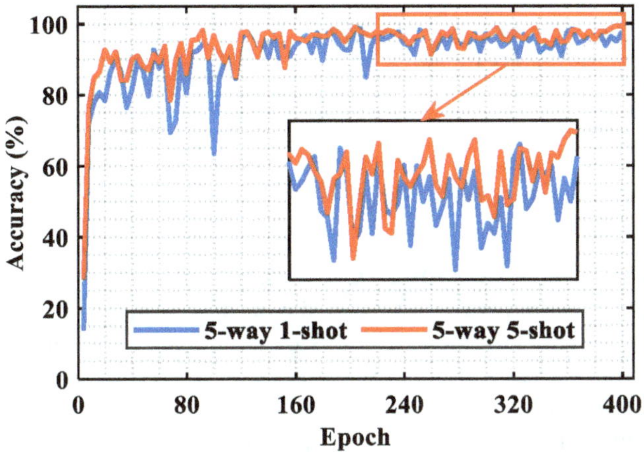

Figure 2.14 *Illustration of fault detection accuracy during the model training process under 5-way 1-shot and 5-way 5-shot conditions [26]*

learning rate is set to 0.1, which allows the model to quickly adapt to various unseen tasks. In contrast to that of meta-learning methods, the training dataset of traditional DL methods is a combination of T_0, T_1, and T_3 datasets.

The fault detection accuracy curves during the model training process under 5-way (randomly sampled 5 out of 10 types of electric machine health states from a task) 1-shot and 5-shot conditions are shown in Figure 2.14. It can be observed that with 400 epochs of iterations, the proposed strategy can achieve promising performance both in 1-shot and 5-shot situations. The results of the training process demonstrate that the proposed electric machine fault diagnosis strategy is effective. Moreover, in contrast to traditional DL models, the accuracy curve of the proposed model produces oscillations during the training process. This is because each epoch of the proposed strategy is trained under different tasks.

Figure 2.15 shows the performance of the model with different steps of adaptation (parameter updating) under 5-way 1-shot and 5-way 5-shot conditions. It can be observed that the fault detection accuracy improves with an increasing number of parameter adaptations in both 1-shot and 5-shot conditions. Moreover, with only five steps of parameter updating, the proposed model can achieve an accuracy of 98.22 ± 0.18% and 98.62 ± 0.14% with respect to 1-shot and 5-shot, respectively. The experimental results show that the proposed strategy is

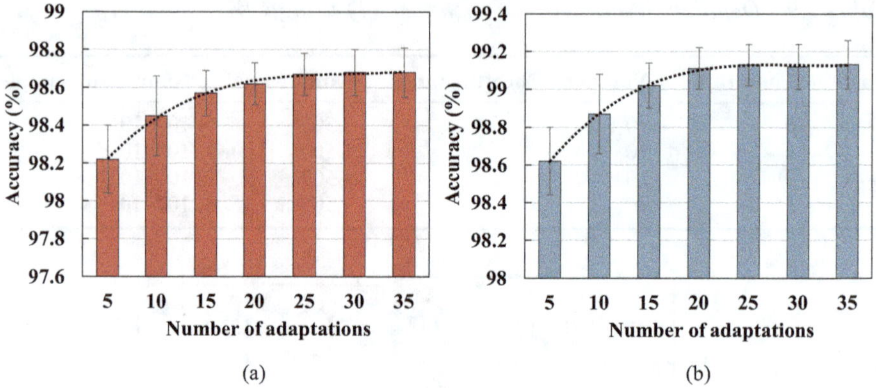

Figure 2.15 Performance of the proposed model with 5 to 35 steps of adaptations under 5-way (a) 1-shot and (b) 5-shot conditions [26]

feasible for fault detection with a small number of samples, as well as under changing working conditions. In addition, the proposed strategy can obtain the best performance with approximately 25 steps of parameter fine-tuning in the 1-shot condition. Moreover, as the number of samples in each training task increases, the performance of the proposed strategy also improves. Figure 2.15 demonstrates that with only 15 adaptation steps, the accuracy of the 5-shot condition rises to greater than 99%. With approximately 20 steps of parameter updating, the proposed model nearly achieves the optimum performance.

The performance of the method proposed in this work and other methods is listed in Table 2.10. The proposed strategy demonstrated a promising performance, achieving $98.68 \pm 0.12\%$, $99.13 \pm 0.11\%$, and $99.38 \pm 0.08\%$ accuracy with respect to 1-shot, 3-shot, and 5-shot conditions, respectively. The results demonstrate that the proposed strategy has the best performance, except that the accuracy of the proposed strategy is slightly lower than that of the WDCNN-few-shot method under the 1-shot condition. In addition, it can be observed that the fault detection accuracy of the three traditional DL-based methods applied for comparison is considerably lower than that of the meta-learning-based methods. This is because the training of the traditional DL method requires a large amount of data, and the limited amount of data used here will cause overfitting. Moreover, the training and testing of these models are carried out on datasets with different loads, and because of the weak generalization ability of these traditional DL models, changes in working conditions will also affect their performance. The experimental results show that meta-learning-based methods can adapt to these problems well.

Effectiveness of the proposed model under various noise levels

The vibration signal measured by CWRU is not a pure fault source signal, as it contains a certain level of noise. However, compared with the actual applications, the signal-to-noise ratio (SNR) of the data obtained in a reasonable laboratory environment is relatively high. In the real world, in the process of collecting data, because of the

Table 2.10 Performance of various fault detection methods

Method	**Fault detection accuracy (%)**		
Meta-learning-based	**10-shot**	**5-shot**	**1-shot**
Proposed	99.78 ± 0.08	99.13 ± 0.11	98.68 ± 0.12
PN	99.11 ± 0.25	98.96 ± 0.21	96.78 ± 0.19
WDCNN-Few-shot	99.25 ± 0.18	98.88 ± 0.13	98.71 ± 0.22
Traditional DL-based	**Fault detection accuracy with all dataset (%)**		
Cap-Net	96.97 ± 0.71		
Pre-trained	97.82 ± 0.47		
WDCNN	96.75 ± 0.56		

Table 2.11 Relationship between SNR and signal-to-noise amplitude ratio

SNR (dB)	**−6**	**−4**	**−2**	**0**	**2**	**4**	**6**
$\frac{A_{signal}}{A_{noise}}$	0.5	0.63	0.79	1.0	1.26	1.58	2.0

influence of different working environments, human intervention, equipment defects, and many other factors, the electric machine data will contain different levels of noise under different conditions. In the real world, the real noise is a sum of a series of independent random variables with different probability distributions [33]. According to the central limit theorem [34], the normalized sum of these random variables approaches a Gaussian distribution with an increase in the number of noise sources. Therefore, to better simulate the unknown real noise and explore the influence of noise on fault diagnosis, white Gaussian noise (WGN) with various SNR levels is added to the vibration signal of the CWRU dataset. The SNR (dB) is defined as follows:

$$SNR_{dB} = 10\log_{10}\left(\frac{P_{signal}}{P_{noise}}\right) = 20\log_{10}\left(\frac{A_{signal}}{A_{noise}}\right), \tag{2.28}$$

where P_{signal} and P_{noise} represent the effective powers of the signal and noise, respectively; and A_{signal} and A_{noise} are the average amplitudes of the signal and noise, respectively. Table 2.11 shows the relationship between the SNR and $\frac{A_{signal}}{A_{noise}}$.

As shown in Table 2.11, WGN with an SNR of -6 dB to 6 dB is added to the vibration signal. To show the noisy signals with various SNRs more intuitively, taking the data with 0.007 inches in diameter to 0.021 inches in diameter IR faults under a 0 hp load as an example, Figure 2.16 shows the bearing signals without noise and with various levels of noise. As shown in Table 2.11 and Figure 2.16, when the SNR is 4 dB, the amplitude ratio of the signal and noise is 1.58. In this condition, the bearing vibration signal can maintain the same shape without noise. However, with a decrease in SNR, it is difficult to see the approximate shape of the original signal at the time the domain diagrams and the low-amplitude vibration signal are submerged into the noise. Therefore, it is essential to diagnose faults accurately under various levels of noise.

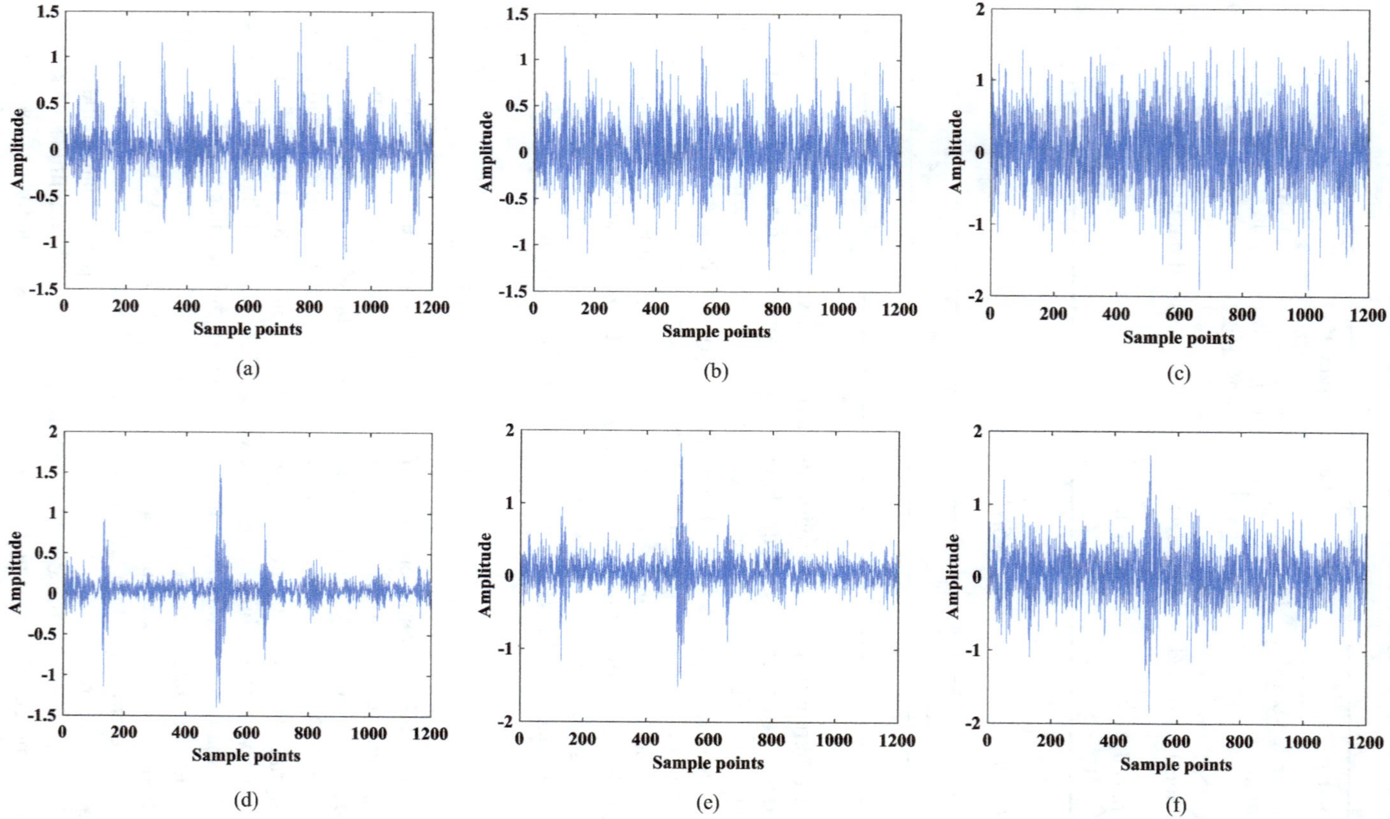

Figure 2.16 *Time domain of signals without noise and with various levels of noise and faults. IR: inner raceway; WGN: white Gaussian noise. (a) 0.007 inches in diameter IR fault, without WGN; (b) 0.007 inches in diameter IR fault, with SNR = 4 dB WGN; (c) 0.007 inches in diameter IR fault, with SNR = 4 dB WGN; (d) 0.014 inches in diameter IR fault, without WGN; (e) 0.014 inches in diameter IR fault, with SNR = 4 dB WGN; and (f) 0.014 inches in diameter IR fault, with SNR = 4 dB WGN [26].*

To explore the influence of various levels of noise on fault detection performance, data with no load is selected for model training and testing. Fault detection under various WGN levels can also be regarded as a different task. Therefore, there are, in total, seven different tasks. Detailed descriptions of each task are provided in Table 2.12. Each task can be regarded as a multiclass classification task. To ensure that the proposed model has a strong generalization ability under various noise environments, it is trained for 500 epochs under different tasks with various levels of noise. In addition, the learning rates of the learner and meta-learner are set to 0.01 and 0.1, respectively.

To verify the performance of the proposed strategy, the models were trained using the original signal without additional noise and then tested using data from various tasks (T_4 to T_{10}). For meta-learning-based methods, the experiment was completed under the 5-way 5-shot condition. The WGN causes a significant deviation in the accuracy and variance among the various fault detection models. Average accuracy was used as the performance evaluation index. Table 2.13 lists the performances of the various methods without parameter fine-tuning.

Figure 2.17 illustrates the maximum performance of the three meta-learning-based methods and the number of parameter adaptation steps required when the model is adapted to different noise environments. As shown in Table 2.13, although the accuracy of the proposed model under the 6 dB WGN condition can reach 98.82%, with a

Table 2.12 Detailed descriptions of each task under CWRU dataset

Task	**Load (hp)**	**Noise level (SNR/dB)**	**Number of health states**	**Sample size**	**Number of samples**
T_4	0	−6	10	1,500	100*10
T_5	0	−4	10	1,500	100*10
T_6	0	−2	10	1,500	100*10
T_7	0	0	10	1,500	100*10
T_8	0	2	10	1,500	100*10
T_9	0	4	10	1,500	100*10
T_{10}	0	6	10	1,500	100*10

Table 2.13 Performance of various methods under various levels of noise

Method	**Accuracy (%)**						
	SNR = −6 dB	**SNR = −4 dB**	**SNR = −2 dB**	**SNR = 0 dB**	**SNR = 2 dB**	**SNR = 4 dB**	**SNR = 6 dB**
Proposed	31.86	40.93	56.51	73.42	87.16	94.73	98.82
PN [30]	29.53	37.28	58.72	71.76	85.17	89.22	97.63
WDCNN-Few-shot [32]	28.19	38.64	53.95	72.38	83.93	92.72	97.52
Cap-Net [15]	29.93	39.95	52.37	71.09	84.57	92.04	97.66
Pre-trained [29]	31.02	37.87	55.74	70.44	86.64	88.91	95.47
WDCNN [31]	30.28	38.79	57.08	69.23	84.94	90.79	96.89

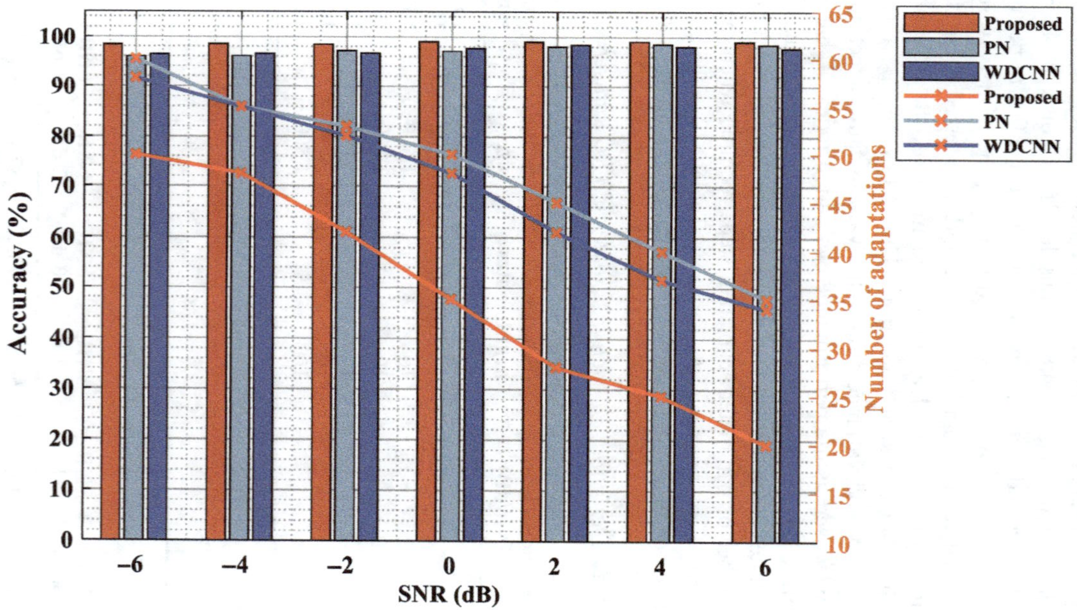

Figure 2.17 Performance of different models under different noise conditions (5-way 5-shot) [26]

decreasing SNR value, the accuracy of these models will decrease. This is because as the SNR decreases, the signal contains more noise, making it more difficult to accurately detect the corresponding health state of the data. Figure 2.17 shows that, compared with the results in Table 2.13, the performance of the meta-learning-based methods is significantly improved after parameter adaptation. With an increase in the SNR, the number of parameter fine-tuning steps required for the various models decreases. Moreover, the strategy proposed in this study can achieve better performance with fewer adaptation steps, which demonstrates that when the model needs to adapt to a new environment, the proposed method can reduce computational costs.

Theoretically, if part of the noisy data is used for training, the model will require fewer steps to adapt to other noisy environments. In this case, the model training is carried out using the datasets of T_4, T_6, T_8, and T_{10}, and the model testing is carried out using the datasets of T_5, T_7, and T_9.

Figure 2.18 shows the relationship between fault detection accuracy and the steps of the parameter adaptations. It can be observed that, regardless of whether it

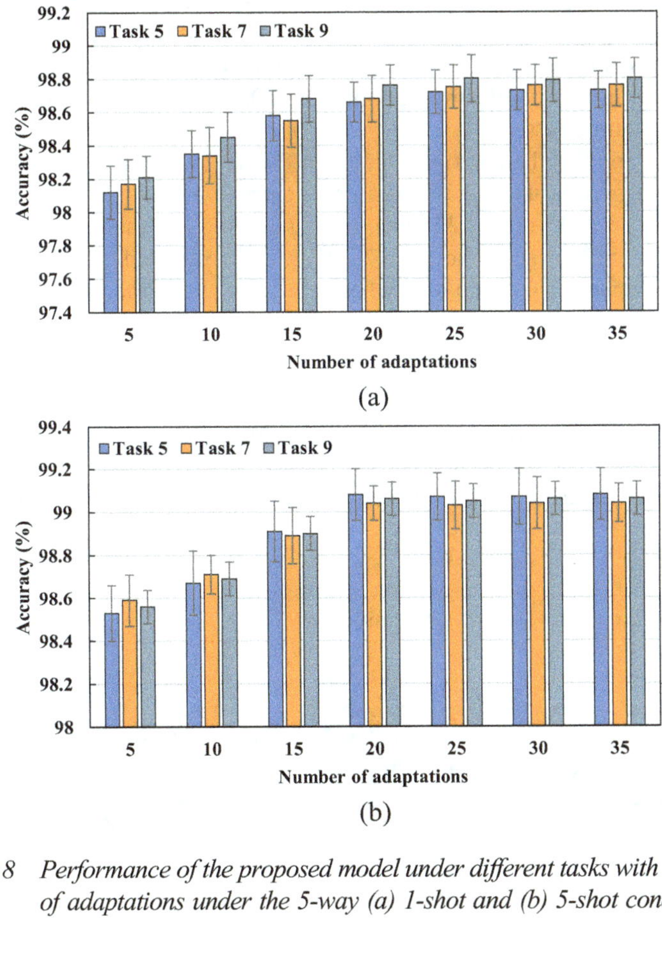

Figure 2.18 Performance of the proposed model under different tasks with 5 to 35 steps of adaptations under the 5-way (a) 1-shot and (b) 5-shot condition [26]

is 1-shot or 5-shot conditions, as the number of parameter fine-tunings increases, the fault detection accuracy will also increase under various noise environments (different tasks). In contrast to the traditional application of meta-learning in fault diagnosis, which is usually trained under signal working conditions, the strategy proposed in this study considers training under different tasks simultaneously. Therefore, the model has stronger adaptability. Examining the results in Table 2.13, the accuracy of the proposed method can reach at least 98.6% in various noise environments. The results demonstrate that the proposed model can adapt better to fault diagnosis in a noisy environment. Moreover, when an unseen working condition appears, such as T_5, T_7, or T_9, the model only requires a few adaptations to achieve a promising performance, which also reduces the computational costs. For instance, compared with the results in Figure 2.17, the proposed model requires only approximately 20 steps to achieve the best performance in T_9 (SNR = 4 dB).

2.3.2.2 Case 2: Paderborn University Rolling Bearing Dataset

Dataset description

The Rolling Bearing Dataset of Paderborn University consists of undamaged and damaged bearings. The damage data consists of two groups: artificial and natural damage. Artificial damage is caused by three methods: electric discharge machining (EDM), drilling, and electric engraving (EN). Natural damage is generated by accelerated lifetime testing.

An illustration of the test rig is shown in Figure 2.19, where it can be observed that the test rig primarily consists of a drive electric motor (Hanning synchronous motor: 425 W, 3,000 r/min, 1.35 Nm), rolling bearing test module, and load motor (Siemens synchronous servomotor: 1.7 kW, 3,000 r/min, and 6 Nm). In addition, a torque measurement shaft and a flywheel are installed in the test rig. Finally, the motor current and vibration signals of 32 different health states (26 damaged bearings and six undamaged bearings) under four different working conditions are collected. The four different working conditions are shown in Table 2.14.

Effectiveness of the proposed model under artificial damage signal

In this study, the vibration signal caused by artificial EDM damage is selected for model learning. Bearing health state detection under different operation parameter

Figure 2.19 Illustration of test rig from Paderborn University [26]

Table 2.14 Parameters of various working conditions

No.	Rotational speed [rpm]	Load torque [Nm]	Radial force [N]
0	1,500	0.7	1,000
1	900	0.7	1,000
2	1,500	0.1	1,000
3	1,500	0.7	400

Table 2.15 Descriptions of the selected artificial damage signal for model learning

Health state	Operation parameters setup (No.)	Descriptions
UD	0, 1, 2, 3	Undamaged bearing
IR	0, 1, 2, 3	Rolling bearing with inner ring artificial damage caused by EDM
OR	0, 1, 2, 3	Rolling bearing with outer ring artificial damage caused by EDM

Table 2.16 The detailed descriptions of each task under Paderborn University Dataset

Task	Operation parameters setup (No.)	Number of health number of health states	Sample size	Number of samples
T_1^a	0	3	2,500	500*3
T_2^a	1	3	2,500	500*3
T_3^a	2	3	2,500	500*3
T_4^a	3	3	2,500	500*3

setups is regarded as a different multiclass classification task. The objective of each task is to diagnose three different types of health states of bearings: undamaged (UD), inner ring (IR), and outer ring (OR) faults. The description of the selected artificial damage-bearing signal for model learning is shown in Table 2.15.

Each dataset under different working conditions with different health states is resampled into 500 subsamples using a sub-sampling window, and each subsample contains 2,500 sample points. Therefore, 1.500 subsamples need to be detected for each task.

Detailed descriptions of each task are given in Table 2.16, which shows that there are four different tasks and each task can be regarded as a 3-way n-shot problem. To verify the effectiveness of the proposed model under various working conditions, a series of cross-validations have been conducted. Three tasks are used for model training, and the remaining task is used for model testing. As a result,

Table 2.17 Detailed descriptions of different task combinations

Combination	Model training tasks	Model testing task
0	T_2^a, T_3^a and T_4^a	T_1^a
1	T_1^a, T_3^a and T_4^a	T_2^a
2	T_1^a, T_2^a and T_4^a	T_3^a
3	T_1^a, T_2^a and T_3^a	T_4^a

four different task combinations are generated. The descriptions of the different task combinations are shown in Table 2.17.

Taking each task in the 3-way 10-shot condition as an example, Figure 2.20 shows the confusion matrices of the proposed model for different task combinations. The bottom row represents the recall rate of each bearing health state, and the right column represents the precision rate of each health state. The figure shows that the proposed model can accurately detect the rolling bearing health states under various working conditions. In the four different task combination cases, the proposed model has promising performance, and the diagnosis accuracy is at least 99%. In addition, it can be observed that the precision rate and recall rate of the UD category are both 100% under four different conditions. This means that despite the diagnosis error of the proposed model primarily coming from the misjudgment of IR and OR faults, the model can still detect whether the bearing is faulty. Specifically, the proposed model will not misjudge healthy bearings and faulty bearings, which ensures that the model will not provide incorrect information when the bearing is in a healthy state in practical applications.

Additionally, Table 2.18 shows the health state diagnosis accuracy of the different methods under a 3-way 10-shot condition. The results show that the proposed method can achieve the best performance among the three methods reported here.

Effectiveness of the proposed model under natural damage signal

The effectiveness and applicability of the proposed model in the real world are also studied. The vibration signal generated by the accelerated lifetime tests is used for model learning. Figure 2.21 illustrates the apparatus used for accelerated lifetime testing and the bearings with real damages. The accelerated lifetime test rig primarily contains an electric motor (Hanning Elektro-Werke: 2.2 kW, 2,841 r/min) and a bearing housing. The real damaged bearings also contain IR and OR faults. Similar to the previous section, the bearing data with real damage under four different operation parameter conditions is selected to verify the performance of the proposed model; the specific descriptions of the natural damage data can be found in [28]. Four different task combinations are taken into account, and the tasks for natural damage fault diagnosis are represented as T_i^n ($i = 1, \ldots, 4$; n represents natural).

Figure 2.22 shows the accuracy of various methods in 3-way 1-shot, 3-shot, 5-shot, 7-shot, and 10-shot conditions, respectively. The proposed model outperforms all of the comparison methods. Under four different task combination setups, the

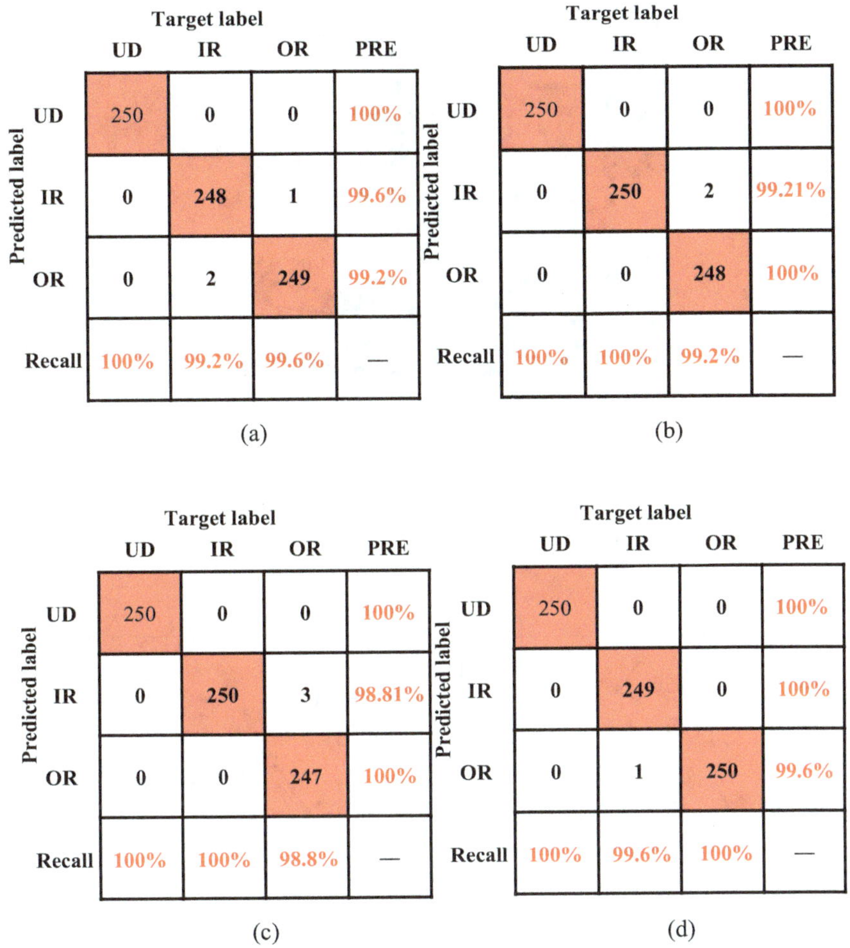

Figure 2.20 Confusion matrices of the proposed model for different task combination cases: (a) Combination 0; (b) Combination 1; (c) Combination 2; and (d) Combination 3 [26]

Table 2.18 Performance of various methods in various task combination cases

Method	**Accuracy (%) of the proposed model under different task combinations**			
	0	**1**	**2**	**3**
Proposed	99.60	99.73	99.60	99.87
PN [30]	98.53	99.33	99.47	99.53
WDCNN-Few-shot [32]	98.67	99.20	99.07	99.47

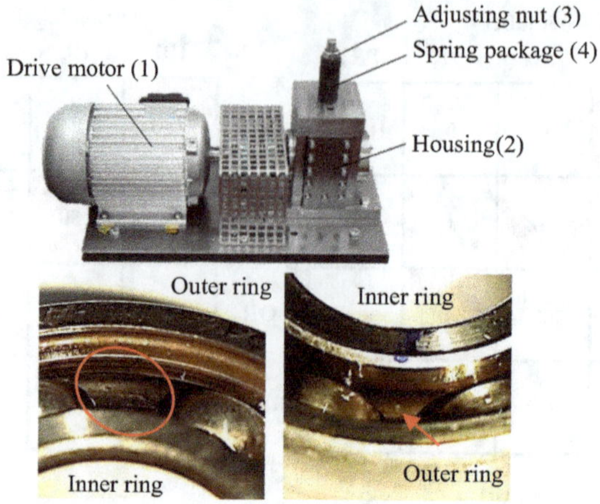

Figure 2.21 Equipment setup for accelerated lifetime testing and bearings with real damages [26]

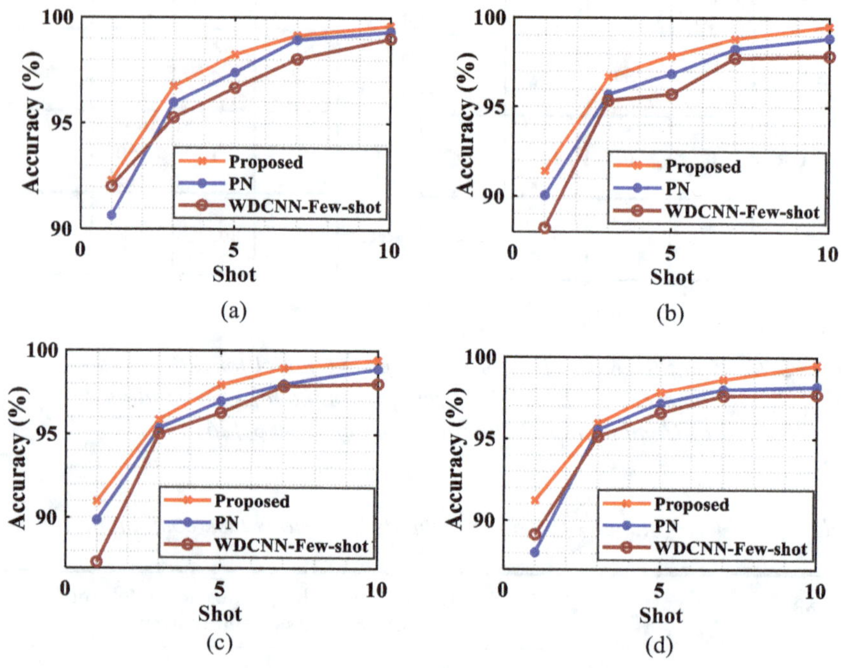

Figure 2.22 Health state detection accuracy of various methods under 3-way 1-shot, 5-shot, and 10-shot conditions: (a) Combination 0; (b) Combination 1; (c) Combination 2; and (d) Combination 3 [26]

diagnosis accuracy rates of the proposed model are greater than 99% in the 3-way 10-shot scenario, which demonstrates that the method can detect the health states of rolling bearings in the real world.

Effectiveness of the proposed model under combination damages

In the field of bearing fault diagnosis research, operation data is usually collected from bearings after artificial damage. In this chapter, the capability of the proposed model to generalize fault diagnosis from artificial damage data to real natural damage data is explored. The vibration signal caused by natural damage for model training and the natural damage signal for model testing is used.

Taking 3-way 10-shot tasks as an example, the tasks (T_1^a to T_4^a) mentioned in Section 2 for model training, and then use the tasks (T_1^n to T_4^n) mentioned in Section 3 to verify the performance of the proposed model. Figure 2.23 shows the relationship between the fault detection accuracy and the number of parameter fine-tuning steps required for various methods.

Compared with other methods, after approximately 20 steps of fine-tuning, the proposed model first achieves the best performance with respect to four unseen tasks, which demonstrates the faster adaptation ability of the proposed model. In

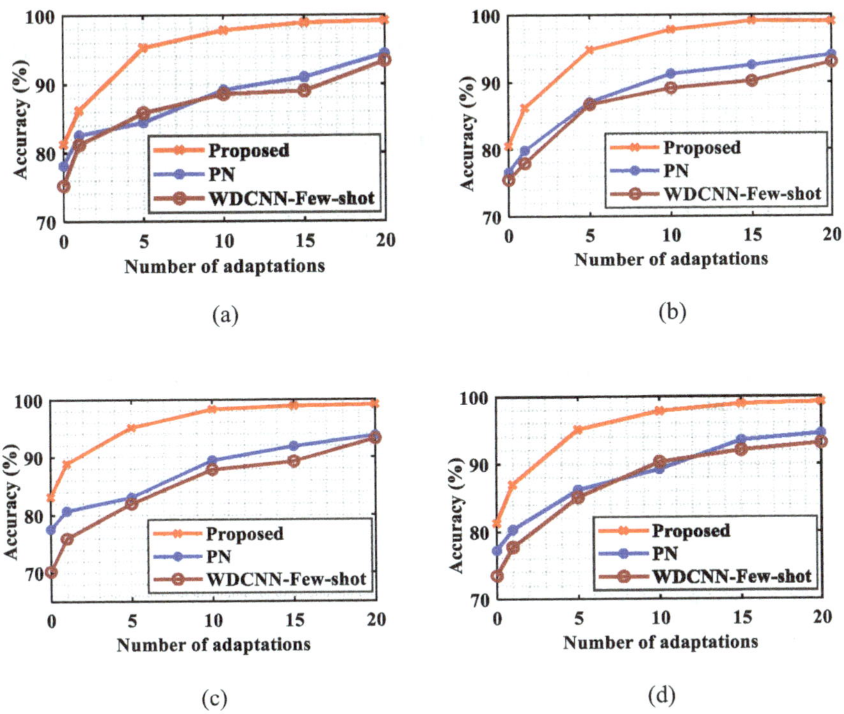

Figure 2.23 Relationship between the fault detection accuracy and the number of parameter fine-tuning steps for various methods in the 3-way 10-shot condition: (a) T_1^n, (b) T_2^n, (c) T_3^n, and (d) T_4^n [26]

addition, the proposed model outperforms other methods under each test task in terms of detection accuracy, which verifies its stronger generalization ability. The experimental results show that it is feasible to make a preliminary exploration of bearing fault diagnosis in the laboratory using artificial damage, and then generalize the model to real-world applications.

In summary, according to the results in this section, the proposed model can detect the bearing health states using a limited number of samples under various working conditions more accurately and effectively. In addition, from the results in Case 1 and Case 2, it can be concluded that the proposed model can effectively address the influence of various noises on the fault diagnosis performance. The results can also prove that the proposed model has potential for practical applications.

2.4 Gaussian-process-enabled method for electric machine fault detection

In this section, a deep Gaussian process (GP) kernel transfer-based few-shot learning method (RNGPT) is proposed for the fault detection of electric machines. First, a deep residual network (ResNet) is used to extract the features of the raw data. Then, the encoded latent feature vector is fed into the GP with kernel transfer ability to make the motor fault detection and classification. The proposed method uses much less data than the traditional AI-based method to achieve fault diagnosis under variable working conditions and does not cause an overfitting problem. Experimental results of two case studies demonstrate that the proposed RNGPT model can accurately and effectively detect motor faults with limited labeled data under different working conditions.

2.4.1 Methodology and proposed framework

2.4.1.1 Few-shot learning

Traditional machine learning technology needs large-scale data to train a model. For example, in a study on the classification of MNIST, there are 10 classes and 6,000 samples in the training set. On average, there are about 600 samples per class. In some practical tasks, machine learning methods face the case of a "small dataset." For example, image data for some rare species are difficult to collect with very few labeled images. In this case, the neural network model needs to be trained based on a small number of sampling dataset. Therefore, it is important to develop a reliable model for a task with only a few labeled training samples. Few-shot learning aims to train a model to identify unseen targets during training with limited labeled examples [35], which could address the previous problem. Few-shot learning is the application of meta-learning in supervised learning, in the case of small sample classification, which is defined as a task with N-class and K-sample (N-way K-shot). In the training phase of few-shot learning, there are many different classification tasks where the purpose is to train a model that can be transferred to a new class with only a few labeled samples.

Let $D_s = \{(x_k, y_k)\}_{k=1}^{K}$ (seen class) be a support set (referred to in conventional deep learning as a training set), which contains input-output pairs. When k is equal to one, this method is called one-shot learning. Let $D_q = \{(x_t, y_t)\}_{t=1}^{T}$ (unseen class) be a query set (referred to in conventional deep learning as a test set), typically T is one order of magnitude greater than K. Let $T = \{D_s, D_q\}$ denote the group of the support and query sets in a training task, $C = \{T_m\}_{m=1}^{M}$ denote a collection of all these tasks. Few-shot learning models are trained on a series of random tasks which are sampled from collection C. In the test task, a new task $T_* = \{D_s^*, D_q^*\}$ is sampled from a test set (which differs from a conventional deep learning test set) for model performance validation. The purpose of the test task is to fine-tune the model based on the support set D_s^*, and then make category predictions on the samples of the query set D_q^*. For a more intuitive comprehension of the difference between few-shot learning and traditional machine learning, a visualized example is presented in Figure 2.24.

Figure 2.24 Difference between the traditional machine learning method and the few-shot learning method. (a) Traditional machine learning method has only one task for model training. (b) Few-shot learning has multiple classification tasks for model training [36].

2.4.1.2 Architecture of the proposed network

The framework of the proposed method is introduced in this section. To establish a strong generalization model for electric machine intelligent fault detection with few signal samples, inspired by the few-shot learning, a GP-based kernel transfer method is applied. The overall framework of the proposed method is illustrated in Figure 2.25. Each task can be regarded as a fault diagnosis task under a specific working condition. It can be observed in Figure 2.25 that a sub-sampling window containing a series of raw data points is collected to be one sample, and there is a sampling interval between each sub-sampling window. A dataset for model training, validation, and testing is a combination of all these samples with different EM health states. The representation layer extracts the discriminative features of the input data (three-phase current and vibration signal). Then the encoded feature maps are fed into a decision-making layer to perform the electric machine health state classification. Finally, the trained base model, which is called the feature fusion model, will be transferred to different electric machine working conditions data for fault diagnosis. The difference between conventional transfer learning methods is that the proposed method uses an identical kernel for all fault diagnosis tasks, which means that the fault diagnosis tasks under different working conditions

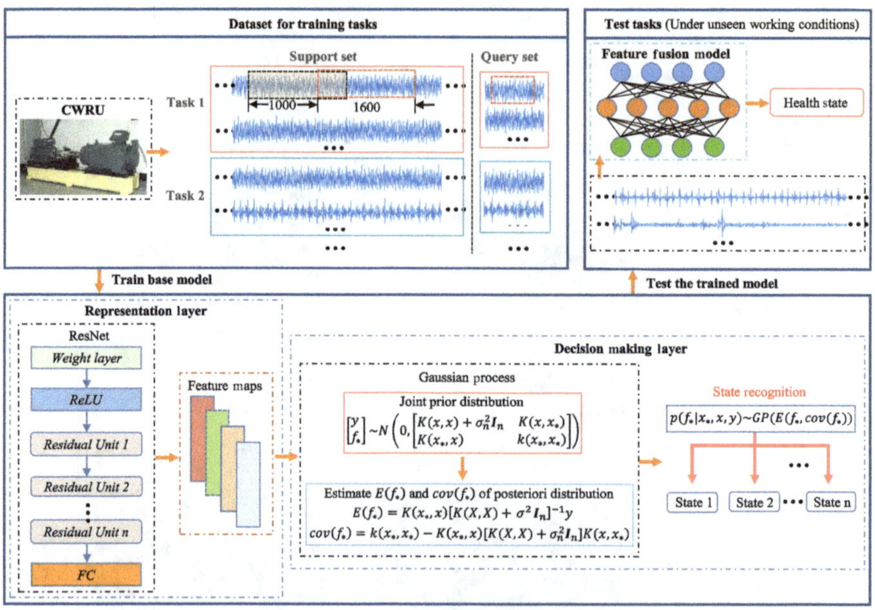

Figure 2.25 The overall framework of the proposed method for intelligent fault detection in electric machines. The few-shot learning-based model uses the current and vibration signal as inputs, extracts the discriminative features, and calculates the GP joint distribution to recognize the electric machine state. CWRU: Case Western Reserve University: FC: Fully Connected [36].

can share a set of "multi-function" hyperparameters. Therefore, there is no additional encoding module required.

Representation layer

Since the characteristics of electric machine signals are different under different fault types, it is reasonable to detect the faults by analyzing the input inner features. The discriminative features extracted from a raw signal play a crucial role in improving the GP joint distribution approximation, which has a great influence on the fault detection ability of the model. A deep neural network is known for its powerful feature extraction ability [37]; it is a machine learning model composed of several stacked nonlinear processing layers and has made bright achievements in the field of smart grid applications [38]. However, the issue of gradient explosion or gradient vanishing will arise as neural network depth grows. This will make it more challenging for the solver to find the best solution and raise the training error of the neural network model. The advantage of deep ResNet is that it cannot only increase the learning efficiency of the deeper network but also increase the accuracy of the model. Moreover, the residual network can avoid the gradient vanishing or gradient explosion problem described previously with the increase of network layers, which makes it possible to develop a very deep neural network. A deep ResNet is applied to extract the raw data features. These encoded features will boost the fault diagnosis performance of the decision-making layer.

In the representation layer of Figure 2.25, a conventional neural network-based ResNet with 10 stacked residual units (ResNet-10) is applied to process the electric machine signals. The difference between a common network and a residual unit is shown in Figure 2.26.

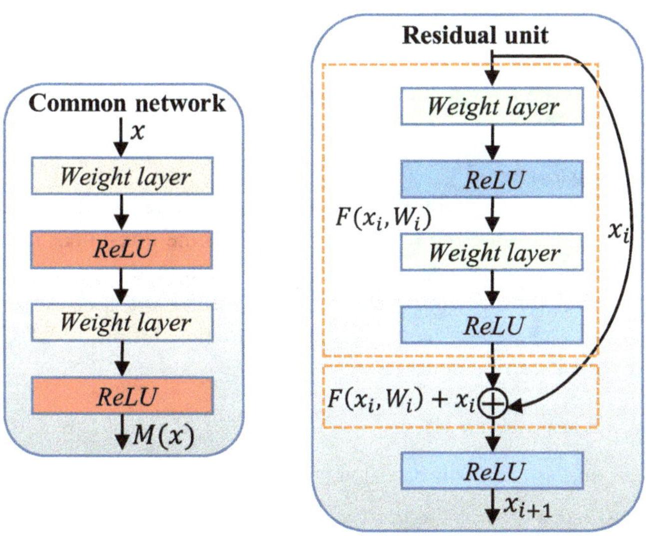

Figure 2.26 The difference between a common neural network and a residual unit. Left: A common neural network; Right: A residual unit of ResNet [36].

It is shown in Figure 2.26 that the output $M(x)$ of a common neural network is a desired mapping of input x, which can be described as follows:

$$M(x) = f(G(x, W)), \tag{2.29}$$

where $G(x, W)$ is a mapping function about input x and weight matrix $W, f(\cdot)$ is the activation function $ReLU$. Assuming that x_i is the input of the i-th residual unit, the output can be described as follows:

$$x_{i+1} = f(F(x_i, W_i) + x_i), \tag{2.30}$$

where the $F(x_i, W_i)$ is the residual function mapping, W_i is the corresponding weight matrix. It should be noted that the dimensions of x_i and $F(x_i, W_i)$ should be equal, otherwise, it could then perform a linear mapping W_s to match the two dimensions

$$x_{i+1} = f(F(x_i, W_i) + W_s x_i), \tag{2.31}$$

The form of residual function $F(\cdot)$ is flexible and changeable. In addition to stacking two convolution layers, it can also stack three convolution layers.

Compared with conventional CNN fitting the underlying mapping directly, the ResNet fits a residual between two adjacent signals, which simplifies the training process of the model. Assuming the underlying feature of non-linear mapping is

$$H(x) = F(x_i, W_i) + x_i, \tag{2.32}$$

the ResNet needs to fit a function mapping of

$$F(x_i, W_i) = H(x) - x_i, \tag{2.33}$$

The results show that the model applying ResNet is easier to train and has better performance. The experimental results will be presented in the next section. After processed by the representation layer, the raw data features will be passed to the decision-making layer for further processing.

Decision-making layer

The decision-making layer is a GP-based electric machine health state classifier. It is shown in Figure 2.25 that the input of this layer is the non-linear mapping based on Eq. (2.32): $H(x) \to \boldsymbol{h}$, where $\boldsymbol{h}$ is a desired feature map.

A GP can be determined by a mean function $\mu(x)$ and a covariance function (kernel function) $k(x, x')$ uniquely. According to the definition of GP, the distribution functions f can be denoted by

$$f(x) \sim GP(\mu(x), k(x, x')), \tag{2.34}$$

where the $\mu(x)$ and $k(x, x')$ can be calculated by

$$\mu(x) = E(f(x)), \tag{2.35}$$

$$k(x, x') = E[\{f(x) - m(x)\}\{f(x') - m(x')\}], \tag{2.36}$$

In general, the value of the mean function $\mu(x)$ is assumed to be 0. This is because, in the application of GP in machine learning, the data preprocessing method, zero-mean normalization is often necessary if any prior knowledge does not exist. The covariance function $k(x,x')$ can capture the relationship between different input points in the input space.

In practical applications, the Gaussian noise should also be considered:

$$\varepsilon \sim N\left(0,\sigma_n^2\right), \tag{2.37}$$

where ε is the Gaussian noise with variance σ_n^2, and it is completely independent of the $f(x)$. Based on Bayesian probability theory, by establishing a priori distribution function on a given training dataset $D=\{(x_n,y_n)\}_{n=1}^N$, the output distribution y_n with Gaussian noise can be obtained as follows:

$$y_n = f(x_n) + \varepsilon \sim N\left(0, K(X,X) + \sigma_n^2 \boldsymbol{I}_n\right), \tag{2.38}$$

where $\boldsymbol{I}_n$ is an $N \times N$ identity matrix; $K(X,X)$ is a Gram matrix:

$$K(X,X) = \left(K_{ij}\right)_{n\times n} = k\left(x_i,x_j\right), i,j = 1,2,\cdots n, \tag{2.39}$$

The joint prior distribution of observations and predictions can be obtained as follows:

$$\begin{bmatrix} y \\ f_* \end{bmatrix} \sim N\left(0, \begin{bmatrix} K(x,x) + \sigma_n^2 \boldsymbol{I}_n & K(x,x_*) \\ K(x_*,x) & k(x_*,x_*) \end{bmatrix}\right), \tag{2.40}$$

where f_* is the predicted value; x_* is a testing point, which has the same Gaussian distribution with the input points x in training set D. $K(x,x_*) = K(x_*,x)^T$ is the $n \times 1$ covariance matrix between test point x_* and input points x. $k(x_*,x_*)$ is the covariance of the point x_*. According to the previous joint prior distribution, the posteriori distribution probability of f_* can be obtained by Bayes' rule as follows:

$$p(f_*|x_*,x,y) \sim GP(E(f_*), \mathrm{cov}(f_*)), \tag{2.41}$$

where the mean $E(f_*)$ and covariance $\mathrm{cov}(f_*)$ of the prediction point f_* can be obtained as follows:

$$E(f_*) = K(x_*,x)\left[K(X,X) + \sigma_n^2 \boldsymbol{I}_n\right]^{-1} y, \tag{2.42}$$

$$\mathrm{cov}(f_*) = k(x_*,x_*) - K(x_*,x)\left[K(X,X) + \sigma_n^2 \boldsymbol{I}_n\right] K(x,x_*), \tag{2.43}$$

The mean $E(f_*)$ can be seen as the estimation value of the prediction point f_* and the $\mathrm{cov}(f_*)$ can reflect the reliability of this value.

The decision-making layer first calculates the joint prior distribution of the underlying features of the training and test sets, the specific calculation process is based on Eq. (2.40). After obtaining the prior distribution, the proposed model calculates the probability corresponding to each electric machine health state according to Eq. (2.41). The specific parameters in Eq. (2.41) can be calculated by combining Eqs. (2.42) and (2.45). The state that corresponds to the highest probability is the motor state predicted by the model.

2.4.1.3 Training of the proposed model

The GP hyperparameters of the proposed model are determined by the maximum likelihood estimation method. In addition, the GP sample function of the model is disconnected from the hyperparameters, which need to be determined. So, the proposed model simply focuses on exploring the hyperparameters of a GP kernel that has the maximum marginal likelihood (marginalizing out the GP sample function). With this approach, it means that it is reasonable to calculate the posteriori distribution probability $P(y_t^*|x, \theta, \varnothing)$ directly and skip a complicated intermediate reasoning step. Let y_t^* denote the target data, D_x denote the input data (a collection of support sets and query sets). The marginal likelihood function $L(\theta, \varnothing)$ can be described as

$$L(\theta, \varnothing) = \log P(y_t^*|D_x, \theta, \varnothing) = \frac{1}{2}(y_t^*)^T C^{-1} y_t^* + \frac{1}{2}\log|C| + \alpha, \tag{2.44}$$

where α is a constant. C is the kernel between all task inputs and this is related to the estimation of hyperparameter θ and neural network weight $\varnothing$. The proposed model is optimized by maximizing the marginal likelihood function.

Algorithm 4 shows the training and test process of the proposed faults detection model.

Algorithm 4: Training and test procedures of the proposed model

---

Training tasks

Input:	Current signal and vibration signal
Output:	The health state of electric machine
Setting:	Parameters of few-shot learning: n and m (n-way, m-shot)
Initialize:	Randomly initialize the ResNet-10's parameters $\varnothing$
1:	**for** epoch=1:stop_epoch **do**
2:	Randomly sample training task $T = \{D_s, D_q\} \sim C$
3:	Get x and y from D_s, get x_* from D_q, then predict signal f_* in D_q
4:	Feature extracting: $H(x) \rightarrow \boldsymbol{h}$ (Eq. (2.32))
5:	Estimate the posteriori distribution probability: $p(f_*|T, \theta, \varnothing)$ (Eq. (2.41)) $\leftarrow E(f_*)$ and $\text{cov}(f_*)$ (Eq. (2.42)) and Eq. (2.43)
6:	Calculate the loss function: $f_{loss} = -L(\theta, \varnothing)$ (Eq. (2.44))
7:	Update parameters of GP: $\theta_{t+1} = \theta_t - \beta\nabla_{\theta,}f_{loss}$
8:	Update parameters of ResNet-10: $\varnothing_{t+1} = \varnothing_t - \gamma\nabla_{\varnothing,}f_{loss}$
9:	**end for**

Test tasks

10:	Randomly sample test task $T_* = \{D_s^*, D_q^*\}$
11:	Get x and y from D_s^*, get x_* from D_q^*, then predict a clean signal f_* in D_q^*
12:	Estimate posteriori distribution probability of f_*: $p(f_*|x_*, x, y)$ (Eq. (2.41)) $\leftarrow E(f_*)$ and $\text{cov}(f_*)$ (Eq. (2.42)) and Eq. (2.43))
13:	**end**

The parameters of the ResNet-10 are denoted as $\varnothing = \{W, b\}$, where $W = \{W_i, W_1, W_2, \cdots, W_{10}\}$ represent the weight parameters of the ResNet-10, and $b = \{b_i, b_1, b_2, \cdots, b_{10}\}$ represent the bias parameters of the ResNet-10. The parameters of the GP are denoted as $\theta = \{\boldsymbol{K}, \sigma_n^2 \boldsymbol{I}_n\}$, where $\boldsymbol{K}$ represents the covariance matrix and $\sigma_n^2 \boldsymbol{I}_n$ represents the noise.

Two stages are included in Algorithm 4: In the training phase, the input of the model is motor current or vibration signal. The data are processed by a neural network and then sent to GP to perform motor state recognition. The loss function of the proposed model can be derived based on Eq. (2.44) as follows:

$$f_{loss} = -L(\theta, \varnothing) = -\log P(y_t^* | D_x, \theta, \varnothing), \tag{2.45}$$

The gradient descent method is applied to update the parameters to minimize the loss function:

$$\theta_{t+1} = \theta_t - \beta \nabla_{\theta_t} f_{loss}, \tag{2.46}$$

$$\varnothing_{t+1} = \varnothing_t - \gamma \nabla_{\varnothing_t} f_{loss}, \tag{2.47}$$

where β and γ are the learning rate, $\beta \nabla_{\theta_t} f_{loss}$ and $\gamma \nabla_{\varnothing_t} f_{loss}$ represent the gradient of the loss function.

In the test phase in Figure 2.25, the input of the model is also the motor current or vibration signal, but its working conditions may be different from those of the conditions in the training phase. In this phase, even if the input never appears in the training phase, the parameters of the fault diagnosis model are not updated anymore. This is because the model uses the identical kernel for all fault diagnosis tasks, even if the input data comes from different working conditions, the model parameters can be transferred.

2.4.2 Results analysis

In this section, the performance of the proposed model with different kernels (RNGPT-BNCosSim, RNGPT-CosSim, and RNGPT-RBF) will be evaluated in two case studies. Accuracy is used as the evaluation criteria for fault diagnosis performance. Case study I is presented in the next section, where the proposed model is applied to the simulation dataset to demonstrate its feasibility in fault diagnosis. Then the model is tested on a real-world dataset from Case Western Reserve University Bearing Data Center Website [24], and the detailed analysis is presented in the next section. Several state-of-the-art few-shot learning algorithms are also presented for comparison: MAML [25], ProtoNet [30,39], and Baseline [35]. Moreover, two conventional DL methods, WDCNN [31] and Capsule-Net [17], are also used to compare with the proposed method.

The learning rate for GP based decision-making layer is $\beta = 10^{-4}$ and for ResNet-based feature extractor is $\gamma = 10^{-3}$. All parameters are optimized and updated by an Adam optimizer. The other detailed parameters of the experiment are introduced in the following subsections. The training process of the proposed model is completed on a 64-bit computer with one Intel(R) Core(TM) i7-9800X

Table 2.19 Detailed hyperparameters of the proposed methods

Hyper-parameters	RNGPT-BNCosSim	RNGPT-CosSim	RNGPT-RBF
Learning rate	10^{-3}	10^{-3}	10^{-3}
Number of layers	10	10	10
Kind of layers	ResNet (CNN)	ResNet (CNN)	ResNet (CNN)
Epoch	400	400	400
Mini-batch	16	16	16
Optimizer	Adam	Adam	Adam
Loss function	Marginal likelihood	Marginal likelihood	Marginal likelihood
Activation function	ReLU	ReLU	ReLU
GP Kernel function	BNCosSim	CosSim	RBF

Table 2.20 Detailed parameters of asynchronous motor

Parameter	Setting value
Rated power	11 kW
Rated speed	1,458 rpm
Rated load torque	75 Nm
Rated voltage	380 V
Rated frequency	50 Hz
Efficiency and power factor	0.85 and 0.86

CPU of 3.8 GHz and one RTX2080 Ti GPU of 11 GB VRAM, besides, it also has one 32.0 GB RAM. The code of the proposed model is written in Pycharm platform with Pytorch of GPU version. The values of the detailed hyperparameters of the proposed methods are summarized in Table 2.19. All these hyperparameters are tuned and determined by the Grid Search method [23].

2.4.2.1 Case study I: simulation dataset

In this case study, the dataset used for the motor fault diagnosis experiment was obtained by simulation. The proposed model is trained to identify different motor health states with only a small number of training samples. A three-phase asynchronous induction motor is simulated under three different sizes of loads (working conditions) by ANSYS Electronics Suite software [24]. The electromagnetic performance and system can be uniquely simulated by this software. Moreover, it can also evaluate other important mechanical effects, such as temperature, vibration, etc. The detailed parameters of the motor are summarized in Table 2.20. The three-phase current signal is collected and used as the input of the proposed fault diagnosis model. Each working condition contains seven motor health states, therefore, a total of 21 (3×7) sets of data are obtained in the simulations. The descriptions of each motor health state with three different sizes of loads are shown in Table 2.21.

It can be observed from Table 2.20 that the motor with 68, 70, and 72 Nm loads is simulated, and the data with seven types of motor health states is obtained.

Table 2.21 Descriptions of motor health state

Motor load (Nm)	Health state	Description
68	Normal	Healthy motor without fault
70	BRB	There are four broken squirrel cage Bars, two adjacent, and the other two adjacent in the opposite pole direction
72	ISC	Inter-turn Short Circuit of a slot in phase A
	LM	Sudden Loss of load
	SPF	Stator Phase Failure
	RDE	Rotor Dynamic Eccentricity, the rotor rotates once, and the trajectory of the rotor center forms a circle with a radius of eccentric displacement
	RSE	Rotor Static Eccentricity, the shaft of the rotor is fixed, and the center of the trajectory of the rotor rotates at one point

The objective is to compile a dataset with limited samples for validating the model's performance. The running time of the motor simulation is 4 s, and the data sampling frequency is 1 kHz. Therefore, the length of the data under each health state is 4,000. It is shown in Figure 2.25 that a sub-sampling window containing 200 data points is collected to be one sample, with 50 data points between each window. As described above, the sample size of each health state is 77. In total, there are 539 sets of samples under each working condition for model training, validation, and testing.

Process of model training

To explore the fault diagnosis model with high accuracy and strong generalization under variable working conditions, two datasets with 68 and 70 Nm loads are selected for model training and 72 Nm loads for model final performance testing. The proposed fault detection model is trained for 400 epochs to get the highest accuracy. During the training phase, the parameters of the model with the highest accuracy in the validation set will also be collected at the same time in each epoch. Therefore, the performance of the model can be guaranteed under different working conditions. All the tasks are 5-way (five of the seven types of health states are randomly selected) with 1-shot or 5-shot (one or five health state samples are randomly selected in the support set). Otherwise, there are 12 samples included in each query set. In the representation layer, a 256-dimensional feature vector is extracted from each three-phase current signal sample. This feature map is passed to the input layer of a decision-making layer to make a 5-class classification (5-way).

Evaluation of model performance

The health states detection performances of the various methods are presented in Table 2.22.

It shows that the proposed RNGPT model has the best fault detection performance compared with other methods on the simulation dataset. RNGPT with the

Table 2.22 Performance of different models in the simulation dataset

Health states detection model	Health states detection accuracy (100%)	
	5-shot	1-shot
MAML [25]	89.88 ± 0.41	88.84 ± 1.43
ProtoNet [30,39]	90.51 ± 1.24	90.28 ± 0.70
Baseline [40]	89.60 ± 0.67	89.19 ± 1.45
RNGPT-BNCosSim	90.68 ± 0.70	90.42 ± 0.68
RNGPT-CosSim	90.35 ± 0.77	90.17 ± 0.74
RNGPT-RBF	96.10 ± 0.29	93.28 ± 0.68
Conventional method	**Health states detection accuracy**	
WDCNN [31]	80.23 ± 0.82	
Capsule-Net [17]	84.20 ± 1.76	

RBF kernel has the highest detection accuracy in 5-shot (96.10 ± 0.29%) and 1-shot (93.28 ± 0.68%). Compared with RNGPT with BNCosSim and CosSim kernel, RNGPT-RBF outperforms these two models and is at least 5% more accurate in 5-shot and 2% more accurate in 1-shot. Compared with three few-shot learning-based methods, the proposed RNGPT model can detect the health states of the motor more accurately, than the RNGPT-CosSim model. This is because the performance of the model is degraded by the CosSim kernel function. Therefore, an appropriate GP kernel can make the proposed model more effective. In addition, compared with Capsule-Net, the detection accuracy of RNGPT-RBF is about 12% higher. There are two main reasons for this result: one is the limited data for model learning, which makes Capsule-Net do overfitting; another is that when the working condition changes during the model test, the traditional DL method has a weak generalization ability, and as a consequence, the detection accuracy of this model will decrease if it does not fine-tune the model parameters.

2.4.2.2 Case study II: Case Western Reserve University Bearing Dataset

In this case study, the dataset used for the motor fault diagnosis experiment was obtained from Case Western Reserve University (CWRU) Bearing Data Center. The detailed descriptions of the fault type and diameter are shown in Table 2.23. It shows that there are three different types of motor bearing faults, including inner raceway, rolling element, and outer raceway. Each type of fault has three different sizes, which are 0.007, 0.014, and 0.021 inches. Therefore, there are a total of 10 kinds of motor bearing health states for each working condition, including nine kinds of fault states and one kind of normal state. The vibration signal is collected and used as the input of the proposed fault diagnosis model, the data sampling frequency is 12 kHz. To construct a small-scale dataset, a sub-sampling window containing 1,600 data points is collected to be one sample every 1,000 points, as shown in Fig. 2.25. In total, there are 1,000 sets of samples under each working condition for model training, validation, and testing.

Table 2.23 Detailed parameters of motor for case study II

Motor load (hp)	**Fault diameter (inches)**	**Fault type**		
1	0.007	Inner Raceway	Rolling Element	Outer Raceway
	0.014			
	0.021			
2	0.007			
	0.014			
	0.021			
3	0.007			
	0.014			
	0.021			

Process of model training

The vibration data of the drive end-bearing fault with three different types of loads (1–3 hp) are selected for model training, validation, and testing. The dataset with a 1 hp load is selected for model training, a 2 hp load for validation, and a 3 hp load for model final performance testing. The proposed fault detection model is trained for 400 epochs to get the highest accuracy. During the training phase, the parameters of the model with the highest accuracy in the validation set will also be collected at the same time in each epoch. Therefore, the performance of the model can be guaranteed under different working conditions. All the tasks are 5-way (5 of the 10 types of health states are randomly selected) with 1-shot or 5-shot (one or five health state samples are randomly selected in the support set). There are 16 samples included in each query set. In the representation layer, a 512-dimension feature vector is extracted from the vibration signal. This feature map is passed to the input layer of the decision-making layer then to make a 5-class classification (5-way).

Evaluation of model performance

The evolution of the loss function value over 400 epochs is illustrated in Figure 2.27. It shows that with the continuous optimization of the model, the value

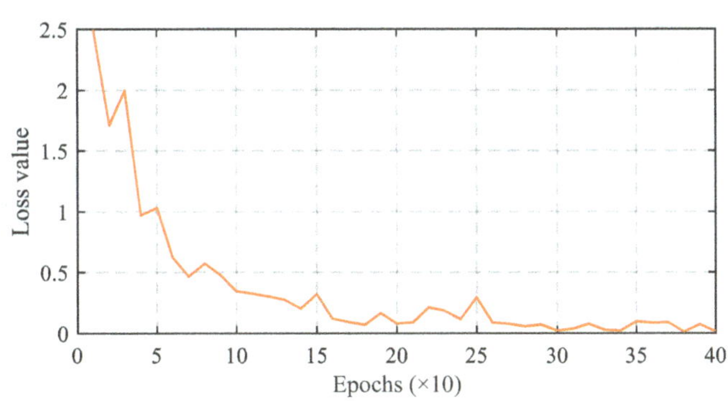

Figure 2.27 The evolution of the loss function value over 400 epochs [36]

of the loss function is constantly decreasing, which demonstrates that the model proposed in this article can be applied to real-world data. The health state detection performances of different methods are presented in Figure 2.28. It can be observed in Figure 2.28 that the proposed RNGPT-RBF model has the best fault detection performance compared with other methods on the CWRU-bearing dataset. RNGPT with the RBF kernel has the highest fault diagnosis accuracy in 5-shot (99.39 ± 0.09%) and 1-shot (98.55 ± 0.16%). In addition, the proposed method has a smaller variance, which indicates that it has a more stable fault diagnosis performance.

The confusion matrices of RNGPT-RBF in the testing phase are presented in Figure 2.29. S_0 to S_4 represent five randomly sampled motor health states. The abscissa represents the motor health states detected by the model, and the ordinate represents the actual states. Figure 2.29(a) is the confusion matrix of RNGPT-RBF

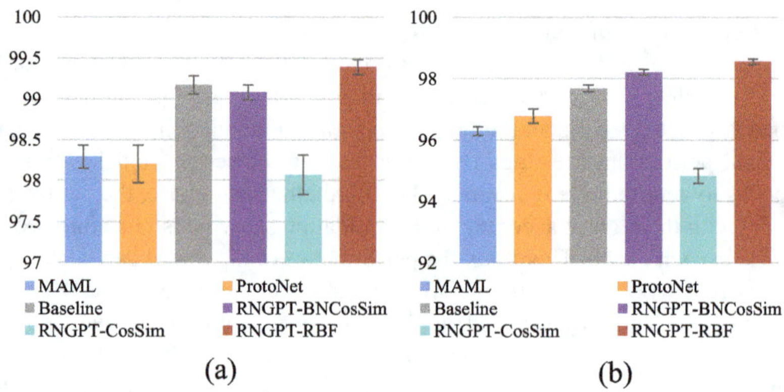

Figure 2.28 Faults detection accuracy of different models in the CWRU dataset: (a) 5-shot and (b) 1-shot [36]

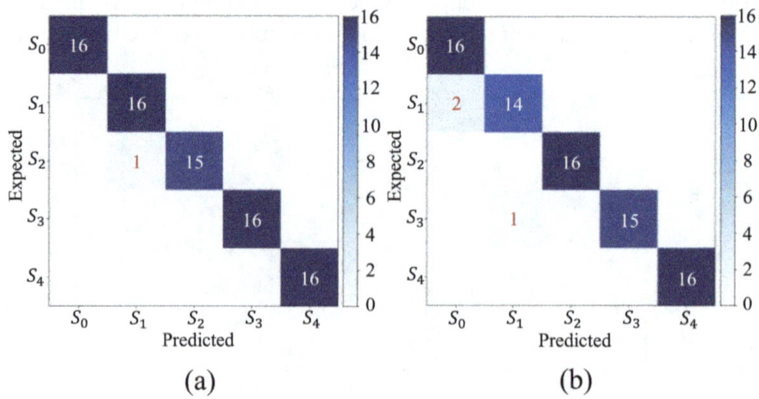

Figure 2.29 Confusion matrices of RNGPT-RBF model health state detection results: (a) 5-shot and (b) 1-shot [36]

in the 5-shot scenario, and it demonstrates that the proposed model can correctly detect the motor states with low error. Figure 2.29(b) is the confusion matrix of RNGPT-RBF in the 1-shot scenario, and it can be seen that the number of misjudgments is higher than that shown in Figure 2.29(a). The reason is that, compared with five input samples in the 5-shot scenario, only one sample is input into the model in the 1-shot scenario.

2.5 Summary and outlook

Motor fault diagnosis research is of great significance in ensuring the safety and stability of industrial production and people's daily lives. In this chapter, research on intelligent fault diagnosis methods for common electrical and mechanical faults in motor systems is carried out. Fault diagnosis strategies under limitations such as changes in motor operating conditions, limited data samples, and environmental noise are studied. The performance of the proposed method has been verified on various datasets.

In Section 2.2, a data-driven multi-fault detection model based on CCaps-Net is proposed for the state detection of EMs under varying working conditions. The CNN without a pooling layer retains the important features of EM data and improves the accuracy of fault identification. The powerful feature learning ability of Caps-Net enables the storage of more data. By combining the capsule network with CNN, the proposed model obtains a higher average multi-fault detection accuracy of at least 98%. Moreover, CCaps-Net has a stronger generalization ability under varying working conditions than other state-of-the-art methods. The EM fault detection accuracy of CCaps-Net under different load size conditions is at least 10% higher than that of other methods.

In Section 2.3, to solve the problems of a limited number of samples and weak model generalization ability, a meta-learning-based electric machine health state detection method with fast adaptation ability is proposed. Experiments are carried out on two datasets, and the comparative results show that the proposed model can achieve state-of-the-art performance. The health state diagnosis accuracy of the proposed model is close to 100% with a limited number of samples under various working conditions. The proposed model overcomes the shortcomings of the traditional DL method, which has the possibility of being overfitted to a small amount of data. In addition, the proposed model can adapt to a new environment faster, which reduces computational costs and illustrates that the model in this study has stronger generalization ability. The proposed model still shows significant fault diagnosis ability, even in various noise environments. The proposed model can effectively detect healthy and damaged bearings, and the results show that fault diagnosis errors primarily arise from misjudgment of the fault condition.

In Section 2.4, a GP kernel transfer-based few-shot learning method is proposed for electric machine fault diagnosis under variable working conditions. Diagnostic knowledge is learned from limited current and vibration signals. The trained model can be transferred to other unseen working conditions without parameter updating and fine-tuning. Two case studies are carried out using two different electric motor health state datasets, including a simulation dataset obtained by ANSYS software and a real-world bearing dataset from CWRU. In the first case, the proposed model has achieved the highest fault detection accuracy of 96.10 ± 0.29% in 5-shot and 93.28 ± 0.68% in 1-shot. In the second case, the proposed model also has the best fault detection performance in 5-shot and 1-shot, the detection accuracy is 99.39 ± 0.09% and 98.55 ± 0.31% respectively. Comparative case studies have shown that the proposed approach can achieve state-of-the-art performance. The results demonstrate that, when limited samples of the electric machine are available and the working conditions are variable, it is possible and significant to explore a fault detection model to exploit the underlying shared features for diagnostics.

This chapter presents a series of diagnostic strategies for common electrical and mechanical faults in motor equipment, proposing three different deep learning-based intelligent fault diagnosis methods and achieving certain research results. However, there are still some shortcomings in this study, and further in-depth research can be conducted on the following aspects:

1. Detection of unknown (unlabeled) motor faults: The research on fault diagnosis methods proposed in this chapter is carried out under the condition that the fault label is known. However, in practical applications, new, unknown faults will likely occur. How to quickly and accurately diagnose unknown motor faults is the next research focus.
2. Life prediction of key components of motors: The research in this chapter focuses on situations where faults have already occurred. If accurate life prediction of key components of motors can be achieved under limited data samples and variable working conditions, the targeted maintenance of motor equipment components can be carried out before faults occur, which has important practical significance for ensuring the safe and stable operation of industrial production. Therefore, life prediction for key motor components is also the next research focus.
3. Motor fault diagnosis under multiple faults: The research in this chapter focuses on the situation of motors under a single fault. However, in the real world, due to the complex operating conditions of motors, multiple different faults may likely occur simultaneously. In this case, compared to a single fault, the fault features of the motor are relatively more complex when multiple faults occur simultaneously, which makes motor state detection more difficult under multiple faults. Therefore, further research can be conducted on the diagnosis of multiple motor faults.

References

[1] P. Gangsar, and R. Tiwari, "Comparative Investigation of Vibration and Current Monitoring for Prediction of Mechanical and Electrical Faults in Induction Motor Based on Multiclass-Support Vector Machine Algorithms," *Mechanical Systems and Signal Process*, vol. 94, pp. 464–481, 2017.

[2] P F Albrecht, J C Appiarius, E P Cornell, *et al.*, "Assessment of the Reliability of Motors in Utility Applications," *IEEE Transactions on Energy Conversion*, vol. 2, no. 3, pp. 396–406, 1986.

[3] C Cempel, "Diagnostically Oriented Measures of Vibroacoustical Process," *Journal of Sound and Vibration*, vol. 73, no. 4, pp. 547–561, 1980.

[4] S Osman, and W Wang, "A Morphological Hilbert-Huang Transform Technique for Bearing Fault Detection," *IEEE Transactions on Instrumentation and Measurement*, vol. 65, no. 11, pp. 2646–2656, 2016.

[5] H Shao, H Jiang, H Zhang, *et al.*, "Electric Locomotive Bearing Fault Diagnosis Using a Novel Convolutional Deep Belief Network," *IEEE Transactions on Industrial Electronics*, vol. 65, no. 3, pp. 2727–2736, 2017.

[6] J Lei, C Liu, and D Jiang, "Fault Diagnosis of Wind Turbine Based on Long Short-Term Memory Networks," *Renewable Energy*, vol. 133, pp. 422–432, 2018.

[7] T Han, and D Jiang, "Rolling Bearing Fault Diagnostic Method Based on VMD-AR Model and Random Forest Classifier," *Shock and Vibration*, vol. 2016, pp. 1–11, 2016.

[8] W Sun, G Yang, Q Chen, *et al.*, "Fault Diagnosis of Rolling Bearing Based on Wavelet Transform and Envelope Spectrum Correlation," *Journal of Vibration and Control*, vol. 19, no. 6, pp. 924–941.

[9] Q Gao, C Duan, H Fan, *et al.*, "Rotating Machine Fault Diagnosis Using Empirical Mode Decomposition," *Mechanical Systems and Signal Processing*, vol. 22, no. 5, pp. 1072–1081, 2008.

[10] Y Wang, Z Pan, X Yuan, *et al.*, "A Novel Deep Learning Based Fault Diagnosis Approach for Chemical Process with Extended Deep Belief Network," *ISA Transactions*, vol. 96, pp. 457–467, 2020.

[11] M Ma, C Sun, and X Chen, "Deep Coupling Autoencoder for Fault Diagnosis With Multimodal Sensory Data," *IEEE Transactions on Industrial Informatics*, vol. 14, no. 3, pp. 1137–1145, 2018.

[12] H Wang, Z Liu, D Peng, *et al.*, "Understanding and Learning Discriminant Features Based on Multi-Attention 1DCNN for Wheelset Bearing Fault Diagnosis," *IEEE Transactions on Industrial Informatics*, vol. 16, no. 9, pp. 5735 – 5745, 2020.

[13] H Wang, Z Liu, D Peng, *et al.*, "Attention-Guided Joint Learning CNN With Noise Robustness for Bearing Fault Diagnosis and Vibration Signal Denoising," *ISA Transactions*, vol. 12, pp. 1–15, 2021.

[14] Z. Liu, Z. Wu, T. Li, J. Li and C. Shen, "GMM and CNN Hybrid Method for Short Utterance Speaker Recognition," *IEEE Transactions on Industrial Informatics*, vol. 14, no. 7, pp. 3244–3252, 2018.

[15] J. Chen, W. Hu, D. Cao, *et al.*, "Novel Data-Driven Approach Based on Capsule Network for Intelligent Multi-Fault Detection in Electric Motors," *IEEE Transactions on Energy Conversion*, vol. 36, no. 3, pp. 2173–2184, 2021.

[16] S Hochreiter and J Schmidhuber, "Long Short-Term Memory," *Neural Computation* vol. 9, no. 8, pp. 1735–1780, 1997.

[17] S. Sabour, N. Frosst, and G. E. Hinton, "Dynamic routing between capsules," in *Proceedings of the Advances in Neural Information Processing Systems*, 2017.

[18] Z. Zhu, G. Peng, Y. Chen and H. Gao, "A Convolutional Neural Network Based on a Capsule Network with Strong Generalization for Bearing Fault Diagnosis," *Neurocomputing*, vol. 323, pp. 62–75, 2019.

[19] R. Liu, G. Meng, B. Yang, C. Sun and X. Chen, "Dislocated Time Series Convolutional Neural Architecture: An Intelligent Fault Diagnosis Approach for Electric Machine," *IEEE Transactions on Industrial Informatics*, vol. 13, no. 3, pp. 1310–1320, 2017.

[20] W. Sun, R. Zhao, R. Yan, S. Shao, and X. Chen, "Convolutional Discriminative Feature Learning for Induction Motor Fault Diagnosis," *IEEE Transactions on Industrial Informatics*, vol. 13, no.3, pp. 1350–1359, 2017.

[21] J. Chen, W. Hu, D. Cao, *et al.*, "An Imbalance Fault Detection Algorithm for Variable-Speed Wind Turbines: A Deep Learning Approach," *Energies*, vol. 12, no. 14, pp. 2764–2778, 2019.

[22] K. He, X. Zhang, S. Ren, and J. Sun, "Deep Residual Learning for Image Recognition," 2016 *IEEE Conference on Computer Vision and Pattern Recognition (CVPR)*, Las Vegas, NV, USA, 2016, pp.770–778.

[23] Hatem A. Fayed, and Amir F. Atiya, "Speed Up Grid-Search for Parameter Selection of Support Vector Machines," *Applied Soft Computing*, vol. 80, pp. 202–210, 2019.

[24] I. M. Abdulbaqi, A. T. Humod, and O. K. Alazzawi, "Application of FEM to Provide the Required Database for MCSA Based On-line Fault Detection System on 3-Phase Induction Motor Using ANSYS Maxwell2D," *Advances in Natural & Applied Sciences*, vol. 10, no. 16, pp. 43–54, 2016.

[25] C. Finn, P. Abbeel, and S. Levine, "Model-agnostic Meta-learning for Fast Adaptation of Deep Networks," arXiv preprint arXiv: 1703.03400v3, 2017.

[26] J. Chen, W. Hu, D. Cao, *et al.*, "A Meta-Learning Method for Electric Machine Bearing Fault Diagnosis under Varying Working Conditions with Limited Data," *IEEE Transactions on Industrial Informatics*, vol. 19(3), pp. 2552–2564, 2023.

[27] Case Western Reserve University Bearing Data Center. [Online]. Available: http://csegroups.case.edu/bearingdatacenter/pages/welcome-case-western-reserve-university-bearing-data-center-website.

[28] C. Lessmeier, J. K. Kimotho, D. Zimmer, and W. Sextro, "Condition Monitoring of Bearing Damage in Electromechanical Drive Systems by Using Motor Current Signals of Electric Motors: A Benchmark Data Set for Data-Driven Classification," In *European Conference of the Prognostics and Health Management Society*, 2016.

[29] S. Shao, S. McAleer, R. Yan and P. Baldi, "Highly Accurate Machine Fault Diagnosis Using Deep Transfer Learning," *IEEE Transactions on Industrial Informatics*, vol. 15, no. 4, pp. 2446–2455, 2019.

[30] Z. Ji, X. Chai, Y. Yu, *et al.* "Improved Prototypical Networks for Few-Shot Learning," *Pattern Recognition Letters*, vol. 140, pp. 81–87, 2020.

[31] W. Zhang, G. Peng, C. Li, Y. Chen, and Z. Zhang, "A New Deep Learning Model for Fault Diagnosis with Good Anti-Noise and Domain Adaptation

Ability on Raw Vibration Signals," *Sensors*, vol. 17, no. 2, pp. 425–446, 2017.

[32] A. Zhang, S. Li, Y. Cui, W. Yang, R. Dong and J. Hu, "Limited Data Rolling Bearing Fault Diagnosis with Few-Shot Learning," *IEEE Access*, vol. 7, pp. 110895–110904, 2019.

[33] A. V. Balakrishnan and R. R. Mazumdar, "On Powers of Gaussian White Noise," *IEEE Transactions on Information Theory*, vol. 57, no. 11, pp. 7629–7634,2011.

[34] D. Dehay, J. Leskow and A. Napolitano, "Central Limit Theorem in the Functional Approach," *IEEE Transactions on Signal Processing*, vol. 61, no. 16, pp. 4025–4037, 2013.

[35] W. Chen, Y. Liu, Z. Kira, *et al.*, "A Closer Look at Few-shot Classification," arXiv preprint arXiv: 1904.04232, 2019.

[36] J. Chen, W. Hu, D. Cao, *et al.*, "Gaussian Process Kernel Transfer Enabled Method for Electric Machines Intelligent Faults Detection with Limited Samples," *IEEE Transactions on Energy Conversion*, vol. 36(4), pp. 3481–3490, 2021.

[37] D. Silver, A. Huang, C. J. Maddison, *et al.*, "Mastering the game of Go with Deep Neural Networks and Tree Search," *Nature*, vol. 529, no. 5787, pp. 484–489, 2016.

[38] B. Huang and J. Wang, "Applications of Physics-Informed Neural Networks in Power Systems - A Review," *IEEE Transactions on Power Systems*, vol. 38, no. 1, pp. 572–588, 2023.

[39] J. Snell, K. Swersky, and R. S. Zemel, "Prototypical Networks for Few-shot Learning," arXiv preprint arXiv: 1703.05175v2, 2017.

[40] W. Chen, Y. Liu, Z. Kira, Y. Wang, and J. Huang, "A Closer Look at Few-shot Classification," In *International Conference on Learning Representations*, 2019.

Chapter 3

Artificial intelligence in power electronic reliability, design, and control

Shuai Zhao[1], *Yi Zhang*[1] *and Frede Blaabjerg*[1]

3.1 Introduction to AI in power electronics

The integration of artificial intelligence (AI) into power electronics represents a transformative leap forward in the field of electrical engineering. Power electronics, a critical component of modern electrical systems, is responsible for the efficient conversion and control of electrical energy, serving as the backbone of various applications, from renewable energy systems to electric vehicles and smart grids. The traditional paradigms in power electronics have primarily relied on deterministic control techniques and rule-based algorithms to optimize energy conversion and distribution. However, the ever-increasing complexity and dynamic nature of today's electrical grids and energy systems demand a more adaptive, intelligent, and responsive approach. The introduction of AI into power electronics signifies a pivotal shift towards enhancing the efficiency, reliability, and sustainability of electrical systems. AI, powered by machine learning algorithms, neural networks, and advanced data analytics, empowers power electronics devices to operate with unprecedented levels of sophistication. This integration allows power converters, inverters, and other key components to learn from real-time data, anticipate fluctuations in power demand, and make dynamic adjustments, optimizing energy conversion and distribution in ways that were once inconceivable.

It is worth mentioning that the power electronics community has been investigating the applications of AI since the 1990s [1]. As shown in Figure 3.1, an overview [1] provides the statistical data that are based on searching the IEEE Xplore from the journals *IEEE Transactions on Power Electronics*, *IEEE Journal of Emerging and Selected Topics in Power Electronics*, *IEEE Transactions on Industrial Electronics*, *IEEE Transactions on Industrial Informatics*, and *IEEE Transactions on Industry Applications*. The data for 2020 is up to May 2020. It can be seen that there has been an exponential increase in AI for power electronics applications in recent decades.

[1]Department of Energy Technology, Aalborg University, Denmark

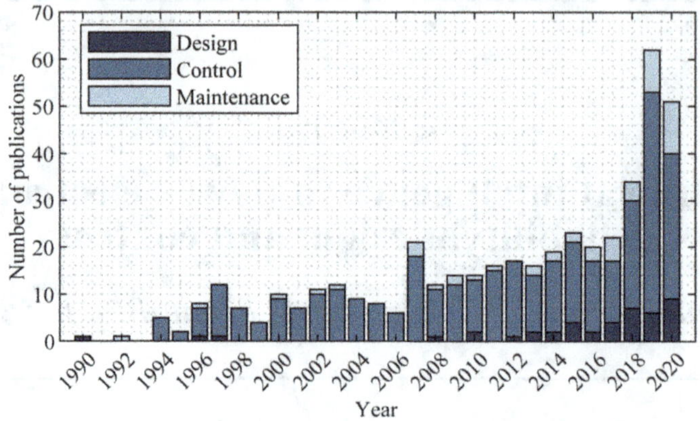

Figure 3.1 Annual number of publications on AI in power electronics since 1990 [1]

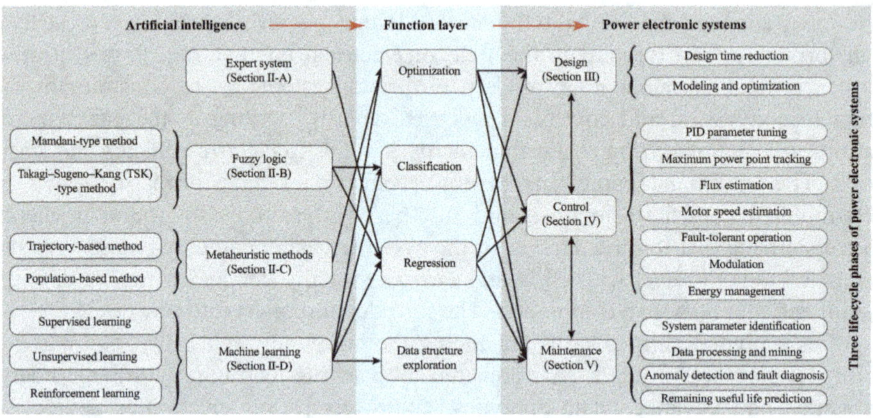

Figure 3.2 Application of AI in the life cycle of power electronic systems [1]

Generally, for AI in power electronics, the three distinct phases of a product's life cycle—design, control, and maintenance—have specific tasks that can be effectively handled by artificial intelligence (AI). These tasks encompass optimization, classification, regression, and the exploration of data structures. Within this context, the tools that enable the applications can be categorized into four main aspects, including expert systems, fuzzy logic, metaheuristic methods, and machine learning. Figure 3.2 illustrates the relationship between the AI tools, the functions, and the relevant applications. More details can be found in [1].

In this era of rapid technological advancement, the synergy between AI and power electronics is poised to redefine the way we generate, distribute, and consume electrical energy. However, it also poses challenges in terms of safety, security, and ethical considerations that must be carefully addressed. As we embark on this journey of incorporating AI into power electronics, it is imperative to strike a balance between innovation and responsibility, harnessing the full potential of this groundbreaking integration while safeguarding the integrity and stability of our electrical systems. In the following sections, we will delve deeper into the various aspects of AI in power electronics, exploring its applications, benefits, challenges, and the promising future it holds for our energy-dependent world.

3.2 Condition and health monitoring in power electronics systems using AI

Power electronics systems, integral to a multitude of applications from renewable energy sources to transportation solutions, play a pivotal role in our modern energy infrastructure. However, as with any sophisticated machinery, these systems are not immune to operational wear and potential failures. This underscores the essentiality of condition monitoring—a systematic approach focused on real-time assessment and analysis of the health and performance of power electronic components. By proactively detecting and diagnosing anomalies, condition monitoring not only enhances system reliability and safety but also aids in preemptive maintenance, reducing costly downtimes. As demand and reliance on power electronics surge, the importance of rigorous condition monitoring becomes ever more paramount, ensuring that these systems operate at their peak efficiency and longevity.

In the meantime, various measures can be applied in the design, production, and operation phases of Power Electronic Systems (PES) to address reliability and availability requirements. Due to the limitations and uncertainties in the models and data used in the design and production phases, the risk of unexpected failure events still exists. It calls for proactive actions that can be facilitated with condition monitoring, where the health status of PES is monitored to identify and mitigate potential risks. Moreover, the obtained in-situ operation status can also be used for operation optimization of PES, if it is not necessarily for preventing unexpected failure.

The health status can be explicitly analyzed with underlying physical principles. The increasing complexity and integration make the PES system structure prohibitively intricate. Deepening into the underlying physical models to facilitate condition monitoring is challenging in most cases. With the implementation of monitoring techniques, the accumulated data provides a new perspective for condition monitoring implementation, i.e., data-driven solutions. The condition monitoring in PES has drastically evolved in the last decade with recent advancements in artificial intelligence (AI), industrial Internet-of-Things (IIoT), and data analytics. Data-driven condition monitoring solutions have been emerging as a promising direction and show significant potential and merits, e.g., flexibility, lower requirements on hardware circuitry and system physical principles, etc. The typical workflow of data-driven condition monitoring is given in Figure 3.3 [2]. It is summarized as six core tasks:

1. Data acquisition: it involves the acquisition of raw physical signals/data from PES. It includes sensor hardware/software, sensor configuration, data

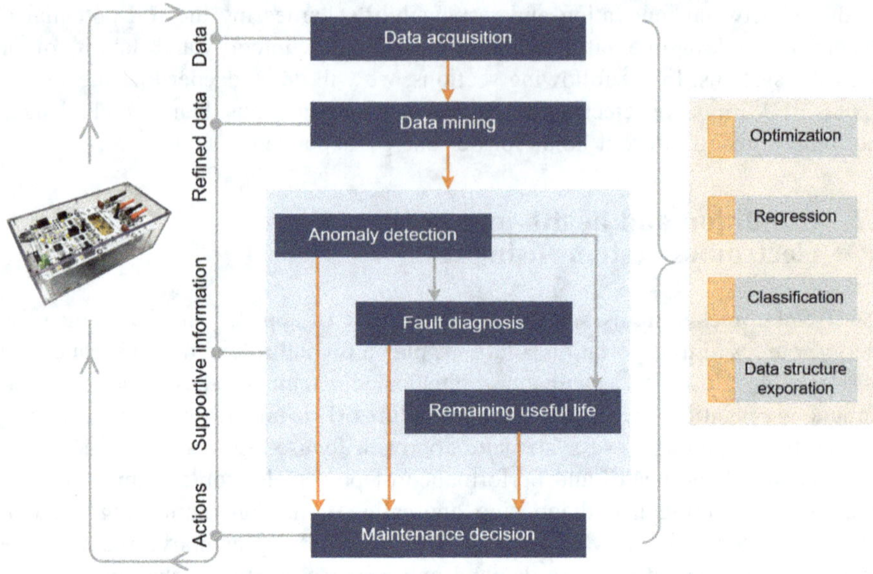

Figure 3.3 The flowchart of data-driven condition monitoring for power electronic systems

transmission, data storage, and data retrieval. For example, a mission profile data logger for power electronic converters is developed in [3], collecting the on-state voltages of power devices, ambient humidity, and temperature data.

2. Data mining: it is concerned with data preprocessing, data compression, and feature extraction [4]. It prepares the data foundation to improve the data suitability and ease the application complexity in terms of the succeeding functional tasks, i.e., anomaly detection, fault diagnostics, and remaining useful life (RUL) prediction. As an example, data mining for fault diagnosis aims to develop features that make different failure modes classifiable as much as possible.
3. Anomaly detection: it aims to determine the off-normal conditions of PES that deviate from the expected system behavior. Essentially, a binary decision needs to be made in this case. For example, an emulator that characterizes the system's normal behavior can be established for a PES. The field system outputs can be compared in real-time with this emulator, which is not subjected to degradation and failure, to determine the abnormal state.
4. Fault diagnosis: it is a process of deciding the possible fault causes/modes, locations, and severity. Compared to the binary tasks in anomaly detection, fault diagnosis is essentially dealing with a multi-class classification task [5].
5. Remaining useful life (RUL) prediction: it aims to quantify the length of the remaining time that the system can safely operate in the future [6]. It is concerned with the degradation modeling of PES dynamic behavior in terms of the operational time. The RUL evaluation considers uncertainty sources

comprehensively, including the current system state, historical workload, system heterogeneity, and expected future mission profiles.

6. Decision-making: it formulates corrective/preventive maintenance actions or operation optimization schemes [7]. One example of decision-making for operation optimization in PES is a power routing concept. The workloads are balanced among components with different degradation levels to extend the converter's lifetime.

Although there is a lot of research in this field, from an algorithm perspective, the above six core tasks essentially deal with one or more of four functions, as shown in Figure 3.3 [1], i.e., optimization, regression, classification, and data structure exploration. Optimization attempts to find optimal solutions that minimize or maximize an objective function from available alternatives given constraints, equalities, and inequalities. It is an essential step to identify unknown model parameters that properly characterize the data or system behavior. It can be completed deterministically with gradient-related methods or heuristically with metaheuristic methods (e.g., genetic algorithm). Regression establishes mapping relationships between the input and output variables in the training dataset. The mapping relationships characterize the static or dynamic (e.g., time-series data) linkage and system functional behavior. Given new input variables, the objective of the regression task is to predict the corresponding output variables. Classification assigns a discrete label for input variables to indicate the associated category as the output, which identifies the categorical relationships and decision boundaries among the dataset. Notably, the outputs of classification are discrete quantities compared to those of continuous quantities in regression tasks. Data structure exploration includes data clustering and data compression, where the input variables are categorized into different clusters considering their similarities. The irrelevant information is to be eliminated for feature reduction through data compression. It is an essential step to make the dataset well-prepared for the subsequent functional tasks and ease algorithm complexity.

In [2], a novel method for estimating health indicators rooted in the digital-twin concept is proposed, specifically for the condition monitoring of power electronic converters, as shown in Figure 3.4. This approach is non-intrusive, eliminating the need for supplementary hardware circuits and calibration. The method is validated with the application using a Buck DC–DC converter, accompanied by theoretical evaluations, pragmatic insights, and empirical validations. Specifically, a digital counterpart of the test prototype is formulated, encompassing the power phase, sampling mechanism, and closed-loop control. To deduce the unknown circuit parameters, an AI tool called the Particle Swarm Optimization algorithm is based on the data of both the digital twin and its real-world counterpart. The gathered data clusters, showcasing health indicators under varied test conditions for the Buck DC–DC converter, facilitate the tracking of degradation patterns in essential components, including capacitors and MOSFETs. The findings are insightful for realizing unobtrusive, economically efficient condition monitoring for power electronic systems. Moreover, the new digital-twin-based condition

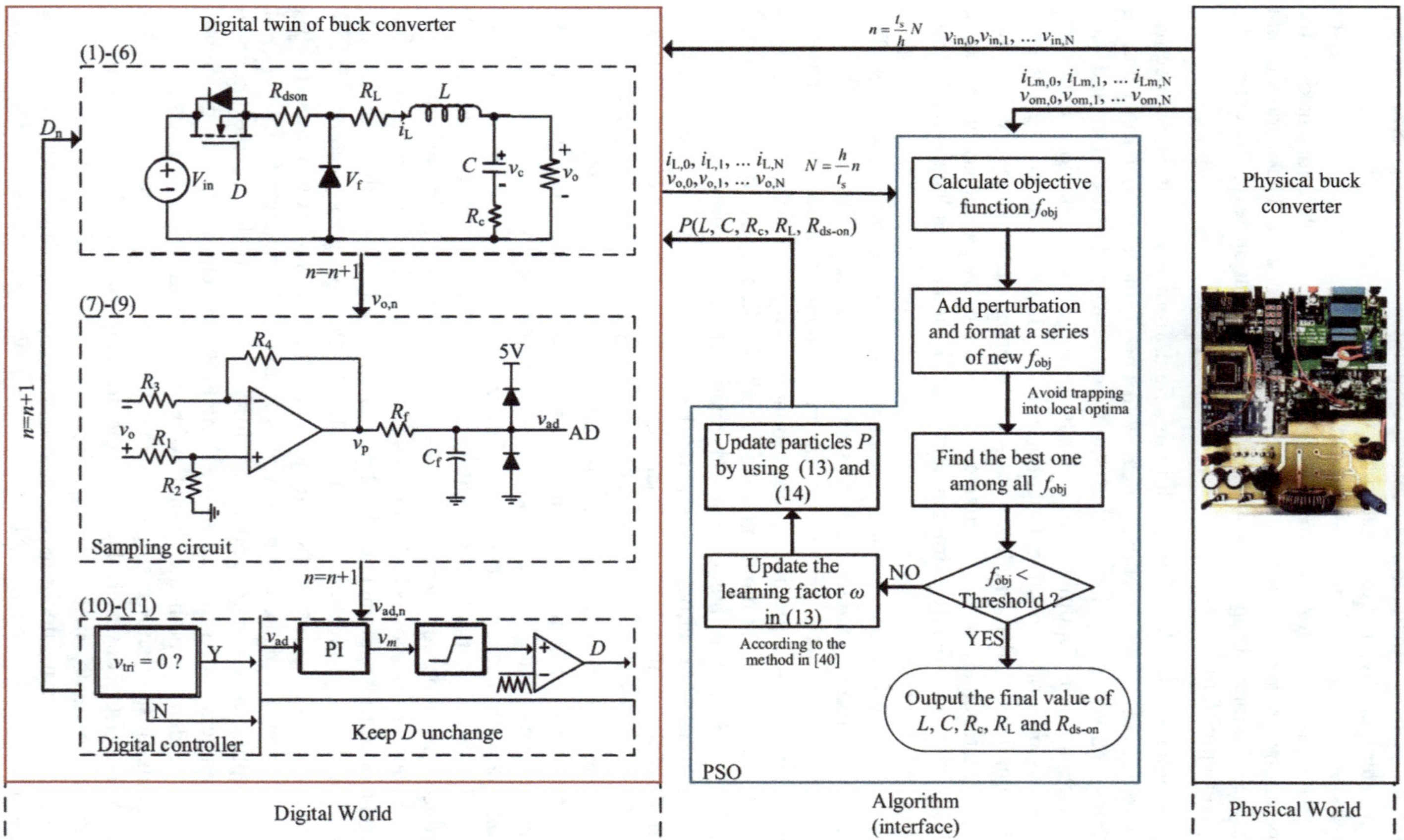

Figure 3.4 Digital twin-based estimation method for health indicators of DC–DC converters [2]

monitoring methods can be extended to other more complex power electronic systems.

In [3], a machine learning tool integrated with physical principles, termed Physics-informed Machine Learning (PIML), is gaining attention as an invaluable direction for engineering domains with well-established physical models and domain-specific knowledge [8]. Venturing into the realm of power electronics, Reference [3] introduces a PIML-based approach for parameter estimation, which has been applied to the condition monitoring of a DC–DC Buck converter. A physics-informed neural network is crafted by coherently merging a deep neural network with the dynamic model of a Buck converter, as shown in Figure 3.5. Its data-efficient nature allows for comprehensive training using a minimal dataset consisting of just 360 samples, a size feasible for practical preparation. Both simulation outcomes and real-world hardware implementations substantiate the accuracy and resilience of the parameter estimation technique. As a result, the proposed methodology addresses several key issues inherent to conventional data-centric models, particularly concerning training data quality, predictive accuracy, and system robustness. With this illustrative implementation, it sheds light on the potential adaptations of established machine learning frameworks specifically tailored for the power electronics sector. The data and code are open-accessed on GitHub at Link*.

Beyond mere data aggregation, it is crucial to emphasize data attributes such as resolution, immunity to noise, diversity, and overall volume. Data mining stands as a pivotal tool, enhancing information quality to bolster subsequent AI algorithmic processes. Given the intricate nature of PES, it's precarious to solely depend on a singular data source or health marker. Rather, garnering data from a broad spectrum of sources proves advantageous, gathering a comprehensive informational perspective from various angles. To optimize this multi-source data, information fusion methods become indispensable, aiming to amplify accuracy while curtailing uncertainties.

Empirical evaluations suggest that several physical parameters correlate with the degradation observed in SiC MOSFETs [9]. Merging these parameters can significantly augment device degradation modeling. Within the RUL prediction

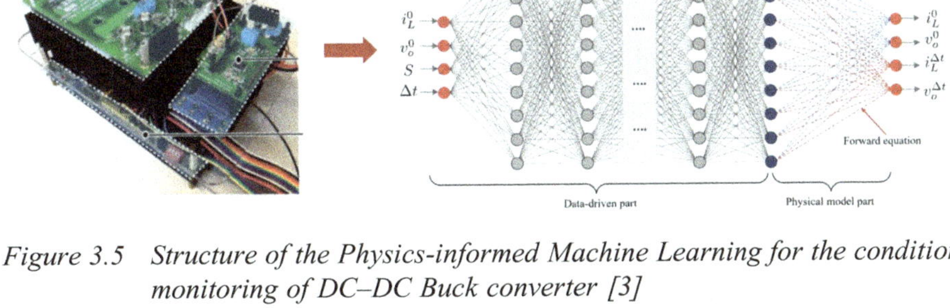

Figure 3.5 Structure of the Physics-informed Machine Learning for the condition monitoring of DC–DC Buck converter [3]

*https://github.com/ms140429/PIML_Converter

exercise presented in [8], genetic programming (GP), a metaheuristic technique, is employed. This method integrates diverse physical attributes of SiC MOSFETs, such as input capacitance (Ciss), output capacitance (Coss), reverse transfer capacitance (Crss), drain-source on-state resistance (Rds(on)), gate threshold voltage (VGS(th)), and diode forward voltage (VSD), forging a unified failure precursor. Leveraging symbolic programming, GP autonomously discovers nonlinear parameter combinations that minimize specific target metrics, like prediction discrepancies. This prospective combination encompasses a plethora of mathematical operations, from basic arithmetic to more intricate functions. Incorporating advanced operations can further enhance GP's versatility and capability. Consequently, a composite failure precursor emerges, as depicted in Figure 3.6. When juxtaposed against the ubiquitously employed parameter Rds(on), the prediction precision employing the synthesized composite failure precursor witnesses a 35.3% enhancement, simultaneously trimming uncertainty by 16.3%. Such outcomes underscore the latent promise of information fusion during data preprocessing. It is noteworthy that the fusion function in Figure 3.6 is not physically explanatory but is rather sculpted from a data-centric viewpoint.

The application of a single AI tool may be subjected to an overfitting effect and limited generalization. Ensemble methods integrating multiple AI tools can be applied to improve robustness and reduce bias and variance. In [10], three typical ensemble methods, i.e., bagging, boosting, and stacking, are used for fault detection of series DC arcs in Buck converters and Boost converters. In the bagging scheme, the training dataset is split into several subsets using the bootstrap method. Based on these subsets, an AI tool will be trained in terms of each subset, resulting in several base learners. Notably, each single base learner is based on the same AI tool, and the training between different learners is independent. Finally, the predictions of all base learners will be averaged as the final output. The bagging scheme is effective in reducing the prediction variance. In the boosting scheme, the

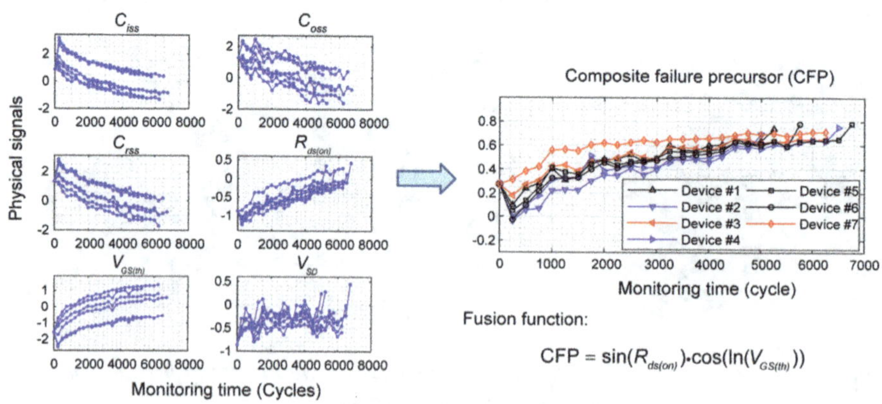

Figure 3.6 Information fusion for composite failure precursor development for SiC MOSFETs [9]

training of base learners is performed in sequence, where the weights of the data in the training set used for the current learner, are dynamically adjusted according to the training results of the previous base learner. The final prediction is obtained by combining each base learner, considering its performance, rather than averaging them uniformly as in the bagging scheme. This ensemble scheme is especially useful when the prediction shows a high bias.

Among the three ensemble schemes, the stacking scheme is superior in terms of accuracy, precision, and recall score for this fault detection task. In the stacking scheme, each base learner is established with a different AI tool. This diversified combination makes the prediction more robust and avoids getting stuck in the possible local minimums. Figure 3.7 shows the flowchart of the stacking method and the integrated various AI tools, including Support Vector Machine (SVM), K-Nearest Neighbors (KNN), Random Forest (RF), Extra Trees (ET). Subsequently, the outputs of these base tools are integrated with Logic Regression (LR). Experimental analyses indicate that the stacking ensemble method can provide reliable fault detection with high-level accuracy under different load types.

One of the PES features is the heterogeneous operational environments, which suggests that the data structure and distribution of the newly obtained condition monitoring data will be different from those in the training dataset. Moreover, other variability sources, including system-to-system variability, temporal variability, and measurement errors, will further increase these differences. As a result, for the model inference, the accuracy of the trained model will be significantly reduced. It calls for the adaptiveness of AI tools that can automatically update the model structure and parameters with the latest condition monitoring data. In this way, the trained AI tool can be dynamically tailored to a particular PES and can deal with unexpected changes.

For AI tools trained by offline data, a model parameter tuning layer is necessary to adjust the tool with the latest condition monitoring data for online

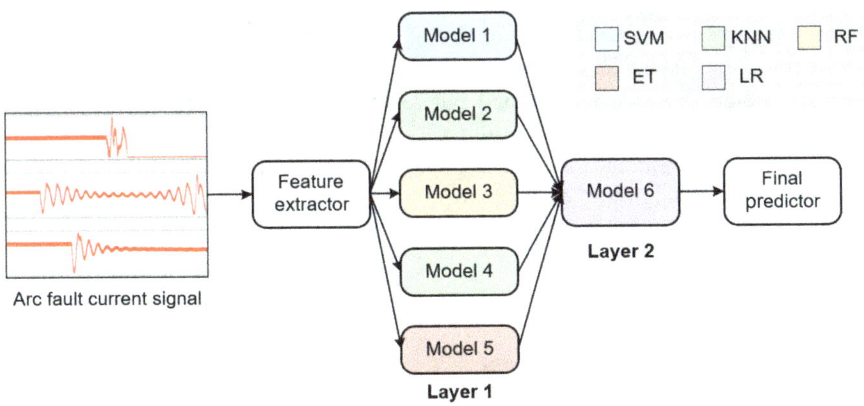

Figure 3.7 The stacking ensemble method integrating multiple AI tools for series DC arc-fault detection in PES [10]

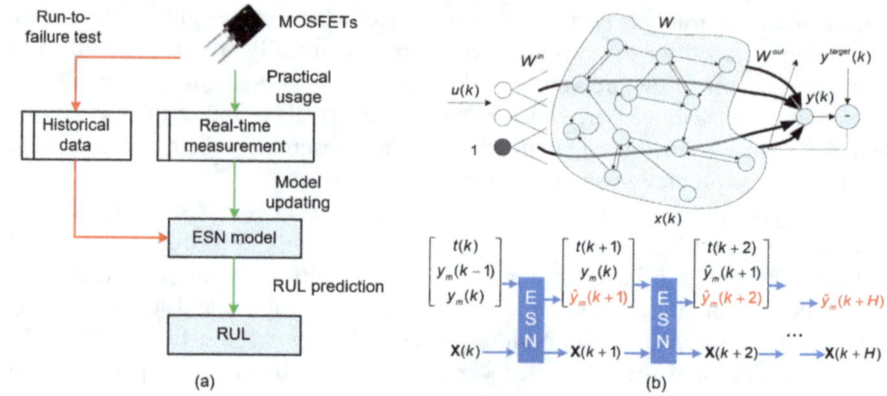

Figure 3.8 RUL prediction of MOSFETs with Echo State Network (ESN) with the adaptiveness enabled by Particle Filter (PF). (a) Flowchart for model training and inference of ESN; (b) dynamic updating mechanism of ESN with PF [11].

applications, enabling the adaptiveness of AI tools. Particle filter (PF), which is in the dynamic Bayesian network category, is an exemplary AI tool with adaptiveness. Its state-space structure is flexible and can be extended with various variants, such as auxiliary PF and unscented PF. For example, an auxiliary PF is applied to the RUL prediction of IGBTs in [11]. With the latest on-state voltage $V_{CE(on)}$ data, the initial PF model is dynamically updated for estimating the degradation state of IGBT. In this way, the initial PF model can be specially tailored to the in-situ IGBT device with the adaptiveness capability. Besides, PF can be incorporated into other AI tools to enable adaptiveness. For the RUL prediction of MOSFETs, PF is applied to update the output weights W^{out} of the trained Echo State Network (ESN) with real-time measurement, as shown in Figure 3.8. The inputs of ESN include the on-state resistance $R_{ds(on)}$ at time k-1 and k (i.e., $y_m(k\text{-}1)$, $y_m(k)$). The output is the predicted $R_{ds(on)}$ at time k+1 (i.e., $\hat{y}_m(k+1)$). The output weights W^{out} is updated with PF once the latest $R_{ds(on)}$ at time k (i.e., $y_m(k)$) becomes available. Such a dynamic prediction process is carried out recursively for predicting $R_{ds(on)}$ in the future to determine the failure occurrence.

Delving deeper into the realm of AI applications, we explore the optimized design of power electronic circuits, where AI not only complements but significantly enhances traditional methods.

3.3 Optimized design of power electronic circuits

Power electronic design usually adjusts the component values based on simulations run iteratively, following an optimization procedure derived from designers' experience. Moreover, the methods for automated synthesis and optimization of circuits are either deterministic or stochastic. Deterministic approaches, like the gradient method, might easily be trapped in a local optima, as high nonlinearity is always involved in PEC optimization. Also, the performance of some deterministic methods is significantly dependent on the initial starting point of a search. On the other hand, stochastic methods tend to be more suitable for PEC optimization since they explore the search space more extensively.

1. Brute-force approach: try out a large number of possible combinations of the available set of component values, and select the one that gives the best simulation results. However, this approach requires large processing resources and/or simulation time.

   In the arena of AI, several innovative approaches like Genetic Algorithms (GA) and Particle Swarm Optimization (PSO) stand out, showcasing the versatility and power of AI in circuit optimization.
2. Genetic Algorithms (GA): The GA is an optimization tool inspired by the process of natural evolution. Basically, GA works with a population whose members are called chromosomes; a chromosome is composed of genes represented as binary strings. Each chromosome encodes a possible solution to the optimization problem. The initial population is randomly generated and evolves from generation to generation over several evolutionary cycles by using the specific phenomena of the evolutions: the crossover, the mutation, and the selection. The best-fit individual is then chosen based on its performance mark, called the fitness function. The fitness is defined so that it accurately and consistently captures the effects of several optimization criteria. This method can provide optimum solutions while using modest amounts of time and resources.

   As shown in Figure 3.9, an optimization tool for the design of analog circuits implemented with commercially available discrete devices using a GA method is presented. The inputs of this optimization are the design specifications of the circuit, the set of component values of the starting solution, the available component libraries, and the inputs for the GA algorithm, including the optimization parameters and their weights, the probabilities of the generic mutation and crossover, the population size, and the number of generations. When the algorithm runs, the lower value of the fitness function corresponds to the best solution, which is an optimal combination of values and options for the circuit components.
3. Particle Swarm Optimization (PSO): PSO is another evolutionary algorithms (EAs), which is a population-based search technique inspired by natural mechanisms or phenomena. Different from the previous GA, PSO solves optimization problems by emulating the flocking behavior of birds. It employs a swarm of particles, which represent feasible solutions to search in a cooperative manner. Each particle has its own position and velocity, which are iteratively updated by tracing the personal best position found so far and the global best position found so far. PSO has also been used in analog circuit synthesis, filter design, etc. Compared to GA, PSO often converges faster, which can considerably reduce the computation time. In [12], a PSO approach

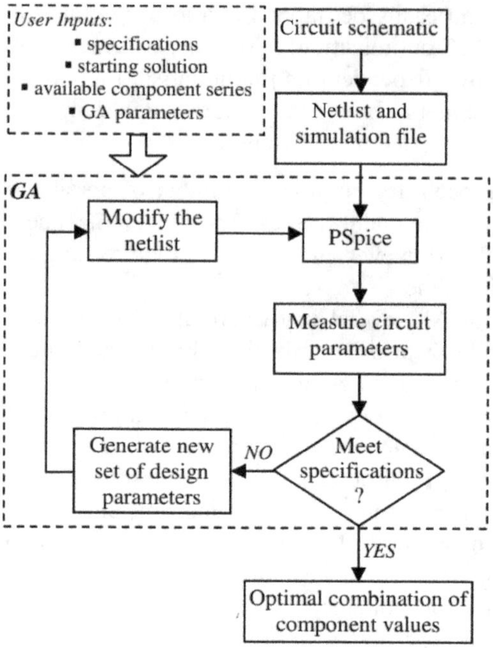

Figure 3.9 The flowchart of the optimization procedure [13]

for the optimization of a buck converter is presented. The power converter is decoupled into two parts, and particles are employed to find the optimal values of the components of each part, aiming to satisfy the specified static and dynamic requirements. To improve the population diversity, a mutation operator is incorporated into the PSO. Their study reveals that the PSO is easy to implement and fast to converge.

4. Artificial neural network (ANN): ANN is a more general method [14] to deal with the complex design process. Li *et al.* in [15] proposed an ANN-based model named Mag-net, which is trained by data from practical experiment measurements, to calculate the magnetic core loss in a fast way. Additionally, an ANN-based inductor evaluation model called AI-Mag [16], which is trained by data from finite element method (FEM) simulation, is proposed to accelerate the inductor design. In [17], a thermal modeling effort considering the cross-coupling effects is trained from the measurement data. In [18], a modeling and parameter design method for DC–DC converter-based artificial intelligence is proposed. A database of switching losses is constructed using Spice simulation data with a single fixed semiconductor switch. Nex, an ANN, is trained by the database. Transfer learning is implemented to train other ANNs for other switches, with much less training data needed. Finally, under the restrictions of current and voltage ripples, a heuristic optimization algorithm is used to obtain the most efficient and optimal design. The design and

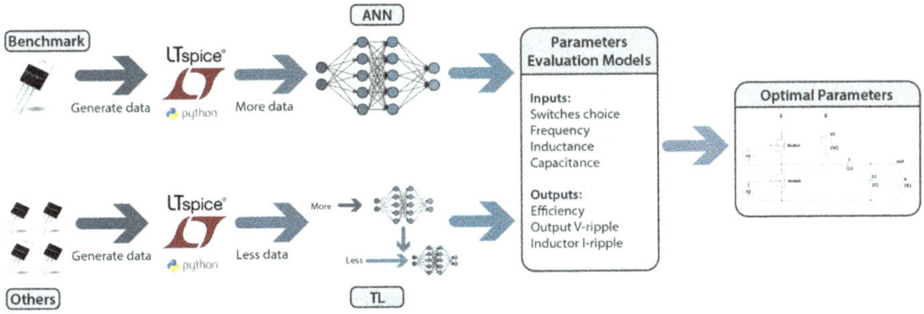

Figure 3.10 Design and optimization workflow using artificial intelligence-based DC–DC converter efficiency modeling [18]

optimization workflow using artificial intelligence-based DC–DC converter efficiency modeling is shown in Figure 3.10. The results show that the ANN gives precise estimates of the converter properties.

5. Other neural networks: In addition to the above artificial intelligence-based methods, deep neural networks (DNN) combined with gate-recurrent units and K-nearest neighbors for designing high-frequency transformers [19], back propagation neural networks (BPNN) combined with multi-objective genetic algorithms (MOGA) for optimizing operational parameters [20], graph isomorphism networks (GIN) for converter inverse design [21], and other neural-network-based methods can be targeted solutions to specific design issues.

Transitioning from design optimization to control systems, the role of AI continues to be pivotal, albeit with unique challenges and applications.

3.4 Control of power electronic systems

In power electronic applications, most applications of artificial intelligence are primarily focused on the design and optimization of components, fault diagnosis, preventive maintenance, and reliability analysis. The applications on control are not very often. The reasons are laid out in the power electronic control, which requires high speed and high accuracy. Compared to conventional model-based control, the black-box nature of artificial intelligence is not very desirable. However, there are still some efforts using AI to achieve better control. AI's adaptability is further exemplified in control strategies like the Inverse Application of Artificial Intelligence (IAAI), Fuzzy Logic Control, and optimizing controllers and modulators [22], each of which unfolds new dimensions of AI's potential in power electronics.

1. An inverse application of IAAI for the control of power electronic converter systems [23]. In this method, the proposed inverse application method of feedforward ANN. Different from the recurrent NN, which stores past outputs as part of inputs, the proposed IA method directly swaps the inputs/outputs in

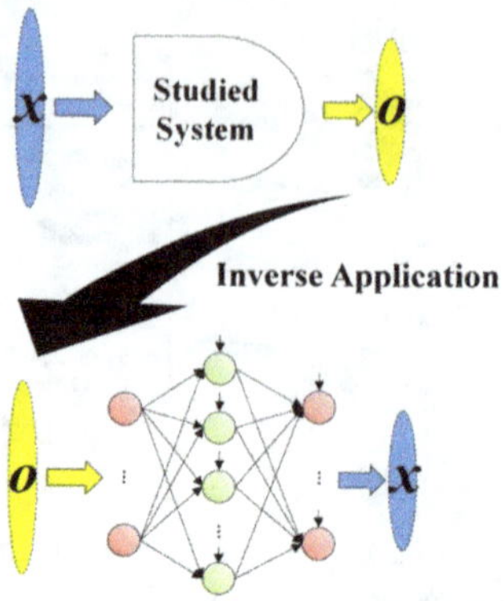

Figure 3.11 Diagram of inverse application of artificial intelligence (IAAI) [23]

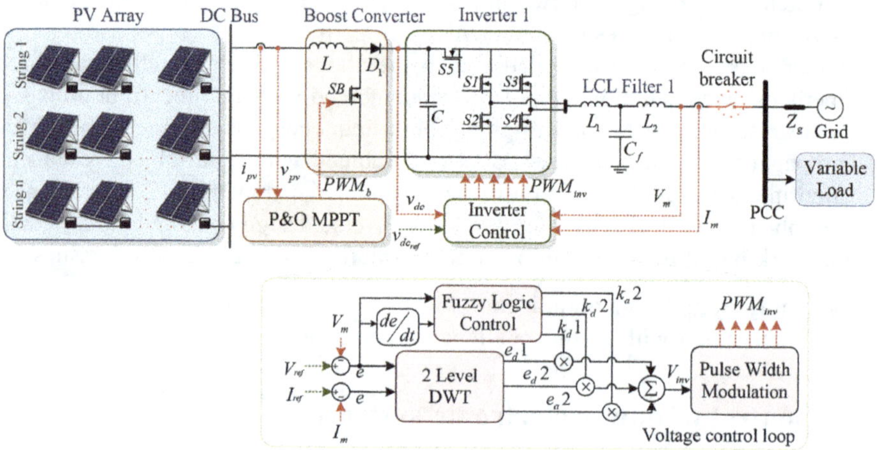

Figure 3.12 A standalone operation of transformerless inverter with proposed fuzzy control [24]

the original system, as shown in Figure 3.11. After ANN training, the trained NN can smoothly provide the desired coefficient or reference for the power converter control. It should be noted that the authors of [23] also mentioned that their proposed method does not use ANN to approximate a power system controller but to provide suitable parameters such as ratios and references for the system control. By running multiple simulations to collect data, this role can be achieved by the inversely designed ANN.

2. Fuzzy logic control: the voltage regulation and harmonic elimination are challenging for photovoltaic systems when the main grid is cut off. The standalone operation must be achieved while operating in voltage control mode. In [24], a wavelet-based fuzzy control for the standalone operation of single-phase inverters is designed as shown in Figure 3.12. The proposed controller

regulates the output voltage by adjusting the fuzzy controller weights, which are operating in a closed loop with the pulse width modulation generator. A hardware prototype of a 1 kWp standalone PV system is developed in their study, and the proposed control method is evaluated to achieve <6% of output voltage regulation under steady-state and step load conditions while reducing total harmonic distortion (THD) for steady-state conditions. The THD of the proposed system operating with linear loads is observed to be 2.18%, and for nonlinear loads, it is around 2.71% under simulation conditions.

3. Optimizing controllers and modulators: As shown in Figure 3.13, another case is using artificial intelligence to improve the conventional PID controller. In [25], an improved metaheuristic algorithm named Runge Kutta optimizer (RUN) is proposed. The improved RUN (iRUN) optimizer is obtained by integrating the pattern search (PS) method into the original form of the RUN optimizer. The constructed iRUN optimizer is proposed as an efficient tool to tune a real proportional-integral-derivative (PID) plus second-order derivative (PIDD2) controller. In the proposed control mechanism, a performance index known as ZLG is used as an objective function. The proposed iRUN optimizer-based real PIDD2 controller is comparatively demonstrated to be superior in terms of transient response and robustness against other available and recently reported methods based on PID, PID with filter (PID-F), PID acceleration (PIDA), fractional order PID (FOPID) and PID plus second-order derivative

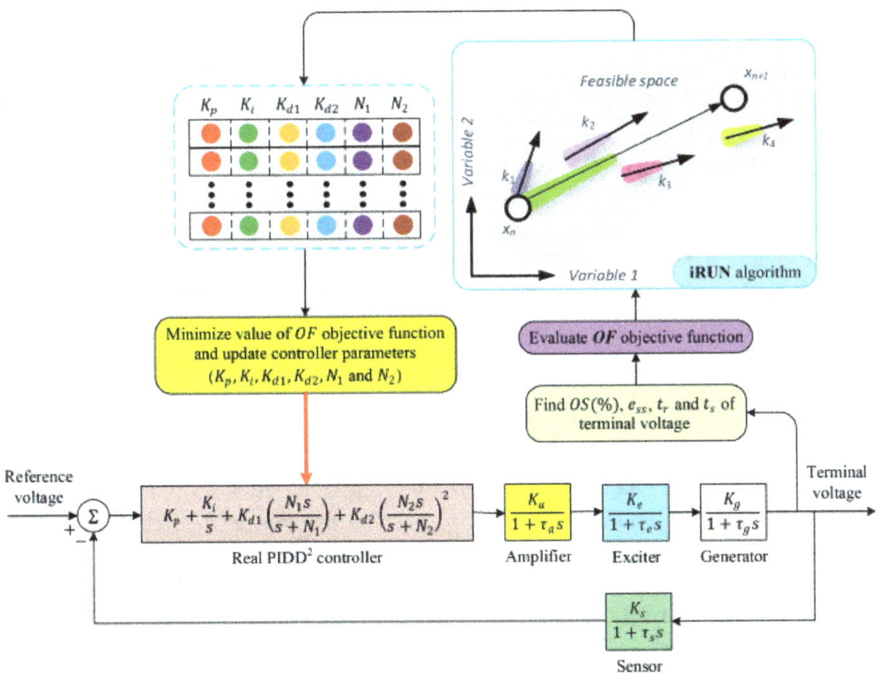

Figure 3.13 The improved PID control [25]

(PIDD2) controllers that are respectively tuned with metaheuristic optimizers of the stochastic fractal search algorithm, Archimedes optimization algorithm, whale optimization algorithm, Henry gas solubility optimization, and particle swarm optimization. The proposed method is shown to achieve no overshot and lower rise and settling times compared to the listed methods. It is also shown to be sufficiently robust in the event of potential variations that may occur in the parameters of the system.

Model predictive controllers (MPC) can also be optimized similarly, as mentioned in [26] and [27]. Besides, complex modulation is a limitation of control performance, so optimizing the modulation process is an effective solution for improving control characteristics, such as optimizing modulation of dual-active-bridge (like in [28,29]) and triple-active-bridge converters (like in [30]) for better power flow control and higher power conversion efficiency.

References

[1] S. Zhao, F. Blaabjerg, and H. Wang, "An Overview of Artificial Intelligence Applications for Power Electronics," *IEEE Transactions on Power Electronics*, vol. 36, no. 4, pp. 4633–4658, 2021, doi:10.1109/TPEL.2020.3024914.

[2] Y. Peng, S. Zhao, and H. Wang, "A Digital Twin Based Estimation Method for Health Indicators of DC–DC Converters," *IEEE Transactions on Power Electronics*, vol. 36, no. 2, pp. 2105–2118, 2021, doi:10.1109/TPEL.2020.3009600.

[3] S. Zhao, Y. Peng, Y. Zhang, and H. Wang, "Parameter Estimation of Power Electronic Converters with Physics-Informed Machine Learning," *IEEE Transactions on Power Electronics*, vol. 37, no. 10, pp. 11567–11578, 2022, doi:10.1109/TPEL.2022.3176468.

[4] A. Moradzadeh, B. Mohammadi-Ivatloo, K. Pourhossein, and A. Anvari-Moghaddam, "Data Mining Applications to Fault Diagnosis in Power Electronic Systems: A Systematic Review," *IEEE Transactions on Power Electronics*, vol. 37, no. 5, pp. 6026–6050, 2022, doi:10.1109/TPEL.2021.3131293.

[5] P. I. Gómez, M. E. L. Gajardo, N. Mijatovic, and T. Dragičević, "A Self-Commissioning Edge Computing Method for Data-Driven Anomaly Detection in Power Electronic Systems," *IEEE Transactions on Industrial Electronics*, doi:10.1109/TIE.2023.3347839.

[6] J. Miao, Y. Liu, Q. Yin, B. Ju, G. Zhang, and H. Wang, "A Novel Soft Fault Detection and Diagnosis Method for a DC/DC Buck Converter Based on Contrastive Learning," *IEEE Transactions on Power Electronics*, vol. 39, no. 1, pp. 1501–1513, 2024, doi:10.1109/TPEL.2023.3320878.

[7] A. Vaccaro, P. Magnone, A. Zilio, and P. Mattavelli, "Predicting Lifetime of Semiconductor Power Devices Under Power Cycling Stress Using Artificial Neural Network," *IEEE Journal of Emerging and Selected Topics in Power Electronics*, vol. 11, no. 6, pp. 5626–5635, 2023, doi:10.1109/JESTPE.2022.3194189.

[8] Y. Fassi, V. Heiries, J. Boutet, and S. Boisseau, "Toward Physics-Informed Machine-Learning-Based Predictive Maintenance for Power Converters—A Review," *IEEE Transactions on Power Electronics*, vol. 39, no. 2, pp. 2692–2720, 2024, doi:10.1109/TPEL.2023.3328438.

[9] S. Zhao and H. Wang, "Enabling Data-Driven Condition Monitoring of Power Electronic Systems with Artificial Intelligence: Concepts, Tools, and Developments," *IEEE Power Electronics Magazine*, vol. 8, no. 1, pp. 18–27, 2021, doi:10.1109/MPEL.2020.3047718.

[10] V. Le, X. Yao, C. Miller, and B. -H. Tsao, "Series DC Arc Fault Detection Based on Ensemble Machine Learning," *IEEE Transactions on Power Electronics*, vol. 35, no. 8, pp. 7826–7839, 2020, doi:10.1109/TPEL.2020.2969561.

[11] Z. Li, Z. Zheng, and R. Outbib, "A Prognostic Methodology for Power MOSFETs Under Thermal Stress Using Echo State Network and Particle Filter," *Microelectron. Rel.*, vol. 88-90, pp. 350–354, 2018.

[12] J. Zhang, Y. Shi, and Z. H. Zhan, Power electronic circuits design: A particle swarm optimization approach, 2008. In *Asia-Pacific Conference on Simulated Evolution and Learning*, pp. 605–614. Berlin, Heidelberg: Springer Berlin Heidelberg.

[13] I. Saracut, M. Neag, R. Onet, and I. Kovacs, "Design Optimization of Analog Active Circuits Using the Genetic Algorithm." *Acta Technica Napocensis*, vol. 53, no. 4:47, 2012.

[14] R. Rajamony, S. Wang, G. Calderon-Lopez, I. Ludtke, and W. Ming, "Artificial Neural Networks-Based Multi-Objective Design Methodology for Wide-Bandgap Power Electronics Converters," *IEEE Open Journal of Power Electronics*, vol. 3, pp. 599–610, 2022, doi:10.1109/OJPEL.2022.3204630.

[15] H. Li, S. R. Lee, M. Luo, C. R. Sullivan, Y. Chen, and M. Chen. MagNet: A Machine Learning Framework for Magnetic Core Loss Modeling, 2020 *IEEE 21st Workshop on Control and Modeling for Power Electronics (COMPEL)*, pp. 1–8.

[16] T. Guillod, P. Papamanolis, and J. W. Kolar. "Artificial Neural Network (ANN) Based Fast and Accurate Inductor Modeling and Design," *IEEE Open Journal of Power Electronics*, vol. 1, pp. 284–299, 2020.

[17] Y. Zhang, Z. Wang, H. Wang, and F. Blaabjerg, "Artificial Intelligence-Aided Thermal Model Considering Cross-Coupling Effects," *IEEE Transactions on Power Electronics*, vol. 35, no.10, 9998–10002, 2020.

[18] F. Tian, D.B. Cobaleda, and W. Martinez. "Artificial-Intelligence based DC-DC Converter Efficiency Modelling and Parameters Optimization." *2022 24th European Conference on Power Electronics and Applications (EPE'22 ECCE Europe)*. IEEE, 2022.

[19] G. S. Lee, S. Kim, and S. Bae, "Efficient Design Method for a Forward-Converter Transformer Based on a KNN–GRU–DNN Model," *IEEE Transactions on Power Electronics*, vol. 38, no. 1, pp. 73–78, 2023, doi:10.1109/TPEL.2022.3203480.

[20] L. Gu, H. Fan, Z. Jin, and D. Xu, "Multi-Objective-Optimized Parameter Design Method for High-frequency Resonant Converter," 2022 *IEEE Energy*

Conversion Congress and Exposition (ECCE), Detroit, MI, USA, 2022, pp. 1–6, doi:10.1109/ECCE50734.2022.9947777.

[21] C. Xue, Y. Li, F. Zargari, and Y. Li, "Graph Isomorphism Network: A Learning-based Workflow for Converter Inverse Design Problem," 2023 *IEEE Energy Conversion Congress and Exposition (ECCE), Nashville, TN, USA, 2023*, pp. 6589–6591, doi:10.1109/ECCE53617.2023.10362399.

[22] Y. Gao, S. Wang, T. Dragicevic, P. Wheeler, and P. Zanchetta, "Artificial Intelligence Techniques for Enhancing the Performance of Controllers in Power Converter-Based Systems—An Overview," *IEEE Open Journal of Industry Applications*, vol. 4, pp. 366–375, 2023, doi:10.1109/OJIA.2023.3338534.

[23] Y. Gao, S. Wang, H. Hussaini, T. Yang, T. Dragičević, S. Bozhko, P. Wheeler, and S. Vazquez, "Inverse Application of Artificial Intelligence for the Control of Power Converters," *IEEE Transactions on Power Electronics*, vol. 38, no.2, 1535–1548, 2022. https://doi.org/10.1109/TPEL.2022.3209093

[24] M. A. Khan, A. Haque, V. S. B. Kurukuru, and F. Blaabjerg, "Optimizing the Performance of Single-Phase Photovoltaic Inverter Using Wavelet-Fuzzy Controller," *e-Prime-Advances in Electrical Engineering, Electronics and Energy*, vol. 3, 100093, 2023.

[25] D. Izci and E. Serdar, "An Improved RUN Optimizer-Based Real PID Plus Second-Order Derivative Controller Design as a Novel Method to Enhance Transient Response and Robustness of an Automatic Voltage Regulator," *e-Prime-Advances in Electrical Engineering, Electronics and Energy*, vol. 2, 100071, 2022.

[26] F. Simonetti, A. D'Innocenzo, and C. Cecati, "Neural Network Model-Predictive Control for CHB Converters with FPGA Implementation," *IEEE Transactions on Industrial Informatics*, vol. 19, no. 9, pp. 9691–9702, 2023, doi:10.1109/TII.2023.3233973.

[27] M. M. Mardani, R. D. Lazar, N. Mijatovic, and T. Dragičević, "Artificial Neural Network-Based Constrained Predictive Real-Time Parameter Adaptation Controller for Grid-Tied VSCs," *IEEE Journal of Emerging and Selected Topics in Power Electronics*, vol. 11, no. 2, pp. 1507–1517, 2023, doi:10.1109/JESTPE.2022.3214342.

[28] X. Li, X. Zhang, F. Lin, C. Sun, and K. Mao, "Artificial-Intelligence-Based Hybrid Extended Phase Shift Modulation for the Dual Active Bridge Converter with Full ZVS Range and Optimal Efficiency," *IEEE Journal of Emerging and Selected Topics in Power Electronics*, vol. 11, no. 6, pp. 5569–5581, 2023, doi:10.1109/JESTPE.2022.3185090.

[29] X. Li, X. Zhang, F. Lin, C. Sun, and K. Mao, "Artificial-Intelligence-Based Triple Phase Shift Modulation for Dual Active Bridge Converter With Minimized Current Stress," *IEEE Journal of Emerging and Selected Topics in Power Electronics*, vol. 11, no. 4, pp. 4430–4441, 2023, doi:10.1109/JESTPE.2021.3105522.

[30] A. A. Ibrahim, A. Zilio, T. Younis, D. Biadene, T. Caldognetto, and P. Mattavelli, "Optimal Modulation of Triple Active Bridge Converters by an Artificial-Neural-Network Approach," *IEEE Transactions on Industrial Electronics*, vol. 71, no. 3, pp. 2590–2600, 2024, doi:10.1109/TIE.2023.3270529.

Chapter 4

Application of artificial intelligence in dual-active-bridge (DAB) converters

Weihao Hu[1], Guozhou Zhang[2], Zhenyuan Zhang[1], Sayed Abulanwar[3,4] and Frede Blaabjerg[5]

4.1 Introduction

In recent years, with the development of power electronics in power systems, DC microgrid systems based on new energy sources have become important for research [1]. For the collection, conversion, and transmission of new energy sources, the Dual-Active-Bridge (DAB) DC–DC converter has become a key energy conversion device. The DAB DC–DC converter was first proposed in the early 1990s by Rik De Doncker and Deepakraj M. Divan *et al.* [2,3]. It contains an isolation transformer T_r, an external series inductor L_r, and two bridge converter units (Full bridge 1 and Full bridge 2) as shown in Figure 4.1. The power delivery of DAB converters can be both bidirectional and unidirectional; thus, they are widely used in renewable energy power generation systems, Solid State Transformer (SST), Energy Storage Systems (ESS), Electric Vehicle (EV), and aerospace because their simple structure, wide range of soft switching, reliable performance, and high power density are achievable [1,4,5].

However, the mathematical model of the DAB DC–DC converter is complicated due to the large number of active and passive components in the converter. In addition, because the relevant parameters and external circuit environment in the DAB DC–DC converter may change with the change in operating conditions, it is difficult to model it accurately using the traditional mathematical modeling approach. Since the DAB DC–DC converter model has many adjustable control variables, it aggravates the computational effort and complexity of the efficiency optimization solution for this type of converter. The existing numerical optimization methods, iterative

[1]School of Mechanical and Electrical Engineering, University of Electronic Science and Technology of China, China
[2]College of Engineering and Technology, Southwest University, China
[3]Electrical Engineering Department, Faculty of Engineering, Mansoura University, Egypt
[4]Faculty of Engineering, Horus University-Egypt, Egypt
[5]Department of Energy Technology, Aalborg University, Denmark

algorithms, and heuristic algorithms have the drawbacks of complex models, time-consuming computation, and difficulty in finding the optimum, which make it difficult to ensure fast and comprehensive optimization for each optimization objective [1,6,7]. Therefore, how to improve the efficiency and performance of DAB DC–DC converters has become important in the field of power electronics.

Due to it being a high-dimensional system for optimization with complex and changing models, the new generation of artificial intelligence techniques has shown their advantages in many systems with large amounts of data, complex modeling, and uncertainty in optimization decisions, and they are well-suited to solve the optimization problems of high-dimensional complex systems [8,9]. Currently, these methods have demonstrated superior performance in finding and making decisions in other areas of electrical engineering (e.g., power systems, and power markets). Currently, reinforcement learning methods are rarely used in the field of power electronics optimization and are therefore well worth exploring [10,11]. For power electronic converters with complex parameters like DAB DC–DC converters, it is of theoretical and practical value to deeply integrate artificial intelligence algorithms and study the key bottlenecks of reinforcement learning to be used in DAB DC–DC converter efficiency optimization.

Based on the above background, this chapter will delve into the problem of optimal modulation of reinforcement learning algorithms in DAB DC–DC converters to improve the performance and transmission efficiency associated with DAB DC–DC converters.

4.2 Modulation modeling analysis for dual active bridge converters

In this section, two basic modulation modeling methods, the linear piecewise time-domain (LPTD) method and the harmonic analysis method, are introduced. A DAB converter based on triple-phase-shift (TPS) modulation is taken as an example here. In TPS modulation, there are three phase shift angles, D_1, D_2, and D_3 for the linear piecewise time-domain method or D_1, D_2, and D_φ for the harmonic analysis method, which provides more scope for optimization than SPS and DPS modulations.

4.2.1 Linear piecewise time-domain (LPTD) method

The circuit configuration of the DAB converter is depicted in Figure 4.1, and the equivalent circuit is shown in Figure 4.2. According to Figure 4.1, it contains two symmetrical full bridges, a high-frequency isolation transformer T_r, and a series inductor L_r, which is applied to facilitate the power transfer. Moreover, the turn ratio of T_r is n: 1, and the corresponding voltage conversion ratio is $k = V_1/nV_2$, where $k > 1$ and $k < 1$ indicate the buck and boost scenarios, respectively. Assuming $k \geq 1$, the power is flowing from the primary to the secondary side. Corresponding key waveforms are depicted in Figure 4.3, where v_{AB} denotes the primary voltage and v'_{CD} represents the equivalent secondary voltage, i_{Lk} represents the current flow through the inductor L_k (where L_k represents the leakage of T_r and the value of L_r), T_s represents the switching period, and T_h denotes the half

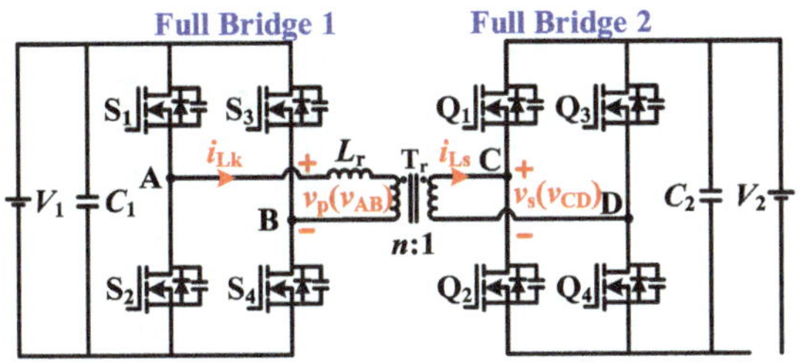

Figure 4.1 Circuit configuration of the DAB converter

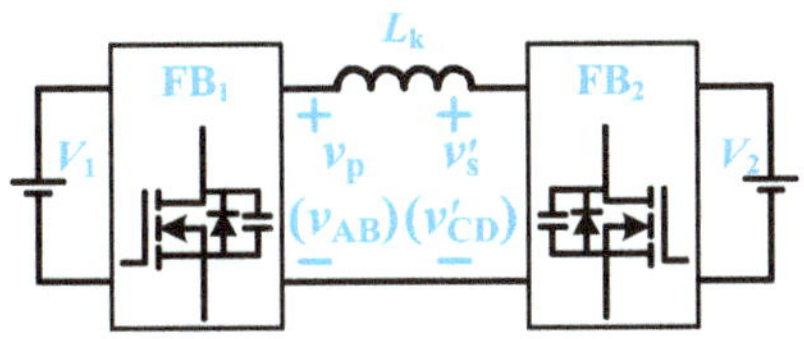

Figure 4.2 The simplified equivalent circuit of the DAB converter

of the switching period. More specifically, each switch is conducted for a half-switching period, and the gate pulse for the switches at the same arm is complementary. By controlling the phase shift between different legs, unique transferred power and corresponding performance can be obtained.

With the help of a TPS control strategy, the DAB converter is controlled by the phase shift angles D_1, D_2, and D_3. Specifically, TPS can be regarded as extended-phase-shift (EPS) once $D_2 = D_3$ or $D_1 = 0$, can be regarded as double-phase-shift (DPS) once $D_1 = D_3 – D_2$, and can be regarded as single-phase-shift (SPS) once $D_2 = D_3$ and $D_1 = 0$. In addition, due to the symmetrical structure characteristics, the phase shifts D_1, D_2, and D_3 can be substituted equivalently by D'_1, D'_2, and D'_3 for reverse conversion, respectively, and the direction of i_{Lk} should also be changed appropriately. Thus, based on the different combinations of D_1, D_2, and D_3 having no overlaps, partial and full of both phase shift ratios, there will be 18 modulation modes. Typically, eight operation modes are suitable for the power transformation of the DAB converter [12], and the main waveforms are shown in Figure 4.3. Specifically, D_1, D_2, and D_3 should be subject to the following constraints:

$$0 \leq D_1 \leq 1, 0 \leq D_3 - D_2 \leq 1, 0 \leq D_2 \leq 2, 0 \leq D_3 \leq 2 \tag{4.1}$$

Aiming to realize ZVS turn-on for all the switches (S_1–S_4, Q_1–Q_4), the output capacitance C_{oss} of each power switch before turning on should be fully discharged. Thus, enough energy from the inductor L_k will be needed to complete this resonance between C_{oss} and L_k [13]. After that, the antiparallel diode of the power switch will conduct naturally, which leads to a near-zero voltage between the drain and the source. Therefore, ZVS turn-on will be obtained if the power switch is turned on at this instant.

According to the principle of ZVS, the energy stored in the inductor L_k should be greater than that in the two junction capacitors C_{oss} of the power switch at the

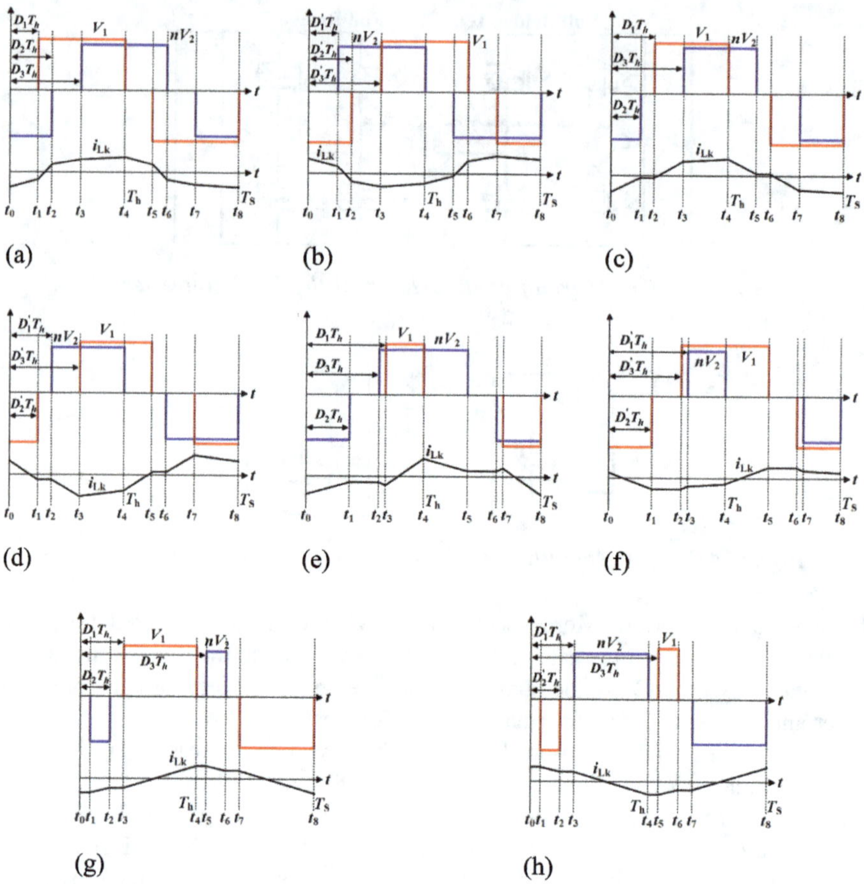

Figure 4.3 Linear piecewise time-domain (LPTD) model based on eight basic modulation modes for the DAB converter. (a) $0 \le D_1 \le D_2 \le D_3 \le 1$, (b) $0 \le D_1' \le D_2' \le D_3' \le 1$, (c) $0 \le D_2 \le D_1 \le D_3 \le 1$, (d) $0 \le D_2' \le D_1' \le D_3' \le 1$, (e) $0 \le D_2 \le D_3 \le D_1 \le 1$, (f) $0 \le D_2' \le D_3' \le D_1' \le 1$, (g) $0 \le D_2 \le D_1 \le 1 \le D_3 \le (1 + D_1) \le 2$, (h) $0 \le D_2' \le D_1' \le 1 \le D_3' \le (1 + D_1') \le 2$ s [1]

bridge arm, for charging/discharging these two capacitors [14,15]. That means the following formula must be satisfied.

$$\frac{1}{2}L_{\mathrm{k}}i_{\mathrm{Lk}}^2 \ge \frac{1}{2}C_{\mathrm{oss}}V_{\mathrm{DS}}^2 + \frac{1}{2}C_{\mathrm{oss}}V_{\mathrm{DS}}^2 \tag{4.2}$$

where V_{DS} indicates the drain-source voltage and the value is equal to the input voltage V_1 for the primary side, and the output voltage V_2 for the secondary side.

Aiming to simplify the analysis, the ZVS turn-on is approximately to be achieved if the current is flowing from source to drain before it is turned on [7,13].

Based on this, the ZVS boundary conditions can be defined as $i_{DS} \leq 0$, where i_{DS} is the current that flows through the drain to the source of the MOSFET. Therefore, the ZVS constraints for switches (S_1–S_4, Q_1–Q_4) can be summarized as follows based on the waveforms illustrated in Figure 4.3.

$$\begin{cases} i_{Lk}|_{t=turn\,on} \leq 0 & \text{For}: S_1, S_2, Q_1, Q_2 \\ i_{Lk}|_{t=turn\,on} \geq 0 & \text{For}: S_3, S_4, Q_3, Q_4 \end{cases} \tag{4.3}$$

It is worth noting that the critical ZVS turn-on will be obtained once the current shown in (4.3) is equal to zero. The critical ZVS turn-on mode may bring certain turn-on losses, while it can provide ZCS turn-off for another switch at the corresponding bridge arm.

4.2.2 Harmonic analysis method

The equivalent circuit of the DAB converter is depicted in Figure 4.2, and the corresponding operation waveforms of the TPS modulation are illustrated in Figure 4.4. Specifically, v_p denotes the AC voltage of the primary side, v'_s is the equivalent AC voltage of the secondary side, L_k is the equivalent inductance of the auxiliary inductance and leakage of the transformer, and the duty ratio of the transformer is N:1. D_1 is the inner phase shift of the full bridge FB_1, D_2 denotes the inner phase shift of the full bridge FB_2, and D_φ is the phase shift between the center points of v_p and v'_s. Moreover, i_{Lk} is the current flowing through L_k. It is assumed that the power flows from the primary to the secondary, and $V_1 \geq V_2$, thus the range of all phase shift angles (D_1, D_2, D_φ) is limited to $[0{\sim}\pi]$.

Based on the harmonic analysis method, the periodic function expression of the TPS can be made up of the combination of a series of odd-order harmonic components, which are expressed by simple sine and cosine functions [7,16]. Based on this, the active power P_o can be expressed as (4.4)

$$P_o = \sum_{n=1,3,5,..} \frac{8V_1V_2}{n^3\pi^2\omega_0 L_k} \cos\left(n\frac{D_1}{2}\right)\cos\left(n\frac{D_2}{2}\right)\sin\left(nD_\varphi\right) \tag{4.4}$$

where $\omega_0 = 2\pi f_s$, and f_s denotes the switching frequency. The reactive power Q can be calculated as (4.5) and (4.6)

$$Q_{n=1,3,5,..} = \sum_{n=1,3,5,..} \frac{8V_1\sqrt{A^2+B^2}}{n^3\pi^2\omega_0 L_k} \cos\left(n\frac{D_1}{2}\right)\sin\left(-\arctan\frac{A}{B}\right) \tag{4.5}$$

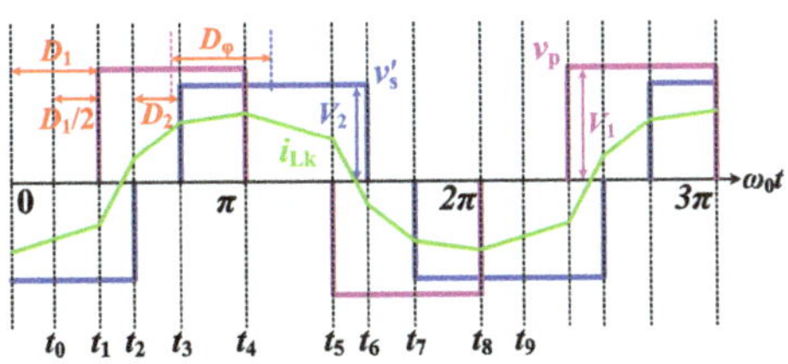

Figure 4.4 Typical equivalent waveforms of the DAB converter based on the TPS [17]

$$Q_{m\neq n=1,3,5,..} = \sum_{m\neq n=1,3,5,..} \frac{8V_1 \cos\left(m\frac{D_1}{2}\right)}{mn^2\pi^2\omega_0 L_k}\sqrt{A^2+B^2} \tag{4.6}$$

where A and B can be expressed as

$$\begin{cases} A = V_2 \cos\left(n\dfrac{D_2}{2}\right)\cos(nD_\varphi) - V_1 \cos\left(n\dfrac{D_1}{2}\right) \\ B = V_2 \cos\left(n\dfrac{D_2}{2}\right)\sin(nD_\varphi) \end{cases} \tag{4.7}$$

Thus, the reactive power is given by $Q = Q_{n=1,3,5,..} + Q_{m\neq n=1,3,5,...}$. Moreover, the current flowing through L_k is described as

$$i_{Lk}(t) = \sum_{n=1,3,5,..} \frac{4}{n^2\pi\omega_0 L_k}\sqrt{A^2+B^2}\sin\left(n\omega_0 t + \arctan\frac{A}{B}\right) \tag{4.8}$$

For ensuring the ZVS performance for all power switches, the current direction should be from the source to drain at the switching-on instants. Thus, the ZVS boundary conditions for the switches of each bridge leg can be expressed as in [10]

$$\begin{cases} i_{Lk}\left(\omega_0 t = \dfrac{D_1}{2}\right) \leq 0 \quad i_{Lk}\left(\omega_0 t = \pi - \dfrac{D_1}{2}\right) \geq 0 \\ i_{Lk}\left(\omega_0 t = D_\varphi + \dfrac{D_2}{2}\right) \geq 0 \quad i_{Lk}\left(\omega_0 t = D_\varphi - \dfrac{D_2}{2}\right) \geq 0 \end{cases} \tag{4.9}$$

In this chapter, the harmonic components are chosen as m, n = 1, 3, 5, 7 to simplify the calculations, as the higher order shows a small influence on the calculation accuracy.

4.3 Loss distribution of DAB converter

Nowadays, one of the most important tasks for power electronic converters is to improve the power conversion efficiency of the DAB converter. Therefore, it is of first interest to quantify the losses of each component. Remarkably, the main power losses are the power-switch losses, the magnetic-component losses, and the other losses [1,18,19]. The magnetic-component losses include the copper losses and the core losses of T_r and L_r. Moreover, the power-switch losses consist of the conducting losses, the switching losses, and the gate driver losses. Furthermore, the most dominant of the unknown losses is the temperature-dependent conduction-loss component related to the MOSFET components [19]. Usually, the unknown losses account for only a small fraction of the overall power losses, which is ignored in the efficiency optimization process to simplify the theoretical analysis [1]. A detailed power loss analysis will be described as follows.

4.3.1 Switching losses

In general, the calculation of the switching losses depends on the switching condition (soft-switching or hard-switching) of the power switches. Thus, the switching losses contain turn-on losses and turn-off losses. Specifically, the turn-on loss and the turn-off loss of the MOSFET can be calculated as

$$\begin{cases} P_{\text{Sw\_on}} = \dfrac{C_{\text{D}} V_{\text{D}}^2}{2} \cdot f_{\text{s}} + \dfrac{1}{2} V_{\text{D}} \cdot I_{\text{on}} \cdot (t_{\text{d\_on}} + t_{\text{r}}) \cdot f_{\text{s}} \\ P_{\text{Sw\_off}} = \dfrac{1}{2} V_{\text{D}} \cdot I_{\text{off}} \cdot (t_{\text{d\_off}} + t_{\text{f}}) \cdot f_{\text{s}} \end{cases} \tag{4.10}$$

where $P_{\text{Sw\_on}}$ indicates the turn-on losses, $P_{\text{Sw\_off}}$ indicates the turn-off losses, f_{s} is the switching frequency, V_{D} is the voltage between the drain-to-source of the MOSFET, C_{D} is the parasitic capacitor of the MOSFET, I_{on} is the turn-on current, I_{off} is the turn-off current. Moreover, $t_{\text{d\_on}}$ denotes the turn-on delay time, t_{r} denotes the rise time, $t_{\text{d\_off}}$ denotes the turn-off delay time, and t_{f} denotes the fall time of the MOSFET, respectively. Thus, the total switching losses can be calculated as

$$P_{\text{Sw}} = P_{\text{Sw\_on}} + P_{\text{Sw\_off}} \tag{4.11}$$

Based on this, the power switching losses in eight power switches (S_1–S_4, Q_1–Q_4) can be further defined as

$$P_{\text{S\_sw}} = P_{\text{Sw,S}_1 \sim \text{S}_4} + P_{\text{Sw,Q}_1 \sim \text{Q}_4} \tag{4.12}$$

More remarkably, the switching losses can be ignored once the soft switching is obtained. Specifically, the turn-on losses can be ignored if the ZVS turn-on is realized, and the turn-off losses can be ignored once the ZCS turn-off is realized.

4.3.2 Conduction losses

The conduction losses of the power switches can be estimated by

$$P_{\text{cond}} = I_{\text{rms}}^2 \cdot R_{\text{ds}} \tag{4.13}$$

where I_{rms} is the RMS current following through the power switch and R_{ds} is the on-resistor of the MOSFET.

Due to each of the power switches in the DAB converter conducting for half switching period, the total conduction losses of the power switches can be calculated as:

$$P_{\text{S\_cond}} = 4R_{\text{ds\_p}} \left(\frac{I_{\text{Lk\_rms}}}{\sqrt{2}} \right)^2 + 4R_{\text{ds\_s}} \left(\frac{n \cdot I_{\text{Lk\_rms}}}{\sqrt{2}} \right)^2 \tag{4.14}$$

where $I_{\text{Lk\_rms}}$ indicates the RMS current flowing through L_{k}, $R_{\text{ds\_p}}$ denotes the on-resistor of the power switches (S_1–S_4), and $R_{\text{ds\_s}}$ denotes the on-resistor of the power switches (Q_1–Q_4).

4.3.3 Gate driver losses

The gate driver loss can be calculated as

$$P_{\text{gate}} = Q_{\text{g}} \cdot V_{\text{gs}} \cdot f_{\text{s}} \tag{4.15}$$

where V_{gs} represents the gate driver voltage and Q_g indicates the gate charge capacitance in the MOSFET, respectively. Based on this, the total gate driver loss is given by

$$P_{\text{S\_gate}} = 4Q_{\text{g\_p}} \cdot V_{\text{gs\_p}} \cdot f_{\text{s}} + 4Q_{\text{g\_s}} \cdot V_{\text{gs\_s}} \cdot f_{\text{s}} \tag{4.16}$$

where V_{gs_p} denotes the gate driver voltage and Q_{g_p} denotes the gate charge capacitance of the power switches (S_1–S_4) in the primary side, while V_{gs_s} denotes the gate driver voltage and Q_{g_s} denotes the gate charge capacitance of the power switches (Q_1–Q_4) in the secondary side.

To sum up, the losses in the power switches can be calculated as

$$P_{\text{S\_loss}} = P_{\text{S\_sw}} + P_{\text{S\_cond}} + P_{\text{S\_gate}} \tag{4.17}$$

The magnetic-component (T_r and L_r) losses are broken down into copper losses and core losses. Assuming that the winding resistors in both T_r and L_r are constant, the copper losses of the magnetic components can be defined as

$$P_{\text{copper}} = I^2_{\text{m\_rms}} \cdot R_{\text{m}} \tag{4.18}$$

where I_{m_rms} represents the RMS current following through the magnetic components and R_m indicates the winding resistor of the magnetic components.

To simplify the calculations of the core losses, assume T_r and L_r are fed the fundamental sinusoidal components. Thus, the Steinmentz equation can be used to calculate the core losses, which is represented as

$$P_{\text{core}} = k \cdot f_{\text{s}}^{\alpha} \cdot \widehat{B}^{\beta} \cdot V_{\text{e}} \tag{4.19}$$

where k, α, β represent the Steinmentz coefficient which can be obtained from the core datasheet. Moreover, $\widehat{B}$ indicates the peak magnetic flux density, and V_e indicates the volume of the magnetic components.

Thus, the losses of the power transformer T_r is described as

$$P_{\text{Tr\_loss}} = I^2_{\text{L}_\text{k}\text{\_rms}} \cdot [R_{\text{T}_\text{r}\text{\_p}} + n^2 \cdot R_{\text{T}_\text{r}\text{\_s}}] + k_{\text{Tr}} F_{\text{s}}^{\alpha\_tr} \widehat{B}_{\text{Tr}}^{\beta_{\text{Tr}}} \tag{4.20}$$

where R_{Tr_p} denotes the winding resistor of T_r on the primary side and R_{Tr_s} denotes the winding resistor of T_r on the secondary side. $\widehat{B}_{Tr}$ can be estimated as

$$\widehat{B}_{Tr} \approx 2V_1/(\pi^2 N_{Trp} A_{Tr}) \tag{4.21}$$

where N_{Trp} is the primary winding turns and A_{Tr} is the effective magnetic cross-section.

Moreover, the losses of the series inductor L_r can be calculated as

$$P_{L_r\_loss} = I^2_{L_k\_rms} \cdot R_{L_r\_p} + k_{L_r} f_s^{\alpha\_L_r} \widehat{B}^{\beta_{Lr}}_{L_r} \tag{4.22}$$

where R_{Lr} denotes the winding resistor of L_r. The $\widehat{B}_{Lr}$ can be estimated as

$$\widehat{B}_{Lr} \approx \mu_{eff} \mu_0 I_{L_k-\max} / l_{L_r} \tag{4.23}$$

where μ_{eff} denotes the equivalent relative permeability of a gapped core, μ_0 represents the winding turns, I_{Lk_max} indicates the maximum current following through L_r, and l_{Lr} represents the effective magnetic path length of the series inductor L_r, respectively.

Finally, the major power losses can be calculated as

$$P_{All\_loss} = P_{S\_loss} + P_{Tr\_loss} + P_{Lr\_loss} \tag{4.24}$$

The efficiency η of the DAB converter can be defined as

$$\eta = \frac{P_o}{P_o + P_{All\_loss}} \tag{4.25}$$

4.4 Deep-reinforcement-learning-aided modulation for power electronic converters

4.4.1 Efficiency optimized modulation for DAB converter based on deep deterministic policy gradient (DDPG) algorithm

In this section, the DDPG algorithm is adopted to improve the efficiency of the DAB converter with the TPS modulation under the ZVS constraints, which implies finding the most suitable phase-shift angles (D_1, D_2, D_3) to minimize the power losses. Moreover, D_1, D_2, and D_3 should be chosen within the eight operation modes as illustrated in Figure 4.3, and subject to the constraints mentioned in Equation (4.1). To avoid an effect on the operation of the physical circuit, offline optimization is adapted in this section.

4.4.1.1 DDPG algorithm

In the optimization control stage of the DAB converter, the main goal is to find the appropriate action that enables the converter to operate under the minimum power losses according to the current operating state (V_1, V_2, P_o). The problem of efficiency optimization can be formulated as a Markov decision process [20].

As the state-of-the-art DRL algorithm, DDPG is very suitable for solving the Markov decision process, which is used for solving the optimal control strategy in this section. In the DDPG algorithm, the policy function maps the state (V_1, V_2, P_o) into the desired output (D_1, D_2, D_3), and the critic function maps the state and the action pair (V_1, V_2, P_o, D_1, D_2, D_3) into the expected outcome R_t. The basic

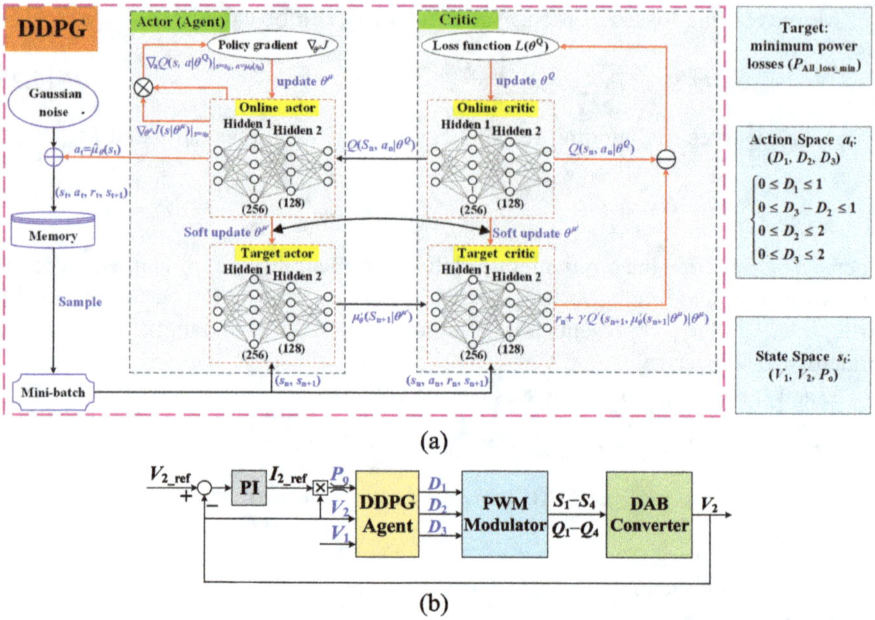

Figure 4.5 The scheme of the DDPG optimization method for the DAB converter. (a) Workflow of the DDPG methodology for the DAB converter [21]. (b) The basic control diagram of a DDPG-based DAB converter.

workflow of the DDPG methodology and the basic control diagram of the DDPG-based DAB converter are depicted in Figure 4.5.

In the DDPG algorithm, the actor function $a_t = \mu_\theta(s_t|\theta^{\mu})$ presents the deterministic policy, which is parameterized by θ^{μ}, while the critic function $Q^{\theta^Q}(s, a)$ approximates the action-value function, which is parameterized by θ^Q. The action-value function $Q^{\theta^Q}(s_t, a_t)$ can be described as

$$Q^{\theta^Q}(s_t, a_t) = \mathbb{E}_{s_{t+1}\sim E}\left[r(s_t, a_t) + \gamma Q^{\theta^Q}(s_{t+1}, \mu_\theta(s_{t+1}))\right] \tag{4.26}$$

where $Q^{\theta^Q}(s_t, \mu_\theta(s_t))$ indicates the expected return by using the policy π to seek the appropriate action in state s. Moreover, the critic network is updated by minimizing the loss function, which can be described as

$$L(\theta^Q) = \mathbb{E}_{\mu'_\theta}[(Q(s_t, a_t|\theta^Q) - y_t)^2] \tag{4.27}$$

where μ'_θ represents a different behavior and y_t indicates the target value generated by the target critic network and target actor network. Moreover, y_t is also dependent on θ^Q, and the target action value y_t is given by

$$y_t = r(s_t, a_t) + \gamma Q'(s_{t+1}, \mu'_\theta(s_{t+1}|\theta^{\mu'})|\theta^{Q'}) \tag{4.28}$$

Furthermore, the parameter of the actor network is updated by the gradient descent $\nabla_{\theta^\mu} J$, where

$$\begin{aligned}\nabla_{\theta^\mu} J &= \mathbb{E}_{\mu'_\theta}\left[\nabla_{\theta^\mu} Q(s,a|\theta^Q)|_{s=s_t,a=\mu_\theta(s_t|\theta^\mu)}\right] \\ &= \mathbb{E}_{\mu'_\theta}\left[\nabla_a Q(s,a|\theta^Q)|_{s=s_t,a=\mu_\theta(s_t)} \cdot \nabla_{\theta^\mu}\mu_\theta(s|\theta^\mu)|_{s=s_t}\right]\end{aligned} \tag{4.29}$$

During the updating process of the action-value function $Q(s_t, a_t|\theta^Q)$, the value y_t keeps changing, which causes the convergence of the critic network to be no longer guaranteed. To solve these problems, the target network is adopted. The "soft update" is used instead of copying the weight parameters into the target network directly. This indicates a new critic network $Q(s_t, a_t|\theta^{Q'})$ and a new actor network $\mu_\theta(s_t|\theta^{\mu'})$ are created to update the appropriate network parameters, where Q' and μ' are the network soft weights, respectively. These soft weights Q' and μ' updated from the main network to the target network, as shown in (4.30).

$$\underset{\alpha=0.001}{\text{softupdate}}\begin{cases}\theta^{Q'} \leftarrow \alpha\theta^Q + (1-\alpha)\theta^{Q'} \\ \theta^{\mu'} \leftarrow \alpha\theta^\mu + (1-\alpha)\theta^{\mu'}\end{cases} \tag{4.30}$$

where α represents the update rate. Based on this, the updating speed of the target values can be restricted effectively, and the stability of the training procedure can be improved significantly.

In the DDPG algorithm, the exploration strategy is used to provide an effective exploration for the continuous action domain. To enhance the exploration ability, the Gaussian noise $\mathcal{N}(\mu_\theta(s_t|\theta_t^\mu), \sigma)$is added to the action strategy μ, and further construct a new exploratory strategy μ' [22,23]. The specific process can be expressed as

$$\mu'_\theta(s_t) = \mu_\theta(s_t|\theta_t^\mu) + \mathcal{N}(\mu_\theta(s_t|\theta_t^\mu), \sigma) \tag{4.31}$$

where $\mathcal{N}(\mu_\theta(s_t|\theta_t^\mu), \sigma)$ is sampled from the noise process and relayed in the practical environment. The parameter σ is constant at the initial time, while σ is attenuated at a constant rate once the memory capacity reaches the upper limit.

4.4.1.2 Reward function for minimizing the power losses

Aiming to solve the minimum operating loss modulation strategy for the DAB converter through the DDPG algorithm, an objective function $f(x)$ can be given by

$$f(x) = \min[P_{\text{All\_loss}}] \tag{4.32}$$

where $P_{\text{All\_loss}}$ denotes the overall power losses of the DAB converter, which is given in (4.24). Besides, there is a nonlinear equality constraint $P_o = P_{ow}$ in the DAB converter, where P_{ow} is the expected transmitted power and P_o indicates the real output power in the training process. To operate as close as possible to the expected P_o after training, a penalty function $\varphi(D_1, D_2, D_3)$ is defined as

$$\varphi(D_1, D_2, D_3) = (P_{ow} - P_o)^2 \tag{4.33}$$

The minimum value zero of $\varphi(D_1, D_2, D_3)$ can be obtained when $P_{ow} = P_o$. Thus, the reward function $F(D_1, D_2, D_3)$ should be constituted by a fitness function P_{loss} and a penalty function $\varphi(D_1, D_2, D_3)$, which is given by

$$F(D_1, D_2, D_3) = -[\omega \cdot P_{loss}(D_1, D_2, D_3) + \alpha \cdot \varphi(D_1, D_2, D_3)] \tag{4.34}$$

where ω and α represent the penalty coefficients. The value of α is chosen as 100. Moreover, the value of ω is chosen as 1 if the soft switching performance is achieved for all power switches (S_1–S_4, Q_1–Q_4) as shown in (7.9), otherwise, the value of ω will be chosen as 10 to avoid switching loss. If the power losses and power errors are decreased, the value of the reward function will be increased appropriately. Thus, the minimum power losses and transmitted power error under soft switching can be acquired by maximizing the reward function $F(D_1, D_2, D_3)$.

4.4.1.3 Training of the DDPG algorithm

The major purpose of the DDPG training is to generate an optimized control strategy with minimum power losses under different operating conditions. Therefore, it is first interesting to select the critical parameters in the procedure of the DDPG algorithm. In this case, the same structure is adopted in the critic networks and actor networks, and each network contains the same two hidden layers, where 256 neurons and 128 neurons are used, respectively, in appropriate hidden layers. Moreover, the appropriate critical parameters can be obtained through the *Grid search* method [24]. More specifically, the max episode N is designed as 20,000, and each episode contains 10 steps. The corresponding training process is depicted in Table 4.1, and the workflow of the DDPG-based agent training process is found in Figure 4.5. As shown in Table 4.1, the states (V_1, V_2, P_o) are sampled

Table 4.1 Training process of the DDPG algorithm

Algorithm: DDPG algorithm
Input parameters: Operation environments (V_1, V_2, P_o)
Output parameters: Phase-shift angles (D_1, D_2, D_3)
1: Randomly initialize the weights of the actor network θ^μ and the critic network θ^Q
2: Initialize target network μ' and Q' by: $\theta^{\mu'} \leftarrow \theta^\mu, \theta^{Q'} \leftarrow \theta^Q$
3: for episode = 1, 2, ... N do
4: Observe current state s_1
5: for actor = 1, 2, ... 10 do
6: Select action $a_t = \mu(s_t|\theta^\mu) + n_t$ according to the deterministic policy μ and exploration rate
7: Execute action a_t, observe the reward r_t and observe new state s_{t+1}
8: Store transition (s_t, a_t, r_t, s_{t+1}) in replay buffer R
9: Sample a random mini-batch of transitions (s_i, a_i, r_i, s_{i+1}) from replay buffer R
10: $y_t = r(s_t, a_t) + \gamma Q(s_{t+1}, \mu(s_{t+1})|\theta^Q)$
11: Update critic net by Equation (4.27)
12: Update actor net by Equation (4.29)
13: Update target net by Equation (4.30)
15: end for
16: end for

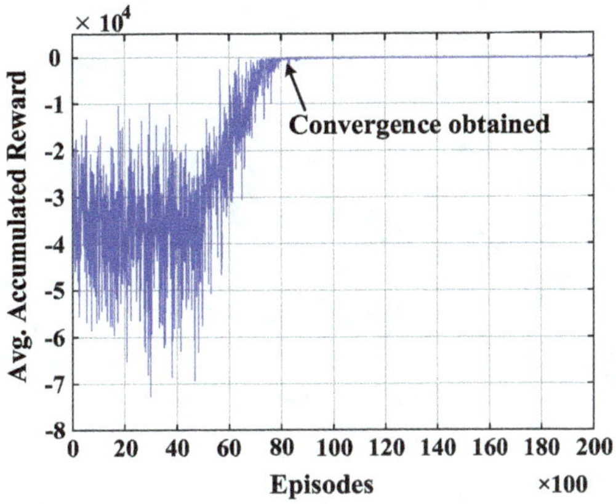

Figure 4.6 Average accumulated reward in the evaluation of policies learned during the DDPG training process [21]

randomly in each episode. In this case, V_1 is varied from 100 to 140 V, V_2 is maintained at 40 V, and P_o is varied from 0 W to 200 W.

Figure 4.6 shows the average accumulated reward in the evaluation of policies learned during the DDPG training process under different operation conditions, where the average accumulated reward of each episode is the average value over 10-time steps. From Figure 4.6, the average accumulated reward is relatively low in the early stages of the learning process (around 8,000 episodes), suggesting that the agent is devoted to gathering experiences by exploring the action space randomly. After 8,000 episodes, the reward is going to keep increasing and reach convergence eventually, which means the optimization control strategy is learned rapidly.

After the whole training process, the trained agent will be stored in a Microprogrammed-Control-Unit (MCU). When the operation environments (V_1, V_2, P_O) are detected in the MCU, the trained agent can be regarded as a fast surrogate predictor that will map the input environments (V_1, V_2, P_O) into the corresponding control decision (D_1, D_2, D_3) immediately.

4.4.1.4 Experimental analysis

Aiming to demonstrate the effectiveness of the proposed DTPS scheme, a 200 W hardware prototype is built, as shown in Figure 4.7. Moreover, SPS, DPS [25], EPS [13], MPS [26], RAPS [27], and the proposed DTPS scheme are implemented in the DSP (TMS320F28335). Relevant design specifications are summarized in Table 4.2.

Figures 4.8 and 4.9 illustrate the steady-state experimental waveforms for the forward power flow of v_{AB}, $v_{CD,}$ and i_{Lk} under different P_o. Solid circles indicate the leading bridge arms, while dotted circles indicate the lagging bridge arms. Figure 4.8 illustrates the steady-state experimental results when V_1 is equal to 100 V and V_2 is equal to 40 V, while Figure 4.9 shows the steady-state experimental results when V_1 is equal to 140 V and V_2 is equal to 40 V. From Figures 4.8(a), 4.8(b), 4.9(a), and 4.9(b),

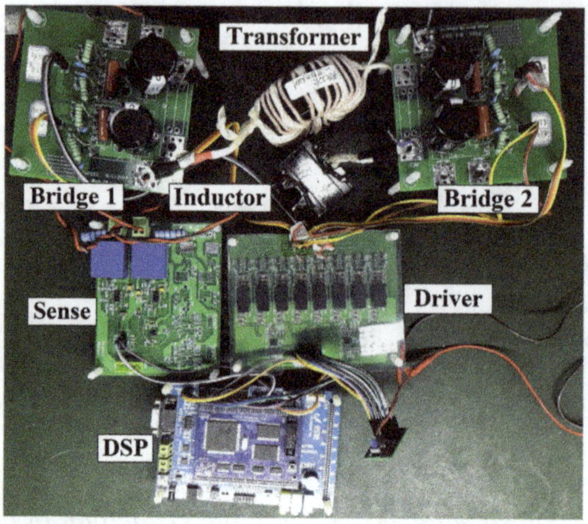

Figure 4.7 Experimental hardware platform for 200 W nominal power dual active bridge converter [21]

Table 4.2 Design specifications of the DAB converter

Symbol	Parameter	Value
P_o	Rated power	200 W
V_1	Input voltage	100 V–140V
V_2	Output voltage	40 V
S_1–S_4	Primary switches	IPB17N25S3-100 (250 VDC, 17 A)
Q_1–Q_4	Secondary switches	IRL530NPBF (100 VDC, 17 A)
n:1	Transformer turn ratio	1:1 (PQ5050+TPW33)
L_k	Series inductor	41 μH (PQ3535+TPW33)
f_s	Switching frequency	50 kHz

it can be observed that all the power switches (S_1–S_4, Q_1–Q_4) can obtain the ZVS turn-on under medium and high-power conditions. Besides, the operation mode is matched to Figure 4.3(a). According to Figures 4.8(c) and 4.9(c), the leading arm of the primary side (S_1/S_2) can realize the ZVS turn-on, while the other three arms can realize the ZCS turn-off, under low power operation. Furthermore, the operation mode is matched to Figure 4.3(c). On this basis, soft switching can be obtained for different operation conditions, because the ZVS constraints are added to the DDPG algorithm.

Figure 4.10 shows the steady-state experimental waveforms for the backward power flow of v_{AB}, v_{CD}, and i_{Lk} under different P_o when V_1 = 100 V, V_2 = 40 V. According to Figure 4.10, all the power switches can realize the soft switching under different load conditions. Furthermore, the experimental waveforms from Figures 4.8 to 4.10 suggest that the oft-switching performances are similar between backward power flow and forward power flow modes. More specifically, the ZVS can be obtained for the leading arm of the primary side, and the ZCS can be

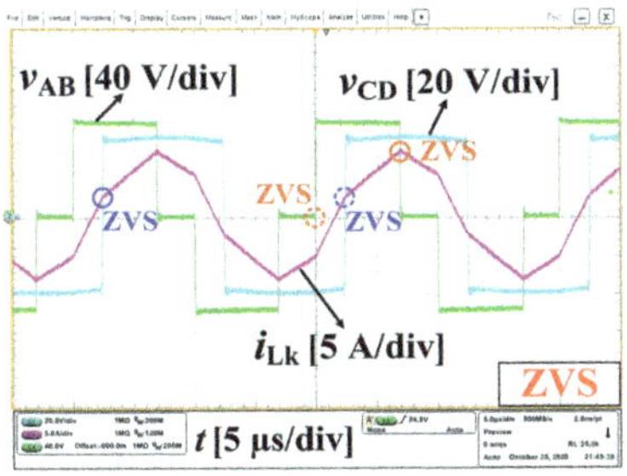

(a)

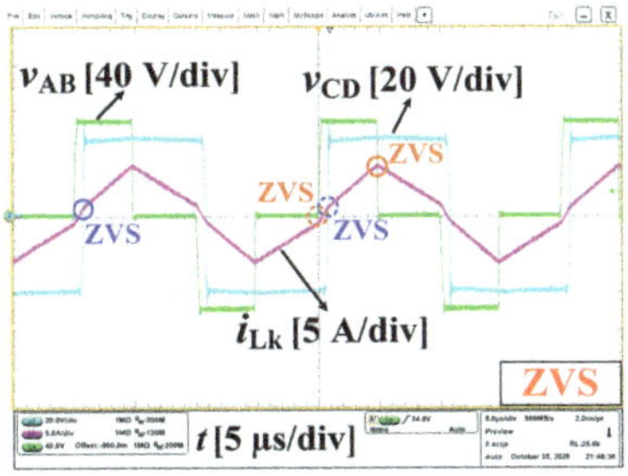

(b)

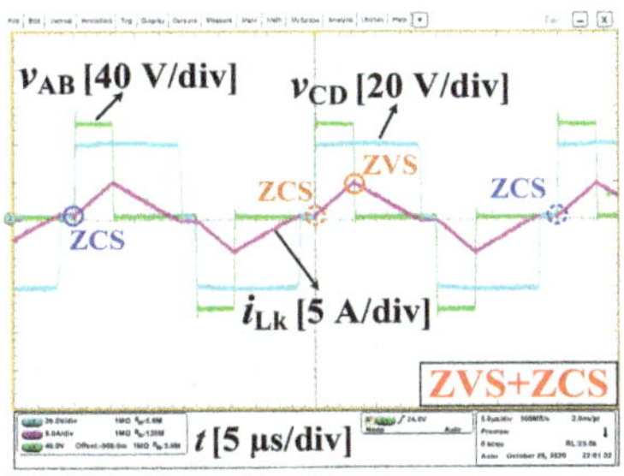

(c)

Figure 4.8 Experimental waveforms of the forward power flow of v_{AB}, v_{CD}, and i_{Lk} under different P_o when $V_1 = 100$ V, $V_2 = 40$ V: (a) $P_o = 200$ W. (b) $P_o = 140$ W. (c) $P_o = 80$ W [21]

obtained for the other three bridge arms under light load conditions, while all the power switches can realize the ZVS under full load conditions.

Dynamic experimental waveforms for a load transition and the corresponding analysis are illustrated in Figure 4.11. From Figure 4.11, the output voltage v_{out} and the current i_{Lk} can quickly restore stability, and slight oscillations are observed during the load-changing condition. Since the trained DDPG agent is used as an implicit function, immediately mapping the input environments into corresponding phase-shift angles (D_1, D_2, D_3), a fast-dynamic response can be obtained.

Figure 4.12 depicts the curves of the efficiency between the theoretical calculations and the experimental measurements, where DTPS', RAPS', and SPS' represent the curves of the theoretically calculated results, and DTPS, RAPS, and SPS represent the curves of the measured results. Figure 4.12 suggests that the measured efficiency is a bit lower than the theoretical measurements because the unknown losses are ignored in the used loss model.

Figure 4.12(a) shows the curves of the efficiencies with respect to P_{o} when V_1 is equal to 100 V, V_2 is equal to 40 V. According to Figure 4.12(a), it is obvious that the efficiency of the RLPS and the proposed DTPS scheme is always higher than the SPS, particularly at low power conditions. Moreover, these efficiency curves are getting closer as the transmitted power increases. Remarkably, the maximum measured efficiency of the proposed DTPS scheme is about 92.5% at 140 W, and

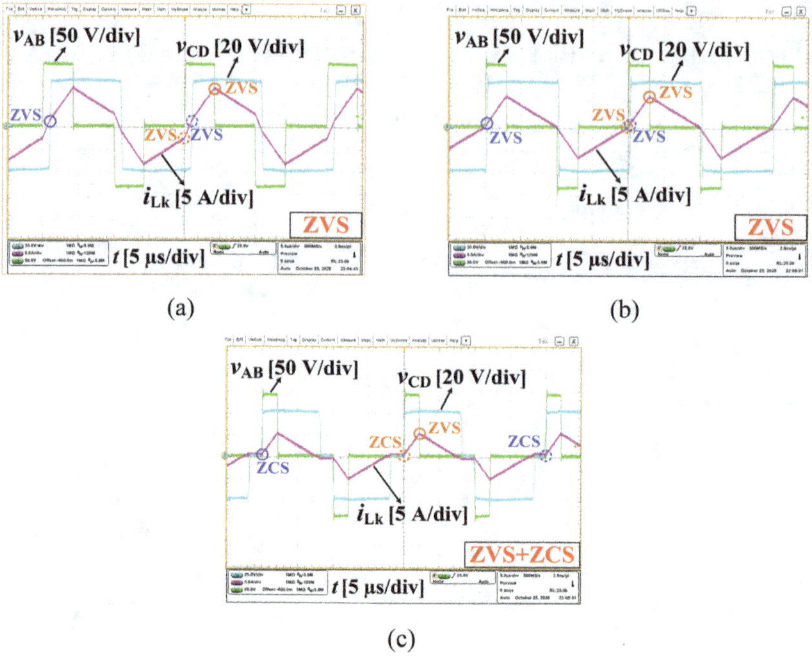

Figure 4.9 Experimental waveforms of the forward power flow of v_{AB}, v_{CD}, and i_{Lk} under different P_o when V_1 = 140 V, V_2 = 40 V: (a) P_o = 200 W. (b) P_o = 140 W. (c) P_o = 80 W [21]

the efficiency is increased by about 6.3% compared to the conventional SPS modulation. Figure 4.12(b) shows the variation of the efficiencies varied with P_o when V_1 is equal to 140 V and V_2 is equal to 40 V. Remarkably, the trends of these efficiency curves shown in Figure 4.12(b) are analogous to Figure 4.12(a), where the maximum efficiency of the proposed DTPS scheme is about 90.6% at 200 W, and the efficiency is increased by about 10.8% compared to the SPS method. Furthermore, the efficiency is improved by 29.4% when the transmitted power P_o = 80 W compared to the SPS. From Figure 4.12, the variation tendency of the proposed DTPS scheme is very close to the RLPS, while the efficiency is slightly higher than the RAPS scheme in most cases.

In summary, the proposed DTPS scheme can reduce power losses and guarantee soft switching performance under all operating conditions. Moreover, the trained agent of the DDPG algorithm provides the phase-shift angles (D_1, D_2, D_3) in real-time online for the continuous operation range without a look-up table, which provides an accurate and rapid dynamic response performance. Moreover, the whole training process of the proposed DTPS scheme can be accomplished at once [28,29], which can reduce the computation complexity and improve the training accuracy compared with the iterative algorithms [7,30,31], RL method [32], and RL-ANN method [27]. Based on the above analysis, the experimental results are in good agreement with the theoretical analysis.

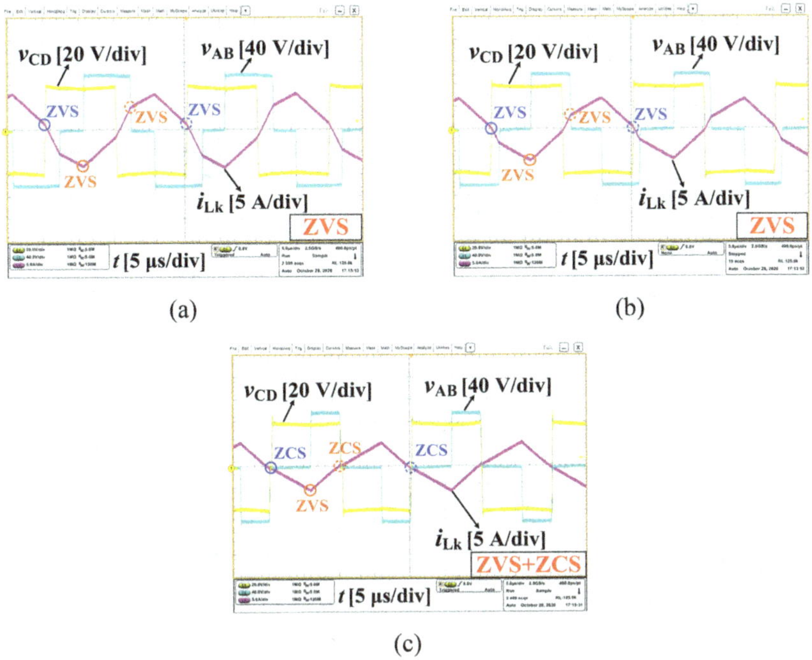

Figure 4.10 Experimental waveforms of the backward power flow of v_{AB}, v_{CD}, and i_{Lk} under different P_o when $V_1 = 100$ V, $V_2 = 40$ V: (a) $P_o = 180$ W. (b) $P_o = 150$ W. (c) $P_o = 100$ W [21]

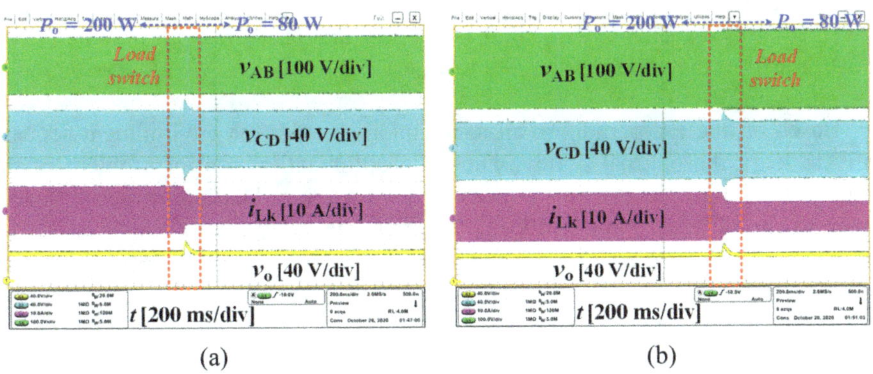

Figure 4.11 Dynamic performance of the DAB converter during a load transition: (a) Experimental waveforms of the load change condition when $V_1 = 100$ V. (b) Experimental waveforms of the load change condition when $V_1 = 140$ V [21]

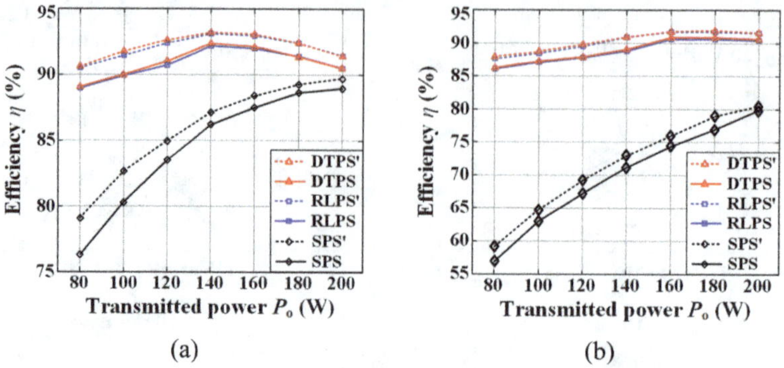

Figure 4.12 Efficiency curves between the theoretical calculations and the experimental measurements. (a) Efficiency with respect to P_o when V_1 = 100 V, V_2 = 40 V. (b) Efficiency with respect to P_o when V_1 = 140 V, V_2 = 40 V [21].

4.4.2 Minimum reactive power control for the DAB converter based on DDPG algorithm and harmonic analysis method

This section introduces a DRL-based method to obtain the optimal control variables for reducing the reactive power flow in the DAB converter. The principle of the introduced DRL method for the DAB converter and the basic control diagram of the DRL-based DAB converter are shown in Figure 4.13.

4.4.2.1 Deep reinforcement learning-based optimization

As an advanced algorithm of the DRL method, DDPG is suitable for solving complex and multi-dimensional optimization problems in the continuous action space [11,29]. The DDPG algorithm is used in this chapter as an off-line policy network to find the control variables for the DAB converter, which correspond to the minimum reactive power.

Based on the RL paradigm, an agent interacts with the environment sequentially to learn the optimal policy. Thus, the optimization control problem of the minimum reactive power can be modeled as a Markov Decision Process (MDP) under stochastic environments [33]. Moreover, the MDP is usually composed of four elements (S, A, P, R), where S is the state space, A denotes the action space, P is the state transition probability function, and R is the reward function. At each step k, an RL agent perceives a state $s_k \in S$, and selects an action $a_k \in S$ according to the policy $\pi(a_k|s_k)$, where $\pi(a_k|s_k)$ indicates the policy function, which maps states s to actions a. After that, the RL agent will receive a corresponding reward $r_k \sim R(s_k, a_k)$ and devise to the next state s_{k+1}. The cumulative discounted reward G_t is used, and it is defined as

$$G_t = \sum\nolimits_{k=0}^{\infty} \gamma^k \cdot R_{k+t} \tag{4.35}$$

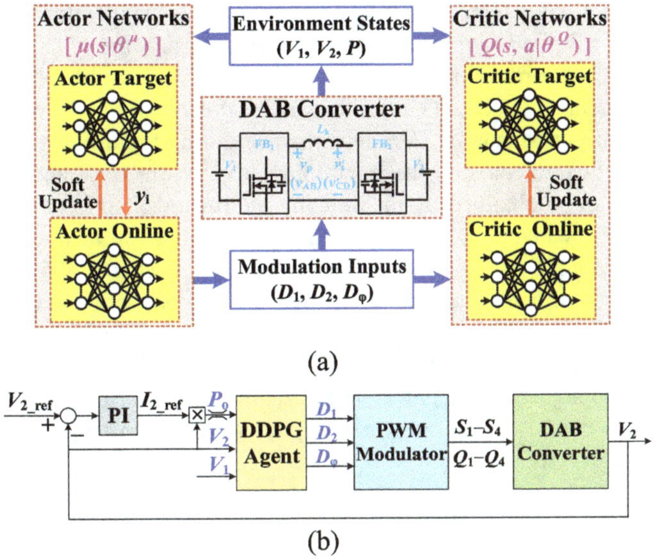

Figure 4.13 The scheme of the DRL optimization method is based on the harmonics analysis method for the DAB converter. (a) Structure diagram of the proposed DDPG method for the DAB converter [17]. (b) The basic control diagram of a DDPG-based DAB converter.

where γ denotes the discount factor and it is limited to [0–1].

Based on this, the aim of maximizing the value of the reward function R is further adjusted to maximize the discounted reward G_t. Moreover, the action-value function $Q^\pi(s_t, a_t)$ is defined in the DDPG algorithm as

$$Q(s_t, a_t) = \mathbb{E}[G(s_t, a_t) + \mathbb{E}_{a_{t+1}}[Q(s_{t+1}, a_{t+1})]] \tag{4.36}$$

In the DAB converter, the environment features consist of the input voltage V_1, the output voltage V_2, and the objective active power P_o'. The active power and operation performance are dependent on the phase-shift angles (D_1, D_2, D_φ) in the harmonic analysis method. Thus, in the proposed DDPG algorithm, the state space S is defined as $S = [V_1, V_2, P_o']$, and the action space A is defined as $A = [D_1, D_2, D_\varphi]$.

Due to the policy agent being improved with trial and error after the explorations [11], it is important to design an appropriate reward function. The aim here is to acquire the optimal actions with respect to the minimum reactive power and the minimum active power error, where the reward function R is set as

$$R(D_1, D_2, D_\varphi) = -[\delta \cdot |Q| + \xi \cdot (P_o - P_o')^2] \tag{4.37}$$

where P_o' is the expected output active power and ξ is the penalty factor of the active power. Remarkably, for the training of the DDPG algorithm, a large value of ξ will cause poor reactive power performance, while a small value of ξ will cause a large active power error. Through trial and error during the simulation, ξ is chosen as 200. δ is set to 1 when the ZVS constraints are satisfied, as mentioned in inequality (4.9). However, a large value of δ should be given once the ZVS is lost, which indicates a small reward value should be given, to avoid such poor action being learned by the agent of the DDPG algorithm. Here, δ is chosen as 10 when inequality (4.9) is not satisfied to ensure a ZVS performance during the training process. Thus, the ZVS can be guaranteed, and the minimum active power error and the minimum reactive power can be obtained by maximizing the reward function R.

As illustrated in Figure 4.13, the DDPG algorithm uses an actor-critic-based framework, which contains two eponymous ingredients (actor network and critic network), where each ingredient is made up of two networks (main network and target network). Moreover, the actor network is used to adjust the weight θ^{μ} in the policy function $\mu(s|\theta^{\mu})$ by fitting a state (V_1, V_2, P_o') to the corresponding action (D_1, D_2, D_{φ}), while the critic network is used to adjust the weight θ^Q in the action-value function $Q(s, a|\theta^Q)$.

The weight θ^Q of the critic network is updated by minimizing the loss function $L(\theta^Q)$, which is given by

$$L(\theta^Q) = E_{(s,a)}[(Q(s_t, a_t|\theta^Q) - y_t)^2] \tag{4.38}$$

where $y_t = r_t(s_t, a_t) + \gamma Q(s_{t+1}, \mu(s_t|\theta^{\mu})|\theta^Q)$. The parameter θ^{μ} of the actor network is updated by using the policy gradient below.

$$\begin{aligned} \nabla_{\theta^{\mu}} J^{\theta^{\mu}} &\approx \mathbb{E}_{s_t \sim \rho^{\beta}}[\nabla_{\theta^{\mu}} Q(s, a|\theta^Q)|_{a=\mu(s|\theta^{\mu})} \nabla_{\theta^{\mu}} \mu(s|\theta^{\mu})] \\ &= \mathbb{E}_{s_t \sim \rho^{\beta}}[\nabla_a Q(s, a|\theta^Q)|_{a=\mu^{\theta}(s)} \nabla_{\theta^{\mu}} \mu(s|\theta^{\mu})] \end{aligned} \tag{4.39}$$

where ρ represents the discounted distribution, and β denotes the specific policy of the current policy π.

Aiming to enhance the stability and reliability of the learning process, two distinct target networks, which are named the actor target network $\mu'(s|\theta^{\mu'})$ and the critic target network Q' $(s, a|\theta^{Q'})$ are added to the actor network and critic network respectively, as depicted in Figure 4.13. In each iteration, these soft weights ($\theta^{\mu'}$ and $\theta^{Q'}$) are updated according to the following formulas.

$$Soft\,update \begin{cases} \theta^{Q'} \leftarrow \tau\theta^Q + (1-\tau)\theta^{Q'} \\ \theta^{\mu'} \leftarrow \tau\theta^{\mu} + (1-\tau)\theta^{\mu'} \end{cases} \tag{4.40}$$

where $\tau << 1$, and τ represents the soft update coefficient. Moreover, Table 4.3 shows the detailed training process of the DDPG algorithm. For the DDPG algorithm, the environments [V_1, V_2, P_o] are used for the training process. During the training of the DDPG algorithm, the training data (State S) is sampled randomly in the range of the environments mentioned above. Besides, the strategy of continuous action space in the DDPG algorithm is learned by a deep neural network. On this basis, after the training of the DDPG algorithm is completed, the trained agent will be downloaded to a digital signal processor (DSP). In practice, once the environments [V_1, V_2, P_o] are detected, such an agent, like an implicit function, will map these input parameters into the corresponding phase shift angles (D_1, D_2, D_{φ}) instantly. Therefore, such a trained agent can dramatically reduce the memory

Table 4.3 Training process of the DDPG algorithm

Algorithm: Offline Learning with the DDPG algorithm
Input: Environments $[V_1, V_2, P_o]$
Output: Phase shift angles (D_1, D_2, D_φ)
1: Randomly initialize the weights of actor network θ^μ and critic network θ^Q
2: Initialize target network μ' and Q' by: $\theta^{\mu'} \leftarrow \theta^\mu, \theta^{Q'} \leftarrow \theta^Q$
3: **for** episode=1 to max-episode **do**
4: Observe current state s_1
5: **for** actor =1, 2, ... 10 **do**
6: Select action $a_t = \mu(s_t|\theta^\mu) + n_t$ according to the deterministic policy μ and exploration rate
7: Execute action a_t, observe the reward r_t and observe new state s_{t+1}
8: Store transition (s_t, a_t, r_t, s_{t+1}) in replay buffer R
9: Sample a random mini-batch of transitions (s_i, a_i, r_i, s_{i+1}) from replay buffer R
10: $y_t = r(s_t, a_t) + \gamma Q(s_{t+1}, \mu(s_{t+1})|\theta^Q)$
11: Update critic net by using the loss function $L(\theta^Q)$ as
(4.38):$L(\theta^Q) = E_{(s,a)}[(Q(s_t, a_t|\theta^Q) - y_t)^2]$
12: Update actor net by using the policy gradient as (4.39):

$$\begin{aligned}\nabla_{\theta^\mu} J^{\theta^\mu} &\approx \mathbb{E}_{s_t \sim \rho^\beta}[\nabla_{\theta^\mu} Q(s, a|\theta^Q)|_{a=\mu(s|\theta^\mu)} \nabla_{\theta^\mu}\mu(s|\theta^\mu)] \\ &= \mathbb{E}_{s_t \sim \rho^\beta}[\nabla_a Q(s, a|\theta^Q)|_{a=\mu^\theta(s)} \nabla_{\theta^\mu}\mu(s|\theta^\mu)]\end{aligned}$$

13: Update the target net by using the policy gradient as (4.39):

$$\begin{aligned}\nabla_{\theta^\mu} J^{\theta^\mu} &\approx \mathbb{E}_{s_t \sim \rho^\beta}[\nabla_{\theta^\mu} Q(s, a|\theta^Q)|_{a=\mu(s|\theta^\mu)} \nabla_{\theta^\mu}\mu(s|\theta^\mu)] \\ &= \mathbb{E}_{s_t \sim \rho^\beta}[\nabla_a Q(s, a|\theta^Q)|_{a=\mu^\theta(s)} \nabla_{\theta^\mu}\mu(s|\theta^\mu)]\end{aligned}$$

15: end for
16: end for

---

allocation and the computational time in comparison to the RL method [32] because the look-up table is not needed in this case. Moreover, due to the strategy of continuous action space in the DDPG algorithm being learned by a deep neural network, the trained agent can provide an optimal control strategy under all continuous operation conditions. Based on this, the proposed DRL method is more suitable for online real-time control. It is worth noting that the trained agent can still provide corresponding phase shift angles (D_1, D_2, D_φ), even if the environments $[V_1, V_2, P_o]$ are out of the previous training range.

4.4.2.2 Experimental analysis

Aiming to verify the feasibility and correctness of the proposed AI-based optimized method (AIO), a 200 W experimental prototype is built, where Figure 4.14 shows a photograph of the experimental prototype. The system specifications are summarized in Table 4.4. The parameters of the built DAB prototype are illustrated in Figure 4.13.

The main hyperparameters of the DDPG algorithm are the number of neural layers, the cell of each hidden layer, and the learning rate. Aiming to find the appropriate hyperparameters for the DDPG algorithm, a *Grid Search* method is adopted [24]. The number of neural layers is chosen from 0 to 5 and the step size defaults to 1, the number of hidden layers is chosen from 10 to 150 and the step size

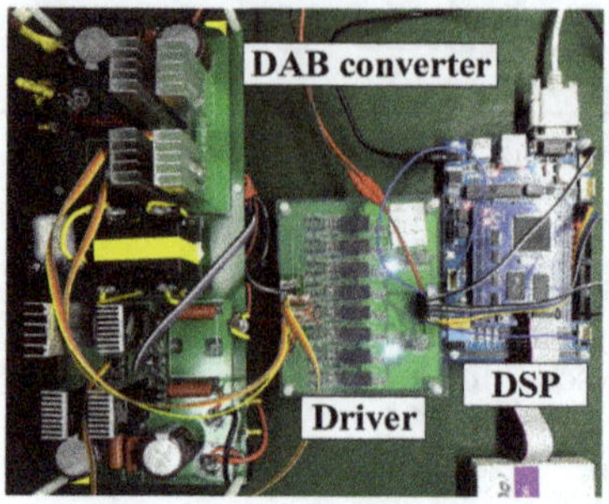

Figure 4.14 Experimental prototype for 200 W nominal power [17]

Table 4.4 System specification of the DAB converter

Parameter	Value
Input voltage V_1 (V)	100–140
Output voltage V_2 (V)	40
Switching frequency f_s (kHz)	50
Inductor L_k (μH)	41 (PQ3535+TPW33)
Turns ratio of the transformer (N:1)	1:1 (PQ5050+TPW33)

Table 4.5 Training parameters adopted in the DDPG algorithm for the DAB converter

Hyperparameter	Value
Learning rate of critic network (λ_c)	0.002
Learning rate for actor network (λ_a)	0.001
Soft update coefficient (τ)	0.001
Replay buffer size (M)	40,000
Max Episode (N)	10,000
Step size of each episode	10
Mini-batch size (m)	32

defaults to 10, and the learning rate is in the range of [10^{-8},0.1]. Similarly, other hyperparameters are also set to a general range. Based on this, two hidden layers with 100 and 100 neurons, respectively, are designed for both the critic and actor networks. Other detailed hyperparameters of the DDPG algorithm are summarized in Table 4.5.

Moreover, the activation functions used in the critic networks are chosen as the rectified linear function for both the hidden layers and the output layer [34,35]. The tanh activation unit and the soft plus activation unit are chosen as the activation functions in the actor network for the output layer.

According to the training parameters adopted in the DDPG algorithm as shown in Table 4.5, the Max Episode N is chosen as 10,000, and the step size of each episode is chosen as 10. Based on hyperparameters shown in Table 4.5, the whole training process can be accomplished in approximately 30 min, where the training of the ANN is running on the *Windows* 10 operating system powered by an Intel(R) Core(TM) i7-8700 CPU @3.20GHz 3.19 GHz.

Based on this, detailed experimental results and corresponding analyses are described as follows.

Figure 4.15 shows the measured experimental waveforms for the forward power flow under different operation conditions when V_2 = 40 V, where purple circles indicate the switching performance of the leading leg in the primary side, purple dotted circles indicate the switching performance of the lagging leg in the primary side, blue circles indicate the switching performance of the leading leg in the secondary side, and blue dotted circles indicate the switching performance of the lagging leg in the secondary side. According to Figures 4.15(a) and 4.15(c), all of the power switches can obtain the ZVS performance. Figures 4.15(b) and 4.15(d) suggest that the leading leg of the primary side can obtain the ZVS turn-on, while

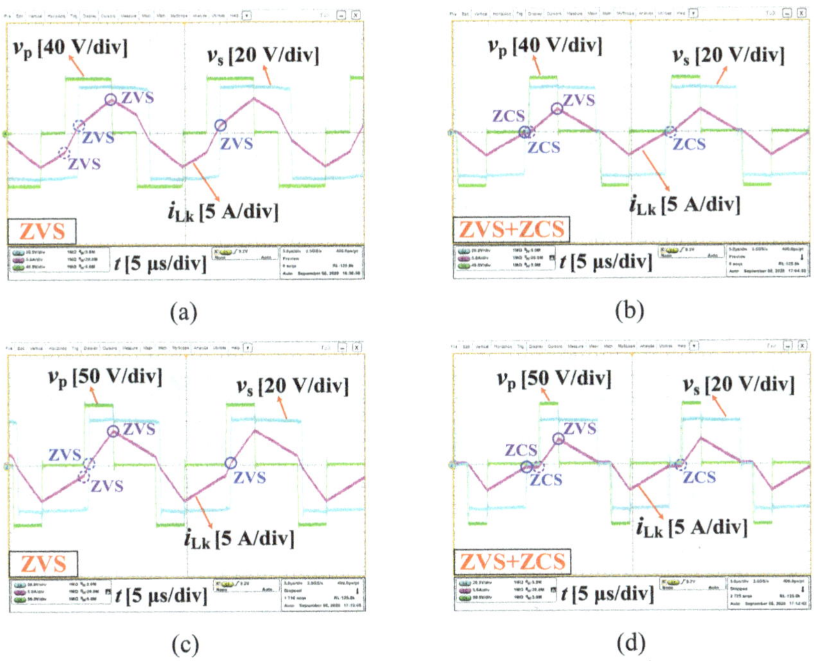

Figure 4.15 Experimental waveforms for the forward power flow of the primary side voltage v_p, secondary side voltage v_s, and current i_{Lk} under different operations with V_2 = 40 V. (a) V_1 = 100 V, P_o = 180 W. (b) V_1 = 100 V, P_o = 100 W. (c) V_1 = 140 V, P_o = 180 W. (d) V_1 = 140 V, P_o = 100 W [17].

the other three legs can obtain the ZCS turn-off. Figure 4.16 shows the measured experimental waveforms for the backward power flow under different operation conditions when $V_1 = 100$ V, $V_2 = 40$ V. Remarkably, the soft switching conditions in Figure 4.16 are analogous to Figure 4.15. The experimental results shown in Figures 4.15 and 4.16 indicate that the soft switching performance can be guaranteed under different operating conditions.

Figure 4.17 illustrates the dynamic performance for the power transition and the input voltage variation when $V_2 = 40$ V. It can be seen from Figure 4.17(a), that the output voltage v_{out} and the current i_{Lk} can quickly return to a steady state after P_o is changed from 180 W to 100 W when $V_1 = 140$ V, $V_2 = 40$ V. Figure 4.17(b)

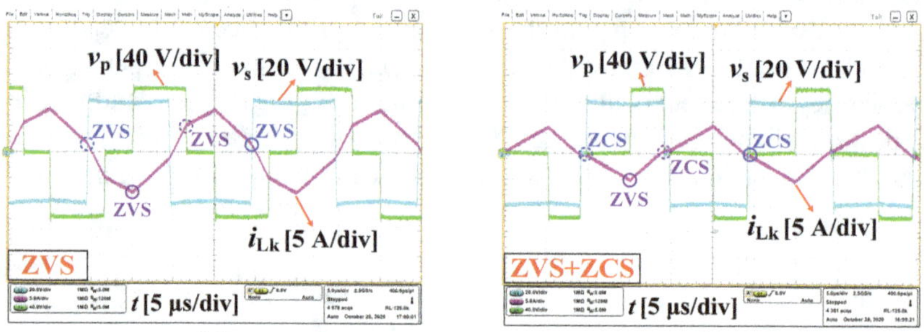

(a) (b)

Figure 4.16 Experimental waveforms for the backward power flow of the primary side voltage v_p, secondary side voltage v_s, and current i_{Lk} under different operations with $V_1 = 100$ V and $V_2 = 40$ V. (a) $P_o = 180$ W. (b) $P_o = 100$ W [17].

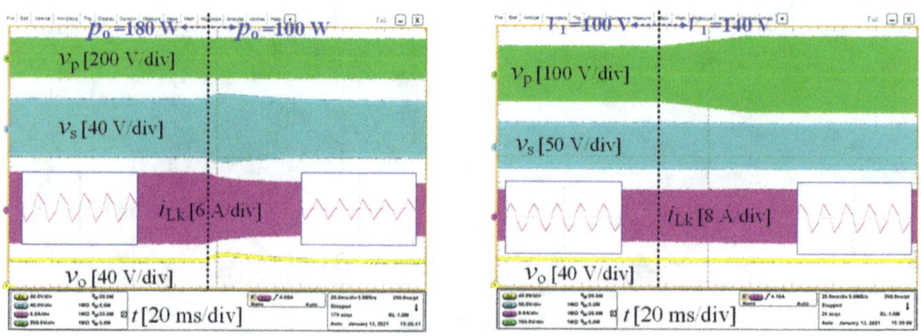

(a) (b)

Figure 4.17 Dynamic performance for the power transition and the input voltage variation. (a) Waveforms of the load transition with $V_1 = 140$ V, $V_2 = 40$ V. (b) Curves of the input voltage variation with $V_2 = 40$ V and $P_o = 180$ W [17].

shows also a good dynamic performance when the V_1 is changed from 100 V to 140 V when $V_2 = 40$ V and $P_o = 180$ W.

Table 4.6 shows the quantitative reactive power comparison between the genetic Algorithm (GA) [30], ant colony optimization (ACO) [36], particle swarm optimization (PSO) [7], minimum current stress phase shift control (MPS) [37], Q-learning optimized triple-phase-shift control (QTPS) [32], and the proposed AIO scheme under different operation conditions when $V_2 = 40$ V. The corresponding histogram of the reactive power comparison between different algorithms is illustrated in Figure 4.18. According to Table 4.5 and Figure 4.18, such

Table 4.6 Reactive power comparison with different optimization methods

Operating conditions	**Algorithms**	**Reactive power (Var)**	
		$P_o = 100$ **W**	$P_o = 180$ **W**
$V_1 = 100$ V $V_2 = 40$ V	GA [30]	115.1	325.1
	ACO [36]	110.3	330.3
	PSO [7]	103.2	320.2
	MPS [37]	98.1	312.4
	QTPS [32]	95.2	310.1
	AIO	101.1	302.1
$V_1 = 140$ V $V_2 = 40$ V	GA [30]	115.3	330.1
	ACO [36]	120.5	320.5
	PSO [7]	105.4	299.3
	MPS [37]	100.1	293.2
	QTPS [32]	98.4	295.4
	AIO	96.0	282.8

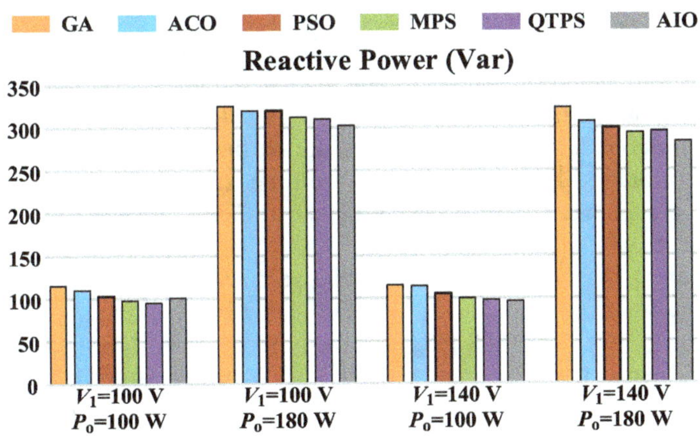

Figure 4.18 Reactive power comparison between different optimization methods with $V_2 = 40$ V at different power loads [17]

six optimization methods have similar reactive power, while the proposed AIO scheme shows lower reactive power at most operating points.

According to the system specifications of the DAB converter mentioned in Table 4.4, the rated power is 200 W, and the input voltage V_1 is varied from 100 V to 140 V. In this application scenario, supposing that the default interval of the V_1 and P_o is set to 0.5 V and 0.5 W in the heuristic algorithm (GA, ACO, PSO) utilization. At least 32,000 optimization processes must be required in these heuristic algorithms, which will cause a heavy calculation burden. Compared to these heuristic algorithms, the QTPS method can complete the training process without such complicated optimization processes, while at least 32,000 training results have to be saved in the look-up table, similar to the heuristic algorithm. Thus, the memory size of such a look-up table is at least 4 MB for the heuristic algorithm (GA, ACO, PSO) and the QTPS method, and such large memory will also slow the query speed down. Remarkably, the generated look-up table is discrete, which cannot provide a continuous control strategy for the DAB converter. In practice, if the detected environments [V_1, V_2, P_o] are not found in such a look-up table directly, the closest environments [V_1, V_2, P_o] and corresponding phase shift angles (D_1, D_2, D_φ) will be selected, which will cause certain power errors and degrade the operation performance of the DAB converter.

However, in the proposed AIO scheme, the whole training process is completed in one go, and the trained agent, like an implicit function, can map the environments [V_1, V_2, P_o] under the whole continuous operation range into the corresponding phase shift angles (D_1, D_2, D_φ) instantly. Moreover, the memory size of the trained agent is less than 200 KB. Based on this, the proposed AIO scheme is more suitable for online real-time control compared to the heuristic algorithms (GA, ACO, PSO) and the QTPS method.

Furthermore, the LPTD model is used in the MPS and the QTPS methods. According to [36], five different switching modes are adopted in the MPS method, and each switching mode should participate in the calculation for specific operating conditions to find the global optimum. Therefore, multiple partial differential calculations will be required for different time intervals and operation modes, which are complex and inconvenient. According to [32], eight different operation modes are considered during the training of the Q-learning algorithm in this method, while a unified harmonic analysis model is used during the training of the DDPG algorithm in the proposed AIO scheme. Thus, the proposed AIO scheme can have less calculation complexity and burden during the training process compared to the QTPS method.

Figure 4.19 illustrates the curve of the reactive power and efficiency with respect to the active power between SPS, DPS [1], EPS [10], PSO-optimized phase shift control (POPS) [7], QTPS [32], and the proposed AIO scheme. According to Figure 4.19, the reactive power and efficiency in the proposed AIO scheme and the QTPS method are obviously better than the other four modulation methods. Besides, the curves of the proposed AIO scheme are similar to those of the QTPS method, while the reactive power and efficiency in the proposed AIO scheme are better than those of the QTPS method for most operating conditions.

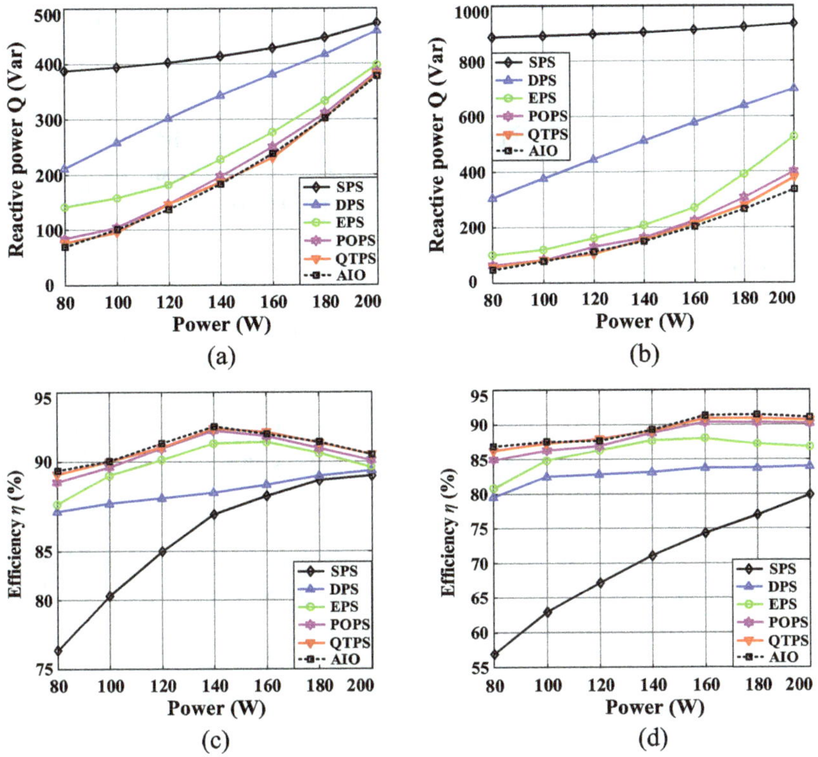

Figure 4.19 Reactive power and corresponding efficiency with respect to the active power with $V_2 = 40$ V measured for different modulation schemes. (a) Reactive power when $V_1 = 100$ V. (b) Reactive power when $V_1 = 140$ V. (c) Efficiency when $V_1 = 100$ V. (d) Efficiency when $V_1 = 140$ V [17].

More specifically, the maximum efficiency in the proposed AIO scheme is high up to 92.6% at $P_o = 100$ W operation, according to Figure 4.19(c), and up to 90.7% when $P_o = 200$ W, according to Figure 4.19(d).

4.4.3 Twin delayed deep deterministic policy gradient (TD3) aided variable-frequency triple phase shift control for DAB converter

In this section, the efficiency optimization of the DAB converter can be equivalent to solving the optimal phase shift angles (D_1, D_2, D_φ) and switching frequency f_s for whole operation conditions with the help of the twin delayed deep deterministic policy gradient (TD3) algorithm. As an improved algorithm built on the DDPG, the TD3 algorithm is more suitable for solving complicated decision-making problems in real-time continuous action space. Here, the TD3 algorithm is adopted to find the optimal control strategy with maximum conversion efficiency and soft switching performance under four control degrees of freedom. A more detailed description of the TD3 algorithm-based optimization is summarized in the following subsections.

4.4.3.1 Basic principle of the TD3 algorithm

A block diagram of the TD3 algorithm and the basic control diagram of the TD3-based DAB converter are shown in Figure 4.20, where an actor-critic-based framework is used [38]. More specifically, one actor network and two critic networks are included in the TD3 architecture, where the actor network is parameterized by ϕ, and two critic networks are parameterized by θ_1 and θ_2, respectively. To ensure stability for the learning of the neural networks, one target actor network (parameterized by ϕ') and two target critic networks (parameterized by θ'_1 and θ'_2) are adopted, by tracking the updates of the (online) actor and the critic networks slowly.

For the TD3 algorithm, these two critic networks are updated simultaneously through the recursive Bellman equation, which can be described as follows:

$$Q^{\pi}(s_t, a_t) = r(s_t, a_t) + \gamma Q^{\pi}(s_{t+1}, \pi(s_{t+1})) \tag{4.41}$$

Aiming to approximate the optimal Q-function, the mean-squared Bellman error function is used to indicate how closely θ_1 and θ_2 satisfy the Bellman

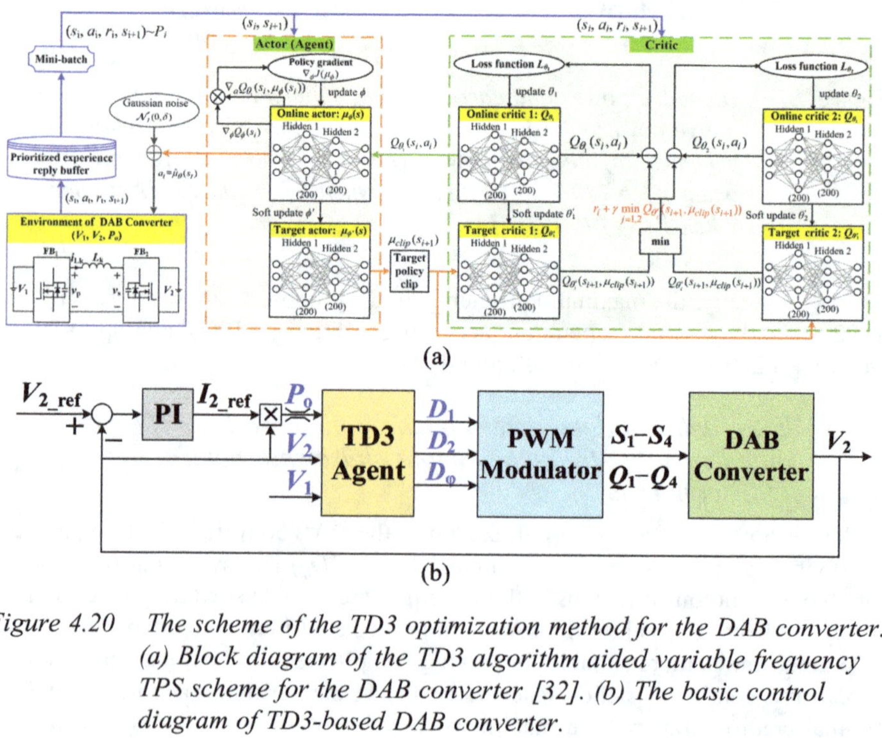

Figure 4.20 The scheme of the TD3 optimization method for the DAB converter. (a) Block diagram of the TD3 algorithm aided variable frequency TPS scheme for the DAB converter [32]. (b) The basic control diagram of TD3-based DAB converter.

equation, which is given by

$$L(\theta_i, \Re) = \mathbb{E}_{(s,a,r,s')\sim\Re}[(Q_{\theta_i}(s,a) - y|(r,s'))^2], i = 1,2 \tag{4.42}$$

Moreover, the minimum value of the two Q-values is used to form the targets in the Bellman error loss functions, which is referred to as the clipped double Q-learning. Thus, the overestimated bias of the Q-values can be reduced effectively, compared with the DDPG algorithm. Besides, a small random noise component ϵ is added to the target policy, which is called target policy smoothing regularization. This added random noise component ϵ is clipped to keep the target close to the original action and avoid over-fitting on the narrow peaks of Q-values. Here, Gaussian noise is adopted. On this basis, the pair of critics are updated as follows [39]:

$$\begin{cases} y(r,s') = r(s,a) + \gamma \min\limits_{i=1,2} Q_{\theta'_i}(s', \pi_{\phi'}(s') + \epsilon) \\ \epsilon \sim \text{clip}(\mathcal{N}(0,\sigma), -c, c) \end{cases} \tag{4.43}$$

The actor network can be updated through the gradient ascent of the expected return $\nabla_\phi J(\phi)$, which is given by

$$\nabla_\phi J(\phi) = \mathbb{E}_{s\sim\mathcal{R}}[\nabla_a Q_{\theta_1}(s,a)|_{a=\pi(s)} \nabla_\phi \pi_\phi(s)] \tag{4.44}$$

Besides, the actor network will be updated after d times updates of the critic network, to reduce the cumulative errors caused by temporal difference learning [40]. Notably, a larger value of d would result in a larger benefit with respect to accumulating errors, while training the actor for too few iterations would cripple learning [38]. After comprehensive consideration, the value of the d is chosen as 2 in this section, in order to reduce the accumulation of errors resulting from temporal difference learning, while ensuring the stability and effectiveness of the training process.

Moreover, a soft update is adopted for the target networks (target actor network, target critic network), which is similar to the DDPG algorithm. Specifically, the soft weights (ϕ', θ'_1, and θ'_2) are updated through the following rules.

$$\textit{Soft update}\begin{cases} \theta'_i \leftarrow \tau\theta_i + (1-\tau)\theta'_i, \text{i} = 1,2 \\ \phi' \leftarrow \tau\phi + (1-\tau)\phi' \end{cases} \tag{4.45}$$

Here, the TD3 algorithm is used to optimize the conversion efficiency under the TPS modulation with varying switching frequency f_s. Thus, the state space can be defined as $S = [V_1, V_2, P_o]$, and the action space can be defined as $A = [D_1, D_2, D_\varphi, f_s]$.

4.4.3.2 Training consideration of the TD3 algorithm

For the TD3 algorithm, the goal of improving conversion efficiency can be equivalent to decreasing power losses and minimizing the power error of the DAB converter. Moreover, the TD3 agent is improved during the exploration process with constant trial and error. Based on this, it is of primary interest to select a suitable objective function $R(D_1, D_2, D_\varphi, f_s)$, which can be defined as follows.

$$R(D_1, D_2, D_\varphi, f_s) = \alpha \cdot \frac{P_o}{P_o + P_{all\_loss}} - \beta \cdot |P_o - P'_o| \tag{4.46}$$

where α indicates the penalty factor for the conversion efficiency, β represents the penalty factor for the power error, P_{all_loss} indicates the total power losses, and P'_o is the target transmitted power for current state s. Notably, a large value of parameter α may increase the power error, while a large value of parameter β may decrease the efficiency improvement capability, for the training of the TD3 algorithm. In practical application, we have tried to train the TD3 agent with different values of parameters α and β through the trial and error method. More specifically, the value of α is limited from 0 to 20, and the value of β is limited from 0 to 2. After each trial, more than 10,000 random sets of data have been selected for analysis, such as the average operation efficiency and average power error. Furthermore, the stability and effectiveness of the training process have also been analyzed. As a result, to balance these three goals mentioned above, the value of α is chosen as 10 if the ZVS constraints are satisfied under current action a, otherwise, it is chosen as 5. Moreover, the value of β is set as 0.1.

In addition, it is also important to choose appropriate hyperparameters and network architecture for the TD3 algorithm. More specifically, the main structure and hyperparameters of the TD3 algorithm are the learning rate, the number of neural layers, and the cell of each hidden layer. First, the *Adam optimizer* [41] is applied to optimize the learning rate of the critic network (λ_c) and actor network (λ_a) during the progress of training, with initial values of 0.001 and 0.0001, respectively. Based on this, stability and convergence can be improved effectively during the training of the TD3 algorithm [42]. After that, other hyperparameters are determined by using the *Grid Search* method [24]. For example, two hidden layers with 200 units and 200 units are used for both actor network and two critic network. Besides, other detailed hyperparameters of the TD3 algorithm are summarized in Table 4.7. Furthermore, both the hidden layers consist of the rectified linear (*ReLu*) activation units, and the *tanh* activation units are adopted for the output of the actor network [34].

Table 4.7 illustrates the training process of the TD3 algorithm. Notably, the state s is selected randomly between the operation environments (V_1, V_2, P_o) for

Table 4.7 Hyperparameters used in TD3 algorithm for optimization of DAB converter

Hyperparameter	**Value**
Initial learning rate of critic network (λ_c)	0.001
Initial learning rate for actor network (λ_a)	0.0001
Soft target update factor (τ)	0.005
Replay buffer size (M)	50,000
Max Episode (T)	12,000
Mini-batch size (m)	32

Table 4.8 Training process of the TD3 algorithm used for the DAB converter

Algorithm Type: TD3 algorithm
Input: Operation environments (V_1, V_2, P_o)
Output: Phase shift angles (D_1, D_2, D_φ)+Switching frequency (f_s)
1: Initialize the actor network μ_ϕ with random parameter ϕ. Initialize the critic networks $Q_{\theta_1}, Q_{\theta_2}$ with random parameters θ_1 and θ_2
2: Initialize target network by: $\phi' \leftarrow \phi, \theta_1' \leftarrow \theta_1, \theta_2' \leftarrow \theta_2$
3: Initialize replay buffer R
4: **for** episode =1, 2, … T **do**
5: Select actions with exploration noise $a \sim \pi_\phi(s) + \epsilon$, $\epsilon \sim \mathbf{N}(0, \delta)$ and observe reward r and new state s'
6: Store transition tuple (s, a, r, s') in replay buffer R
7: Sample mini-batch of N transitions from replay buffer R
8: Compute greedy actions for the next states through: $\tilde{a} \leftarrow \pi_{\phi'}(s') + \epsilon$, where $\epsilon \sim \text{clip}(\mathbf{N}(0, \tilde{\sigma}), -c, c)$
9: Compute targets through: $y \leftarrow r(s, a) + \gamma \min_{i=1,2} Q_{\theta_i'}(s', \tilde{a})$
10: Update critics through: $\theta_i \leftarrow \text{argmin}_{\theta_i} N^{-1} \sum (y - Q_{\theta_i}(s, a))^2, i = 1, 2$
11: **if** Episode mod 2=0 **then**
12: Update actor network parameters ϕ by deterministic policy gradient:
$$\nabla_\phi J(\phi) = \frac{1}{N} \sum_{s \sim N} [\nabla_a Q_{\theta_1}(s, a)|_{a=\pi(s)} \nabla_\phi \pi_\phi(s)]$$
13: Update target networks:
$$\textit{Soft update} \begin{cases} \theta_i' \leftarrow \tau\theta_i + (1 - \tau)\theta_i', \text{i} = 1, 2 \\ \phi' \leftarrow \tau\phi + (1 - \tau)\phi' \end{cases}$$
14: **end if**
15: **end for**

each episode. Furthermore, action a can be selected from four consecutive degrees of freedom (D_1, D_2, D_φ, f_s). Specifically, the value of the (V_1, V_2, P_o, D_1, D_2, D_φ, f_s) chosen in the TD3 algorithm should be limited in Table 4.8.

Figure 4.21 maps the average accumulated reward over 100 independent iterations during the TD3 algorithm training process, where the solid line indicates the average value and the shaded area represents the standard deviation. According to Figure 4.21, the whole training process can be divided into three stages, namely the exploring stage, the learning stage, and the converging stage. Initially, in the exploring stage (about 0 to 1,000 episodes), the TD3 agent makes attempts randomly to find a suitable strategy. Second, in the learning stage (about 1,000 to 6,000 episodes), the TD3 agent learns fast and achieves a high reward soon. Finally, in the converging stage (about 6,000 to 12,000 episodes), the training process gradually comes to stability to find the optimal strategy.

After the whole training process, the trained TD3 agent can produce optimal control strategies (D_1, D_2, D_φ, f_s) in real-time for all continuous operating conditions with soft switching performance and maximum conversion efficiency. In practical applications, such trained TD3 agents can be downloaded to a Microprogrammed Control Unit (MCU), e.g., a Digital Signal Processor (DSP). Once the operation environments (V_1, V_2, P_0) are detected by the MCU, such a TD3

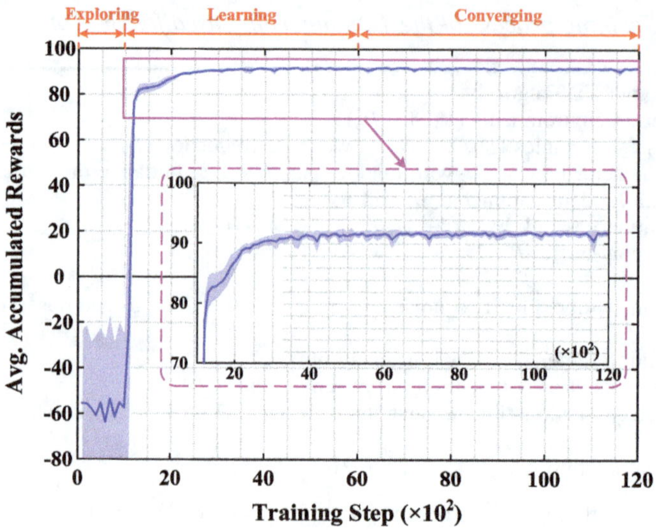

Figure 4.21 Average accumulated rewards over 100 independent iterations during the TD3 algorithm training process [32]

agent acts as a fast surrogate predictor, which can produce the mapped output with optimal control parameters (D_1, D_2, D_φ, f_s) instantly.

4.4.3.3 Experimental analysis

To verify the viability and effectiveness of the proposed TD3 algorithm-aided variable-frequency triple phase shift (TVTPS) scheme, the detailed experimental results are presented and analyzed through a laboratory prototype. Moreover, Figure 4.22 shows the picture of the experimental platform for the DAB converter. In this platform, an oscilloscope is used to monitor the operating state of the prototype, a DC power supply is used to support DC power for the prototype, and a DC electronic load is used to support load current for the prototype. It is worth mentioning that the source and the load are programmable and communicate with the PC for training the AI agent.

The corresponding system specifications are illustrated in Table 4.9.

According to the system specification illustrated in Table 4.9, the input voltage V_1 ranges from 100 V to 140 V, the output voltage V_2 is kept at 40 V, and the rated power is chosen as 200 W. Assuming the power is flowing from the primary side to the secondary side, the ranges of the phase shift angles (D_1, D_2, D_φ) should be restricted to $[0, \pi]$. Moreover, the switching frequency f_s is chosen from 20 kHz to 60 kHz, which is obtained with the help of design considerations related to the magnetic components and power losses [43,44] in the power devices.

Figure 4.23 shows the surface diagram of the four control variables for various input voltages V_1 and transmitted power P_o after training the TD3 algorithm, where Figure 4.23(a) shows the three-phase shift angles (D_1, D_2, D_φ) and Figure 4.23(b) shows the switching frequency f_s. According to Figure 4.23(a), the TPS modulation

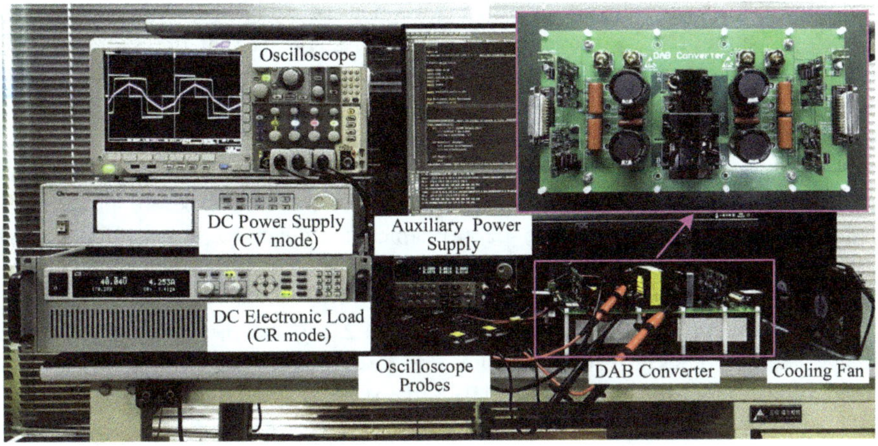

Figure 4.22 Experimental platform for the DAB converter [32]

Table 4.9 System specifications of DAB converter

Parameter	**Value**
Input voltage V_1 (V)	100–140
Output voltage V_2 (V)	40
Switching frequency f_s (kHz)	20–60
Phase shift angles for D_1, D_2, D_φ (rad)	0–π
Rated power (W)	200
Inductor L_k (μH)	41 (PQ3535+TPW33)
Turns ratio of the transformer (N:1)	1:1 (PQ5050+TPW33)

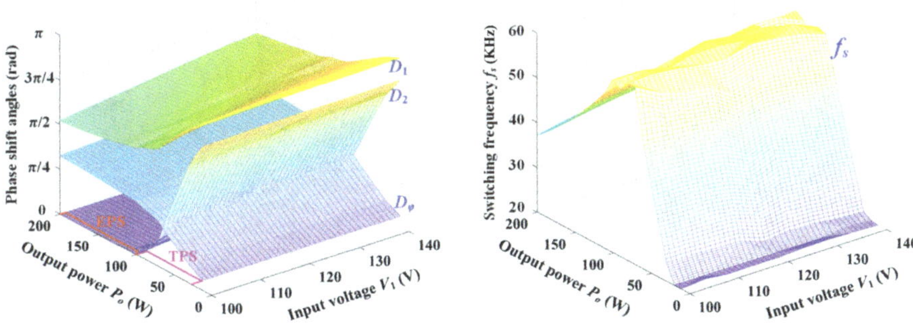

Figure 4.23 A surface diagram of the four control variables under different operation conditions after training the TD3 algorithm. (a) Phase shift angles (D_1, D_2, D_φ). (b) Switching frequency f_s [32].

is adopted at low power levels, and the extend-phase-shift (EPS) modulation is used at high power levels. Moreover, Figure 4.23(b) suggests that the switching frequency f_s is increased with the increase of the output power P_o at low power levels, and vice versa. Based on the above, the proposed TVTPS scheme overcomes the defect of the fixed frequency modulation method.

Figures 4.24 and 4.25 show the forward dynamic experimental waveforms and the corresponding enlarged steady-state experimental waveforms before and after the load variation. Specifically, Figure 4.24(a) depicts the dynamic response for the load changed from 80 W to 200 W, when V_1 = 100 V, V_2 = 40 V. Figure 4.25(a) shows the dynamic response for the load changed from 80 W to 200 W when V_1 = 140 V, V_2 = 40 V. According to Figures 4.24(a) and 4.25(a), the measured actual output voltage v_{out} can quickly recover to 40 V and keep stable after the load is changed, which indicates that the proposed TVTPS scheme has a good dynamic performance.

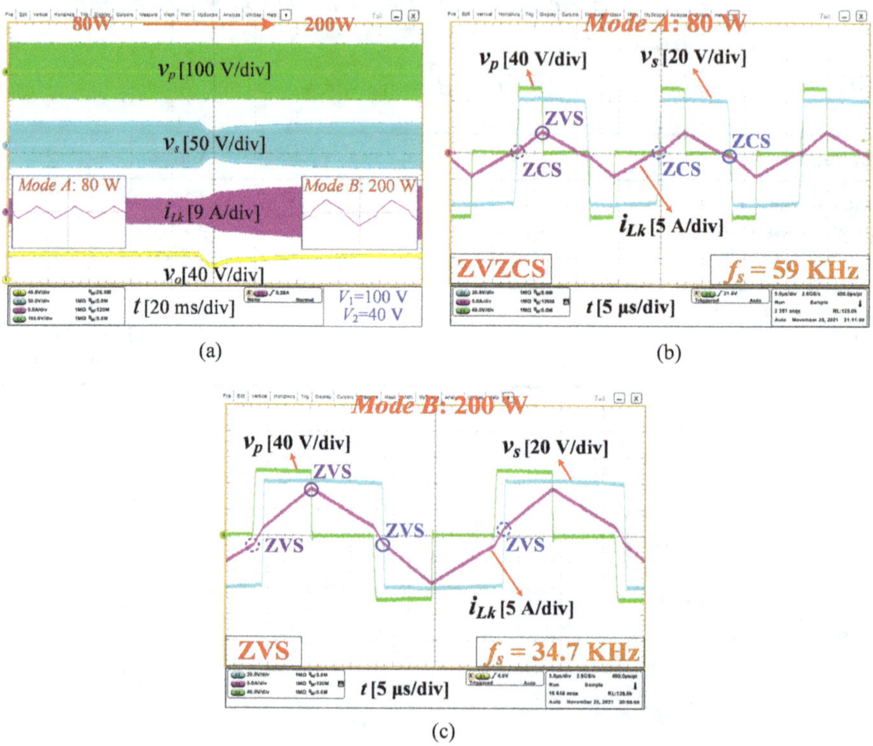

Figure 4.24 Forward dynamic experimental waveforms and enlarged steady-state experimental results before and after load variation. (a) The dynamic response for the load changed from 80 W to 200 W, when V_1 = 100 V, V_2 = 40 V. (b) Enlarged time-base steady-state experimental waveforms before load change (80 W). (c) Enlarged steady-state experimental waveforms after load change (200 W) [32].

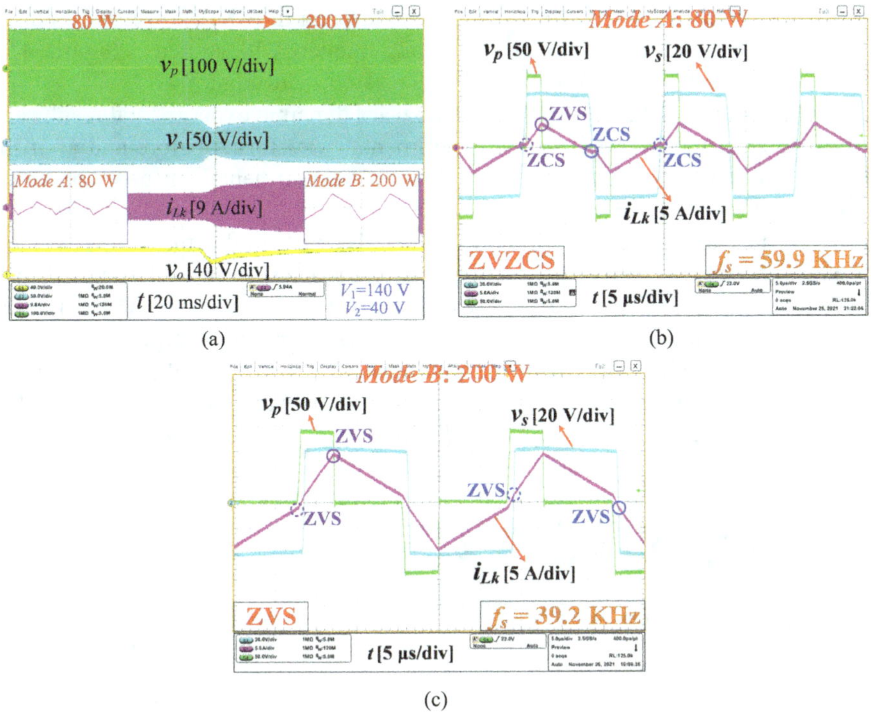

Figure 4.25 Forward dynamic experimental waveforms and enlarged steady-state experimental results before and after load variation. (a) The dynamic response for the load changed from 80 W to 200 W, when V_1 = 140 V, V_2 = 40 V. (b) Enlarged steady-state experimental waveforms before the load change (80 W). (c) Enlarged steady-state experimental waveforms after the load change (200 W) [32].

Figures 4.24(b), 4.24(c), 4.25(b), and 4.25(c) show the corresponding enlarged steady-state experimental waveforms of primary voltage v_p, secondary voltage v_s, inductor current i_{Lk}, and switching frequency f_s. Specifically, *Mode A* maps the experimental results before the load variation, while *Mode B* maps the experimental results after the load variation. Moreover, in the waveforms of the enlarged steady-state experimental waveforms, the purple solid circles display the soft switching performance of the first bridge arm (S_1, S_2), the purple dotted circles display the soft switching performance of the second bridge arm (S_3, S_4), the blue solid circles display the soft switching performance of the third bridge arm (Q_1, Q_2), and the blue dotted circles display the soft switching performance of the fourth bridge arm (Q_3, Q_4). It can be seen from Figures 4.24(b) and 4.25(b), that the Zero-Voltage Zero-Current Switching (ZVZCS) performance can be obtained for light load conditions. Moreover, Figures 4.24(c) and 4.25(c) suggest that ZVS performance can be acquired for heavy load conditions. Therefore, soft switching performance can be obtained for all power switches, due to the soft switching constraints considered for the TD3 algorithm, as aforementioned in (4.9) and (4.46). In addition, the modulation strategy and the switching frequency illustrated in Figures 4.24 (b), 4.24(c), 4.25(b), and 4.25(c) are also consistent with Figure 4.23.

Figures 4.26 and 4.27 illustrate the backward dynamic experimental results and corresponding enlarged steady-state experimental waveforms before and after the variation, which corresponds to the forward experimental results shown in Figures 4.24 and 4.25, respectively.

According to Figures 4.26(a) and 4.27(a), the measured actual output voltage v_{out} and the inductor current i_{Lk} can quickly return to stability after the load is changed from 80 W to 200 W, which is matched with the waveforms in Figures 4.24(a) and 4.25(a). Furthermore, the corresponding enlarged steady-state experimental waveforms of the primary voltage v_p, secondary voltage v_s, inductor current i_{Lk}, and switching frequency f_s before and after the variation are also illustrated in Figures 4.26(b), 4.26(c), 4.27(b), and 4.27(c). More specifically,

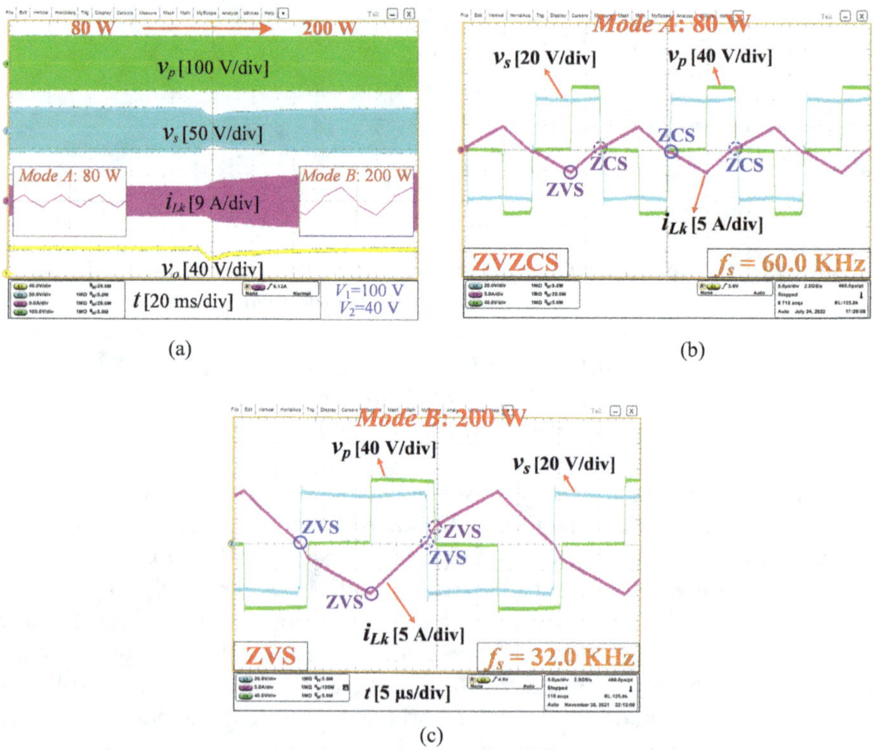

Figure 4.26 Backward dynamic experimental waveforms and enlarged steady-state experimental results before and after load variation. (a) The dynamic response for the load changed from 80 W to 200 W when V_1 = 100 V, V_2 = 40 V. (b) Enlarged time-base steady-state experimental waveforms before the load change (80 W). (c) Enlarged steady-state experimental waveforms after the load change (200 W) [32].

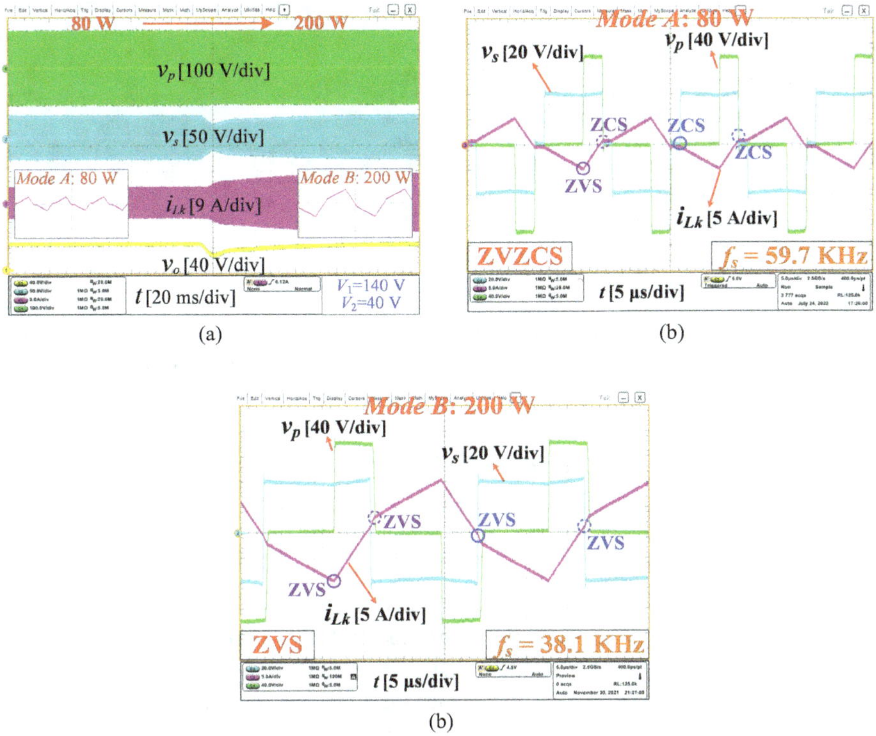

Figure 4.27 Backward dynamic experimental waveforms and enlarged steady-state experimental results before and after load variation. (a) The dynamic response for the load changed from 80 W to 200 W when V_1 = 140 V, V_2 = 40 V. (b) Enlarged steady-state experimental waveforms before the load change (80 W). (c) Enlarged time-base steady-state experimental waveforms after the load change (200 W) [32].

Figure 4.26(b) shows the experimental results at V_1 = 100 V and P_o = 80 W, Figure 4.26(c) illustrates the experimental results at V_1 = 100 V and P_o = 200 W, Figure 4.27(b) depicts the experimental results at V_1 = 140 V and P_o = 80 W, and Figure 4.27(c) demonstrates the experimental results at V_1 = 140 V and P_o = 200 W. It is clearly observed that the soft switching performance and switching frequency shown in Figures 4.26 and 4.27 are well matched with the forward steady-state experimental waveforms in Figures 4.24 and 4.25. More specifically, the ZVS can be guaranteed under high-output power conditions, while the ZVZCS can be obtained under low-output power conditions.

Figure 4.28 shows the curves of conversion efficiency varied with the transmitted power P_o between the EPS [13], particle-swarm-optimization aided TPS (PTPS) modulation method [7], *Q*-learning algorithm based TPS (QTPS) modulation method [32], DDPG algorithm aided TPS (DTPS) modulation method [17], constrained-nonlinear-optimization based variable-frequency triple phase shift

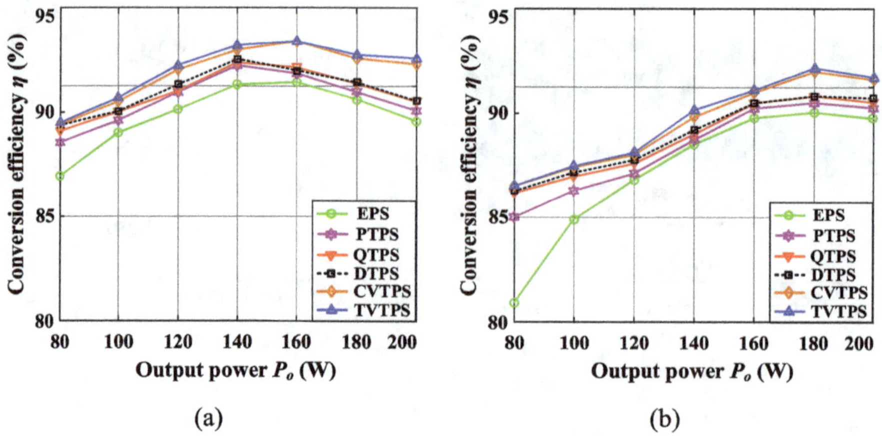

Figure 4.28 Optimized conversion efficiency changed with the transmitted power P_o for different control strategies when V_2 = 40 V. (a) V_1 = 100 V. (b) V_1 = 140 V [32].

(CVTPS) scheme [43], and the proposed TD3 algorithm aided variable-frequency triple phase shift (TVTPS) scheme. More specifically, Figure 4.28(a) shows the curves of the measured efficiency when V_1 = 100 V and V_2 = 40 V, while Figure 4.28(b) shows the curves of the measured efficiency when V_1 = 140 V and V_2 = 40 V.

As can be seen from Figure 4.28, the conversion efficiency of the CVTPS scheme and the proposed TVTPS scheme is higher than the other four constant frequency control methods, particularly for heavy load conditions. Notably, the DTPS scheme has been proven to obtain superior efficiency improvement capability compared with the existing phase shift modulation methods under constant frequency, while the CVTPS scheme and the proposed TVTPS scheme can further enhance the conversion efficiency. This is due to the variable-frequency TPS modulation having four control degrees of freedom, which provides a great potential for efficiency improvement of the DAB converter. As illustrated in Figure 4.28(a), the maximum conversion efficiency of the proposed TVTPS scheme is up to 93.4% at 140 W and has improved by about 1.3% compared with the DTPS scheme. Moreover, the conversion efficiency of the proposed TVTPS scheme is improved by about 2% compared with the DTPS scheme at 200 W. As shown in Figure 4.28(b), the maximum efficiency of the proposed TVTPS scheme is up to 92.2% at 180 W and has improved by about 1.8% compared to the DTPS scheme.

As depicted in Figure 4.28, it is worth noting that the efficiency curves of the proposed TVTPS scheme get close to the CVTPS scheme. However, the conversion efficiency of the proposed TVTPS scheme is slightly higher than the CVTPS scheme in most situations. Notably, the CVTPS scheme belongs to a constrained numerical minimum search method, which indicates a separate optimization

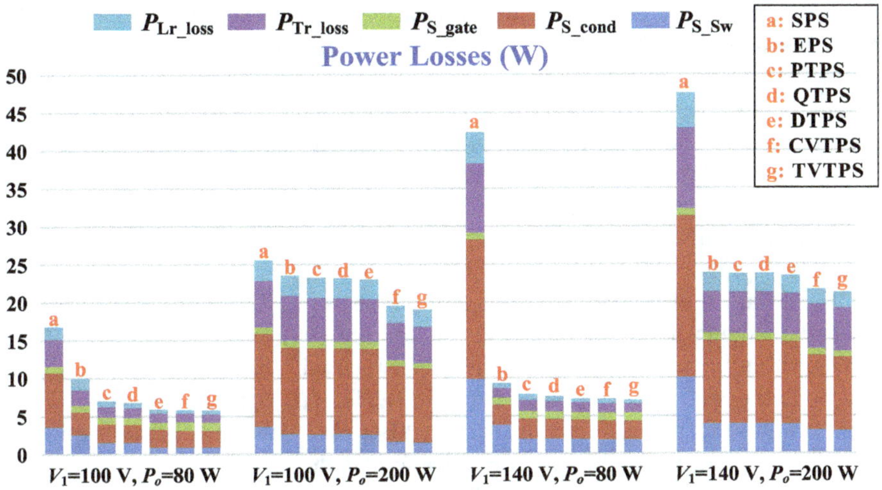

Figure 4.29 Power losses comparison for different control strategies at V_2 = 40 V [32]

process will be needed for each operation condition. Based on this, such a CVTPS scheme possesses relatively high computational complexity, which is time-consuming for implementation. Moreover, an extra look-up table will be needed to save optimized results in practical applications. Thus, such a look-up table cannot provide continuous control in practical applications and may cause corresponding power errors and decrease the conversion efficiency. By comparison, the proposed TVTPS scheme can complete the whole training process at one time, and the trained TD3 agent just acts as a fast surrogate predictor, which can produce optimal control strategies in real-time for all continuous operating conditions. Thus, the proposed TVTPS scheme has a great advantage over the CVTPS scheme.

The detailed power losses can be calculated through the formulas from (4.10) to (4.25). Specifically, the corresponding key parameter values can be extracted from the experimental waveforms in one switching cycle under current operation conditions. Figure 4.29 shows the histogram of the detailed loss distribution between different modulation methods. More specifically, P_{Lr_loss} indicates the power losses caused by the series inductor L_r, P_{Tr_loss} indicates the power losses caused by the power transformer T_r, P_{S_gate} represents the gate driver losses in all power switches, P_{S_cond} is the condition losses in all power switches, and P_{S_Sw} indicates the switching losses in all power switches. Together it can be seen that the proposed TVTPS scheme can reduce power losses effectively, which is matched with the conversion efficiency curves in Figure 4.28.

4.5 Summary

In recent years, with the development of the power electronics-dominated power system, distributed renewable energy generation has become an important research topic. To realize the collection, transformation, and transmission of renewable energy resources, the Dual-Active-Bridge (DAB) DC–DC converter has become a key energy conversion device.

Due to the many nonlinear parameters being included in the DAB DC–DC converter, it is difficult to model accurately using traditional mathematical modeling methods. Moreover, this converter contains many switching devices, and its operating conditions are very complex in practical applications, which leads to a heavy calculation burden, high complexity, and optimization difficulties when solving the optimization modulation strategy.

Faced with such a high-dimensional optimization solution system with complex and changing modeling, artificial intelligence (AI) technology has shown its advantages in many system optimization decisions with large data volumes, complex modeling, and uncertainty, and is suitable for fast optimization solving problems in high-dimensional complex systems. Based on this, this chapter applies AI technology to the comprehensive optimization of the modulation strategy of the DAB DC–DC converter, focusing on the following main contents of this chapter.

1. The Linear Piecewise Time-Domain (LPTD) model of the DAB DC–DC converter is analyzed in detail, and the key waveforms and inductor current expressions are described. Moreover, the unified harmonic analysis model of the DAB DC–DC converter is analyzed in detail, and the current, voltage, power, etc. in the circuit are sorted out into a unified harmonic form for analysis by using the Fourier decomposition method. Based on the above analysis method, the ZVS characteristics and power loss distributions are analyzed for further optimization methods in the DAB converter.
2. To reduce the reactive power of the DAB DC–DC converter, this chapter proposes a deep reinforcement learning (DRL)-aided minimum reactive power modulation scheme. Specifically, as an advanced algorithm of the DRL method, the Deep-Deterministic-Policy-Gradient (DDPG) is used to train an agent offline to solve the optimized modulation strategy. During the training of the DDPG algorithm, the TPS modulation is adopted and the ZVS constraints are considered. The proposed DDPG algorithm-aided TPS modulation strategy can effectively reduce the reactive power and achieve soft-switching performance over the entire operating range of the DAB DC–DC converter.
3. Aiming to further promote the conversion efficiency of the DAB DC–DC converter, this chapter demonstrates a variable-frequency phase shift-based minimum power loss modulation scheme. As an improved algorithm of the DDPG, the twin delayed deep deterministic policy gradient (TD3) algorithm is adopted to solve the optimized modulation strategy under the variable-frequency TPS modulation with the aim of minimizing power losses and keeping the ZVS constraints. The proposed variable-frequency TPS modulation strategy based on the TD3 algorithm using four optimization degrees of freedom can improve the conversion efficiency and steady-state performances of the DAB DC–DC converter using soft switching performance compared with other optimization strategies.

References

[1] N. Hou and Y. W. Li, "Overview and Comparison of Modulation and Control Strategies for a Nonresonant Single-Phase Dual-Active-Bridge DC–DC Converter," *IEEE Transactions on Power Electronics*, vol. 35, no. 3, pp. 3148–3172, 2020.

[2] R. W. De Doncker, D. M. Divan and M. H. Kheraluwala, "A Three-Phase Soft-Switched High Power Density DC/DC Converter for High Power Applications," *Conference Record of the 1988 IEEE Industry Applications Society Annual Meeting*, vol.1, Pittsburgh, PA, USA, 1988, pp. 796–805.

[3] M. N. Kheraluwala, R. W. Gascoigne, D. M. Divan and E. D. Baumann, "Performance Characterization of a High-Power Dual Active Bridge DC-to-DC Converter," *IEEE Transactions on Industry Applications*, vol. 28, no. 6, pp. 1294–1301, 1992.

[4] Z. Zhang and K.-T. Chau, "Pulse-Width-Modulation-Based Electromagnetic Interference Mitigation of Bidirectional Grid-Connected Converters for Electric Vehicles," *IEEE Transactions on Smart Grid*, vol. 8, no. 6, pp. 2803–2812, 2017.

[5] N. Hou, Y. Zhang and Y. W. Li, "A Load-Current-Estimating Scheme with Delay Compensation for the Dual-Active-Bridge DC–DC Converter," *IEEE Transactions on Power Electronics*, vol. 37, no. 3, pp. 2636–2647, 2022.

[6] B. Zhao, Q. Song, W. Liu and Y. Sun, "Overview of Dual-Active-Bridge Isolated Bidirectional DC–DC Converter for High-Frequency-Link Power-Conversion System," *IEEE Transactions on Power Electronics*, vol. 29, no. 8, pp. 4091–4106, 2014.

[7] H. Shi, H. Wen, Y. Hu and L. Jiang, "Reactive Power Minimization in Bidirectional DC–DC Converters Using a Unified-Phasor-Based Particle Swarm Optimization," *IEEE Transactions on Power Electronics*, vol. 33, no. 12, pp. 10990–11006, 2018.

[8] J. C. Peterson, D. D. Bourgin, M. Agrawal, D. Reichman and T. L. Griffiths, "Using Large-Scale Experiments and Machine Learning to Discover Theories of Human Decision-Making," *Science*, vol. 372, no. 6547, pp. 1209−1214, 2021.

[9] M. Haenlein and A. Kaplan, "A Brief History of Artificial Intelligence: On the Past, Present, and Future of Artificial Intelligence," *California Management Review*, vol. 61, no. 4, pp 5−14, 2019.

[10] S. Zhao, F. Blaabjerg and H. Wang, "An Overview of Artificial Intelligence Applications for Power Electronics," *IEEE Transactions on Power Electronics*, vol. 36, no. 4, pp. 4633–4658, 2021.

[11] D. Cao, W. Hu, J. Zhao, *et al.*, "Reinforcement Learning and Its Applications in Modern Power and Energy Systems: A Review," *Journal of Modern Power Systems and Clean Energy*, vol. 8, no. 6, pp. 1029–1042, 2020.

[12] F. Krismer and J. W. Kolar, "Accurate Power Loss Model Derivation of a High-Current Dual Active Bridge Converter for an Automotive Application,"

IEEE Transactions on Industrial Electronics, vol. 57, no. 3, pp. 881–891, 2010.

[13] B. Liu, P. Davari and F. Blaabjerg, "An Optimized Hybrid Modulation Scheme for Reducing Conduction Losses in Dual Active Bridge Converters," *IEEE Journal of Emerging and Selected Topics in Power Electronics*, vol. 9, no. 1, pp. 921–936, 2021.

[14] G. Xu, D. Sha, J. Zhang and X. Liao, "Unified Boundary Trapezoidal Modulation Control Utilizing Fixed Duty Cycle Compensation and Magnetizing Current Design for Dual Active Bridge DC–DC Converter," *IEEE Transactions on Power Electronics*, vol. 32, no. 3, pp. 2243–2252, 2017.

[15] G. G. Oggier, G. O. García and A. R. Oliva, "Switching Control Strategy to Minimize Dual Active Bridge Converter Losses," *IEEE Transactions on Power Electronics*, vol. 24, no. 7, pp. 1826–1838, 2009.

[16] B. Zhao, Q. Song, W. Liu, G. Liu and Y. Zhao, "Universal High-Frequency-Link Characterization and Practical Fundamental-Optimal Strategy for Dual-Active-Bridge DC–DC Converter Under PWM Plus Phase-Shift Control," *IEEE Transactions on Power Electronics*, vol. 30, no. 12, pp. 6488–6494, 2015.

[17] Y. Tang, W. Hu, D. Cao, *et al.*, "Artificial Intelligence-Aided Minimum Reactive Power Control for the DAB Converter Based on Harmonic Analysis Method," *IEEE Transactions on Power Electronics*, vol. 36, no. 9, pp. 9704–9710, 2021.

[18] L.-C. Shih, Y.-H. Liu and H.-J. Chiu, "A Novel Hybrid Mode Control for a Phase-Shift Full-Bridge Converter Featuring High Efficiency Over a Full-Load Range," *IEEE Transactions on Power Electronics*, vol. 34, no. 3, pp. 2794–2804, 2019.

[19] H. Akagi, T. Yamagishi, N. M. L. Tan, S.-I. Kinouchi, Y. Miyazaki and M. Koyama, "Power-Loss Breakdown of a 750-V 100-kW 20-kHz Bidirectional Isolated DC–DC Converter Using SiC-MOSFET/SBD Dual Modules," *IEEE Transactions on Industry Applications*, vol. 51, no. 1, pp. 420–428, 2015.

[20] Y. Ye, D. Qiu, M. Sun, D. Papadaskalopoulos and G. Strbac, "Deep Reinforcement Learning for Strategic Bidding in Electricity Markets," *2020 IEEE Power & Energy Society General Meeting (PESGM)*, Montreal, QC, Canada, 2020, pp. 1–1.

[21] Y. Tang, W. Hu, B. Zhang, *et al.*, "Deep Reinforcement Learning-Aided Efficiency Optimized Dual Active Bridge Converter for the Distributed Generation System," *IEEE Transactions on Energy Conversion*, vol. 37, no. 2, pp. 1251–1262, 2022.

[22] A. L Maas, A. Y. Hannun, and A. Y. Ng, "Rectifier Nonlinearities Improve Neural Network Acoustic Models," *Proceedings of the 30th International Conference on Machine Learning*, 2013, p. 3.

[23] Y. Guo, B. Jiang and Y. Zhang, "A Novel Robust Attitude Control for Quadrotor Aircraft Subject to Actuator Faults and Wind Gusts," *IEEE/CAA Journal of Automatica Sinica*, vol. 5, no. 1, pp. 292–300, 2018.

[24] H. A. Fayed, and A. F. Atiya, "Speed Up Grid-Search for Parameter Selection of Support Vector Machines," *Applied Soft Computing* vol. 80, pp. 202–210, 2019.

[25] B. Zhao, Q. Song and W. Liu, "Efficiency Characterization and Optimization of Isolated Bidirectional DC–DC Converter Based on Dual-Phase-Shift Control for DC Distribution Application," *IEEE Transactions on Power Electronics*, vol. 28, no. 4, pp. 1711–1727, 2013.

[26] A. Tong, L. Hang, G. Li, X. Jiang and S. Gao, "Modeling and Analysis of a Dual-Active-Bridge-Isolated Bidirectional DC/DC Converter to Minimize RMS Current with Whole Operating Range," *IEEE Transactions on Power Electronics*, vol. 33, no. 6, pp. 5302–5316, 2018.

[27] Y. Tang, W. Hu, J. Xiao, *et al.*, "RL-ANN-Based Minimum-Current-Stress Scheme for the Dual-Active-Bridge Converter with Triple-Phase-Shift Control," *IEEE Journal of Emerging and Selected Topics in Power Electronics*, vol. 10, no. 1, pp. 673–689, 2022.

[28] G. Zhang, W. Hu, D. Cao, Q. Huang, Z. Chen, and F. Blaabjerg, "A Novel Deep Reinforcement Learning Enabled Sparsity Promoting Adaptive Control Method to Improve the Stability of Power Systems with Wind Energy Penetration," *Renewable Energy*, vol. 178, pp. 363–376, 2021.

[29] T. P. Lillicrap, J. J. Hunt, A. Pritzel, N. Heess, T. Erez, Y. Tassa, D. Silver, and D. Wierstra, "Continuous Control with Deep Reinforcement Learning," *Proceedings of the International Conference on Learning Representations (ICLR)*, pp. 1–14, 2016.

[30] L. Meng, T. Dragicevic, J. C. Vasquez and J. M. Guerrero, "Tertiary and Secondary Control Levels for Efficiency Optimization and System Damping in Droop Controlled DC–DC Converters," *IEEE Transactions on Smart Grid*, vol. 6, no. 6, pp. 2615–2626, 2015.

[31] Z. Du, L. M. Tolbert, J. N. Chiasson and B. Ozpineci, "Reduced Switching-Frequency Active Harmonic Elimination for Multilevel Converters," *IEEE Transactions on Industrial Electronics*, vol. 55, no. 4, pp. 1761–1770, 2008.

[32] Y. Tang, W. Hu, J. Xiao, *et al.*, "Reinforcement Learning Based Efficiency Optimization Scheme for the DAB DC–DC Converter with Triple-Phase-Shift Modulation," *IEEE Transactions on Industrial Electronics*, vol. 68, no. 8, pp. 7350–7361, 2021.

[33] D. Cao, W. Hu, J. Zhao, Q. Huang, Z. Chen and F. Blaabjerg, "A Multi-Agent Deep Reinforcement Learning Based Voltage Regulation Using Coordinated PV Inverters," *IEEE Transactions on Power Systems*, vol. 35, no. 5, pp. 4120–4123, 2020.

[34] G. Zhang, W. Hu, D. Cao, *et al.*, "Deep Reinforcement Learning-Based Approach for Proportional Resonance Power System Stabilizer to Prevent Ultra-Low-Frequency Oscillations," *IEEE Transactions on Smart Grid*, vol. 11, no. 6, pp. 5260–5272, 2020.

[35] M. Gheisarnejad, H. Farsizadeh and M. H. Khooban, "A Novel Nonlinear Deep Reinforcement Learning Controller for DC–DC Power Buck

Converters," *IEEE Transactions on Industrial Electronics*, vol. 68, no. 8, pp. 6849–6858, 2021.

[36] Z. Yin, C. Du, J. Liu, X. Sun and Y. Zhong, "Research on Autodisturbance-Rejection Control of Induction Motors Based on an Ant Colony Optimization Algorithm," *IEEE Transactions on Industrial Electronics*, vol. 65, no. 4, pp. 3077–3094, 2018.

[37] J. Huang, Y. Wang, Z. Li and W. Lei, "Unified Triple-Phase-Shift Control to Minimize Current Stress and Achieve Full Soft-Switching of Isolated Bidirectional DC–DC Converter," *IEEE Transactions on Industrial Electronics*, vol. 63, no. 7, pp. 4169–4179, 2016.

[38] S. Fujimoto, H. Hoof, and D. Meger, "Addressing Function Approximation Error in Actor-Critic Methods," *International Conference on Machine Learning*, pp. 1587–1596, 2018.

[39] D. Hong, S. Lee, Y. H. Cho, D. Baek, J. Kim and N. Chang, "Energy-Efficient Online Path Planning of Multiple Drones Using Reinforcement Learning," *IEEE Transactions on Vehicular Technology*, vol. 70, no. 10, pp. 9725–9740, 2021.

[40] Y. Wang, Z. Gao, J. Zhang, *et al.*, "Trajectory Design for UAV-Based Internet of Things Data Collection: A Deep Reinforcement Learning Approach," *IEEE Internet of Things Journal*, vol. 9, no. 5, pp. 3899–3912, 2022.

[41] D. P. Kingma, J. Ba, "Adam: A Method for Stochastic Optimization," *Proceedings of the International Conference on Learning Representations*, pp. 1–15, 2015.

[42] A. Abdalrahman and W. Zhuang, "Dynamic Pricing for Differentiated PEV Charging Services Using Deep Reinforcement Learning," *IEEE Transactions on Intelligent Transportation Systems*, vol. 23, no. 2, pp. 1415–1427, 2022.

[43] J. Everts, F. Krismer, J. Van den Keybus, J. Driesen and J. W. Kolar, "Optimal ZVS Modulation of Single-Phase Single-Stage Bidirectional DAB AC–DC Converters," *IEEE Transactions on Power Electronics*, vol. 29, no. 8, pp. 3954–3970, 2014.

[44] X.-F. He, Z. Zhang, Y.-Y. Cai and Y.-F. Liu, "A Variable Switching Frequency Hybrid Control for ZVS Dual Active Bridge Converters to Achieve High Efficiency in Wide Load Range," *2014 IEEE Applied Power Electronics Conference and Exposition - APEC 2014*, Fort Worth, TX, USA, 2014, pp. 1095–1099.

Chapter 5

An active distribution network voltage control using artificial intelligence

Weihao Hu[1], Guozhou Zhang[2], Zhenyuan Zhang[1], Sayed Abulanwar[3,4] and Frede Blaabjerg[5]

5.1 Introduction

The power grid's distribution network plays a crucial role in facilitating access to renewable energy generation (REGS). However, REGS are characterized by randomness, volatility, and intermittency. The access of many distributed REGS will bring large challenges to the operation of the distribution network, which may cause a series of problems such as harmonics, voltage out of limit, and network loss increases, thus reducing the security and economy of distribution network operation [1,2]. Among them, the voltage problem caused by access to renewable energy is more prominent.

In a traditional distribution network, the voltage is regulated by adjusting the tap position of the transformer or controlling the switching of the reactive power compensator. However, the traditional control measures are slow in response, low in adjustment accuracy, and unable to operate frequently. Therefore, it is difficult to deal with the voltage fluctuation caused by the intermittent nature of renewable energy. In recent years, with the gradual application of advanced measuring devices, intelligent monitoring, power electronics, and information communication technology, the distribution network has transitioned from the traditional mode with low automation and backward dispatching methods to a more intelligent mode. In this context, the concept of an active distribution network (ADN) has become standard [3].

Figure 5.1 shows the general structure and composition of a 33-bus active distribution network. The ADN can regulate the reactive power output of distributed

[1]School of Mechanical and Electrical Engineering, University of Electronic Science and Technology of China, China
[2]College of Engineering and Technology, Southwest University, China
[3]Electrical Engineering Department, Faculty of Engineering, Mansoura University, Egypt
[4]Faculty of Engineering, Horus University-Egypt, Egypt
[5]Department of Energy Technology, Aalborg University, Denmark

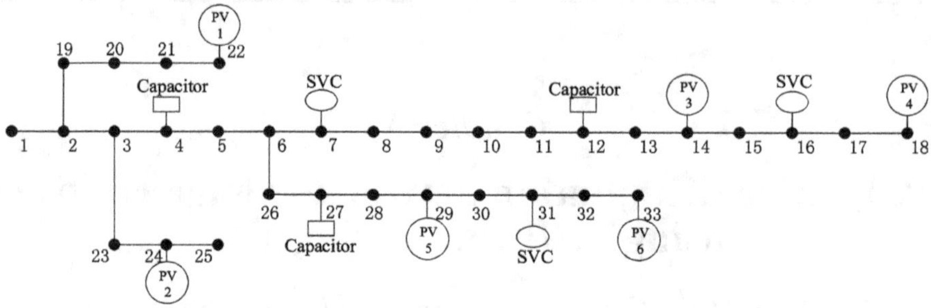

Figure 5.1 An example of the active distribution network

Table 5.1 Comparison of different methods

SP-based method	Require vast computational ability Rely on accurate distribution of random variables Uneasy to realize to practice
RO-based method	Conservative solution Unable to optimize online
AI-based method	Data-driven and Model-free method Able to deal with the rapid voltage fluctuation

generation devices through advanced power electronics and participate in the reactive voltage control of the system to optimize the power flow. Besides, the ADN can also deal with the intermittent power output of new energy generation by regulating the flexible load to offset the adverse effects of the randomness and intermittency of new energy generation and improve the utilization rate and the level of consumption of new energy generation [4]. Operation optimization is the key feature of ADN and the core of active control of the distribution network. Furthermore, voltage control technology is crucial to improving the utilization rate and consumption level of new energy generation, as well as the safety of distribution network operation.

When solving an optimization control problem in ADN, the nonlinear problems caused by power flow need to be addressed, and we have to deal with the uncertainties due to the stochastic nature of REGS and loads. Currently, the methods for dealing with uncertainties are mainly of two types: stochastic programming (SP) and robust optimization (RO) [5,6]. However, the SP-based method requires both vast computational ability and an accurate distribution of random variables that may not be realized in practice [7]. Furthermore, the optimization results calculated by the RO method are usually conservative and utilize only one dispatch solution to deal with all uncertainties of the whole dispatch period. Besides, the RO method cannot be optimized online, and it is difficult to deal with the rapid voltage fluctuation caused by the intermittent nature of PV generation [8]. To this end, artificial intelligence (AI)-based methods can be implemented to address the quality of the solution model. The comparison of these three methods is shown in Table 5.1.

Based on the above background, this chapter will focus on achieving optimal voltage control of ADN with high penetration of renewable energy by implementing AI-based methods.

5.2 Attention-enabled multi-agent DRL for decentralized voltage control

In this section, an attention-enabled multi-agent deep reinforcement learning (MADRL) framework is proposed for active distribution network decentralized Volt-VAR control. Using unsupervised clustering, the whole distribution system can be decomposed into several sub-networks according to the voltage and reactive power sensitivity relationships. Then, the distributed control problem of each sub-network is modeled as a Markov game and solved by the improved MADRL algorithm, where each sub-network is modeled as an adaptive agent. An attention mechanism is developed to help each agent focus on specific information that is mostly related to the reward. All agents are centrally trained offline to learn the optimal coordinated Volt-VAR control strategy and executed in a decentralized manner to make online decisions with only local information. Compared with other distributed control approaches, the proposed method can effectively deal with uncertainties, e.g., renewable generation, achieve fast decision-making, and also significantly reduce communication requirements.

5.2.1 The description of integrated electrical and heating system

Consider an ADN with $N+1$ buses, whose lines and buses are denoted as L and N, respectively. For $i \in \mathcal{N}$, define v_i as the voltage magnitude, and $p_i + jq_i$ as the injected complex power. The injected active power can be divided into PV generation p_i^g and load demand p_i^l, i.e., $p_i := p_i^g - p_i^l$. The injected reactive power is $q_i := q_i^g + q_i^s - q_i^l$, where q_i^g, q_i^s, and q_i^l represents the reactive power of the PV inverter, static reactive power compensation (SVC), and load demand, respectively. Stacking the load demand and generations into vectors, we get $\mathbf{p}^l, \mathbf{q}^l, \mathbf{p}^g, \mathbf{q}^g$, and $\mathbf{q}^s$. Note that on-load tap-changers (OTLCs) and capacitor banks may also exist in distribution feeders, but they typically react in a slow timescale. Furthermore, they are usually scheduled offline due to their limited allowable number of daily switches. This makes it difficult for them to address violent voltage fluctuations caused by the rapid changes in, e.g., PV generations. The SVCs and PV inverters can provide high-speed reactive power support, especially in distribution networks with a high level of PV penetration. This chapter aims to achieve fast voltage control by using SVCs and PV inverters in a scenario of high penetration of DERs. Let π_i denote the parent bus of the non-root bus i. $(\pi_i, i) \in \mathcal{L}$ represents the line between the two nodes, which has impedance $r_i + jx_i$. Let $P_i(t) + jQ_i(t)$ be the complex power that flows from node π_i to node i. Then, the DistFlow equations to model the power flows of the ADN for all buses $i \in N$ at t are given as follows [9]:

$$p_i(t) = \sum_{j \in X_i} P_j(t) - \left(P_i(t) - r_i \frac{P_i^2(t) + Q_i^2(t)}{v_{\pi_i}(t)} \right) \tag{5.1}$$

$$q_i(t) = \sum_{j \in \mathcal{X}_i} Q_j(t) - \left(Q_i(t) - x_i \frac{P_i^2(t) + Q_i^2(t)}{v_{\pi_i}(t)} \right) \tag{5.2}$$

$$v_i(t) = v_{\pi_i}(t) - 2(r_i P_i(t) + x_i Q_i(t)) + (r_i^2 + x_i^2) \frac{P_i^2(t) + Q_i^2(t)}{v_{\pi_i}(t)} \tag{5.3}$$

where χ_i represents the set of children nodes. The task aims to minimize the voltage deviation utilizing the reactive power of SVCs and PV inverters. The voltage regulation problem can be described as:

$$\underset{\mathbf{q}^s, \mathbf{q}^g}{\text{minimize}} \, ||\mathbf{v}(t) - v_0 \mathbf{1}||_1 \tag{5.4}$$

subject to (5.1), (5.2), (5.3) and

$$\underline{\mathbf{v}} \leq \mathbf{v}(t) \leq \bar{\mathbf{v}} \tag{5.5}$$

$$\underline{q}_i^s \leq q_i^s(t) \leq \bar{q}_i^s, \forall i \in \mathcal{N}_s \tag{5.6}$$

$$|q_i^g(t)| \leq \sqrt{(s_i^2) - (p_i^g(t))^2}, \quad \forall i \in \mathcal{N}_g \tag{5.7}$$

where (5.4) is the objective function and it aims to minimize the sum of the voltage deviation of each node; $\mathbf{q}^s$ and $\mathbf{q}^g$ are control variables; (5.5) is the voltage constraint for each node, where $\underline{\mathbf{v}}$ and $\bar{\mathbf{v}}$ are the lower and upper limits; (5.6) describes the reactive power range of SVC, where $\underline{q}_i^s$ and $\bar{q}_i^s$ are its lower and upper limits; $\mathcal{N}_s$ denotes the set of nodes connected with SVCs; from (5.7), the reactive power of the PV inverter at node i depends on the active power of PV at time t, where s_i denotes the rated apparent power of PV connected to the node i; $\mathcal{N}_g$ denotes the set of nodes connected with PVs.

Existing centralized volt-var control methods suffer from a heavy computational burden [10], and they are susceptible to a single point of failure [11]. They also lead to a communication bottleneck since the centralized controller must collect information from the whole system [12]. To this end, this chapter describes a distributed attention-based multi-agent twin delayed deep deterministic policy gradient (MATD3) algorithm to deal with them.

5.2.2 Proposed distributed MATD3 voltage control method

The proposed distributed control method contains three main components, namely (i) network partition via the clustering algorithm; (ii) formulation of the decomposed sub-networks as Markov games; and (iii) voltage control optimization via the proposed attention-based MATD3 algorithm.

5.2.2.1 Network partition

The objective of clustering is to split the ADN into several sub-networks to identify the voltage control areas. Via the network partition, the centralized optimization problem is divided into several small problems, such that the communication requirements and negative impacts on control performance caused by the communication delay are reduced and the robustness to cyber and physical attacks may be improved.

Spectral clustering, an unsupervised learning method derived from spectral graph theory, is used to search for the optimal partition results of ADN. The aim is to reduce voltage deviation by utilizing reactive power from multiple resources, and a voltage-reactive power sensitivity matrix is used to represent the electrical distance. An affinity matrix is derived based on the sensitivity matrix via

$$w_i j = w_{j,i} = \sum_{i=1,j=1}^{N} \exp\left(\frac{-||x_i - x_j||^2}{2\sigma^2}\right) \tag{5.8}$$

where $w_{i,j}$ is the component of the similarity matrix $\boldsymbol{W}$; x_i represents the ith row of the sensitivity matrix; σ is the coefficient that controls the width of the neighborhood. The diagonal degree matrix $\mathbf{D}$ can be obtained by $d_i = \sum_{j=1}^{n} w_{ij}$, where d_i denotes the ith diagonal element of D. After that, the Laplacian matrix L is calculated via $\mathbf{L} = \mathbf{D} - \mathbf{W}$. The clustering problem can be transferred into a graph partition problem and the objective function is as follows [13]:

$$\mathbf{F}_{Ncut}(\mathbf{A}_1, \mathbf{A}_2, ..., \mathbf{A}_k) = \frac{1}{2}\sum_{i=1}^{k} \sum_{m\in A_i, n\in \bar{A}_i} \frac{w_{m,n}}{vol(\mathbf{A}_i)} \tag{5.9}$$

where k is the total number of clusters; $\mathbf{A}_i$ denotes the ith group of the clustering results; $\bar{\mathbf{A}}_i$ is the complement set of $\mathbf{A}_i$; $vol(\mathbf{A}_i)$ represents the weighted sum of all edges in A_i. The objective of (5.9) is to maximize the internal similarity of subgraphs. According to [13], the optimization of (5.9) can be reformulated as:

$$\begin{aligned} &\arg\min tr(\mathbf{F}^T\mathbf{D}^{-1/2}\mathbf{L}\mathbf{D}^{-1/2}\mathbf{F}) \\ &s.t.\ \mathbf{F}^T\mathbf{F} = \mathbf{I} \end{aligned} \tag{5.10}$$

The optimal partition results can be obtained by constructing a space using the eigenvectors corresponding to the first k_1 eigenvalues of the matrix $\mathbf{D}^{-1/2}\mathbf{L}\mathbf{D}^{-1/2}$, and clustering the eigenvectors in the space by the K-means algorithm.

Remark: The key factor that affects the partition result of the distribution system is the network reconfiguration. The topology of the network changes after reconfiguration, leading to a change in the voltage-reactive power sensitivity matrix as well as the partition result. However, the topology changes typically have local impacts on voltage-reactive power sensitivity. Thus, only the local sensitivity indices corresponding to the topology changes need to be updated, which can be done quickly.

5.2.2.2 Formulation of Markov games

After the network partition, the whole network is divided into several sub-regions. Then, the decentralized voltage regulation of multi sub-networks is formulated as Markov Games (MGs), a multi-agent extension of the Markov decision process. In the MGs, each sub-network is modeled as an adaptive agent. At each time step, each agent observes the regional system state $\mathbf{s}_j$, including the active and reactive power of loads, and the active power generation of PVs. Based on the observed information, each agent makes decisions $\mathbf{a}_j$ to schedule the reactive power of SVCs and PV inverters. After all agents' actions are executed, each agent obtains a reward, which represents the total voltage deviation of the whole system. Then, the system transfers to the next state. The MGs provide a mathematical representation for modeling the distributed decision-making process. The key components for an MG include a state set $\mathcal{S}$, an action set $\mathcal{A}$, and a reward function $\mathcal{R}$. They are described as follows:

- $\mathcal{S}$: $\mathbf{s} \in \mathcal{S}$ contains the states for all agents. For agent j, the state $\mathbf{s}_j \in \mathbf{s}$ is the local observation of sub-network j, including (p_i^l, q_i^l, p_i^g), where i is the index of the node in sub-network j.
- $\mathcal{A}$: $\mathbf{a} \in \mathcal{A}$ contains the actions for all agents. For agent j, the action $\mathbf{a}_j \in \mathbf{a}$ includes (α_i^g, α_i^s). The control variables in (5.4) can be derived as:

$$q_i^g = \alpha_i^g \sqrt{(s_i)^2 - (p_i^g)^2}, -1 < \alpha_i^g < 1 \tag{5.11}$$

$$q_i^s = \alpha_i^s \bar{q}_i^s, -1 < \alpha_i^s < 1 \tag{5.12}$$

- $\mathcal{R}$: $r \in \mathcal{R}$ is the immediate reward the agents obtain after the action $\mathbf{a}$ is executed. In this context, all agents share the same reward: $r = -||\mathbf{v}(t) - v_0\mathbf{1}||_1 + \beta$, where $||\mathbf{v}(t) - v_0\mathbf{1}||$ represents the total voltage deviation of all nodes in the DN; β is the penalty term when the voltage constraint in (5.5) is not satisfied.

At each time step, agent j makes a decision $\mathbf{a}_j$ based on the local observation $\mathbf{s}_j$ in the sub-network j. When all agents complete actions, they obtain a shared reward r, and then the system transfers to the next state. This is an MG and the aim of each agent is to learn a policy, which maps its local observation $\mathbf{s}_j$ to action $\mathbf{a}_j$ to maximize the discounted cumulative reward from the current time step onward, $\sum_{k=t}^{T} \gamma^{k-t} r(k)$, where $\gamma \in [0, 1]$ is the discount factor that balances the importance between the future and immediate reward.

Remark: Some distribution systems may not have SVCs, DSTATCOM (distribution static compensator), or fixed/switched capacitors for reactive power control. For a high penetration of the PV integration system, SVC may not be necessary if the PV reactive power capability is properly utilized. The proposed method is general enough to address that.

5.2.2.3 Attention-based MATD3 algorithm for voltage control

The proposed attention-based MATD3 algorithm is developed to solve the formulated MGs. MATD3 is one of the MADRL algorithms. To enhance its scalability when dealing with more sub-networks, the original MATD3 algorithm is improved using the attention model. Each sub-network is modeled as a twin delayed deep deterministic policy gradient (TD3) agent, which is composed of the actor and critic networks. The actor maps the local observation $\mathbf{s}_j$ to action $\mathbf{a}_j$ and it is the policy function. The critic maps the global information $(\mathbf{s}, \mathbf{a})$ to a scalar, which is a judgment of action $\mathbf{a}_j$ considering the impact on other agents. The coordinated control strategy is achieved by adopting a centralized training framework, among which the Actor and Critic functions of each agent are trained against each other iteratively until the critic provides better judgment and the actor can make decisions with reduced voltage deviation. To further enhance the capability of MATD3 in dealing with many agents, an attention mechanism is developed. It allows each agent to focus on the specific information that is mostly related to the reward. The architecture of the proposed method is shown in Figure 5.2 and the details are elaborated below.

Attention-critic functions

For agent j, the critic function $Q_j(\cdot)$ takes the global state $\mathbf{s}$ and actions of all agents $\mathbf{a}$ as inputs and outputs the action value of agent j. To address the uncertainties of ADN, a deep neural network (DNN) is advocated to approximate the critic functions as follows:

$$Q_j(\mathbf{s}, \mathbf{a}) = g_j(\mathbf{s}, \mathbf{a}) = g_{j,l}[...g_{j,1}(\mathbf{s}, \mathbf{a})] \tag{5.13}$$

$$g_{j,i} = \sigma(\mathbf{W}_i^c * \mathbf{o}_{i-1} + \mathbf{b}_i^c), i = 2, 3, ..., l \tag{5.14}$$

where $g_j(\cdot)$ represents the parameterized critic function of agent j approximated by DNN; $g_{j,i}$ represents the function map of the ith layer NN; $\mathbf{W}_i^c$ and $\mathbf{b}_i^c$ represent the weight matrix and bias vector of the ith layer, respectively; σ represents the

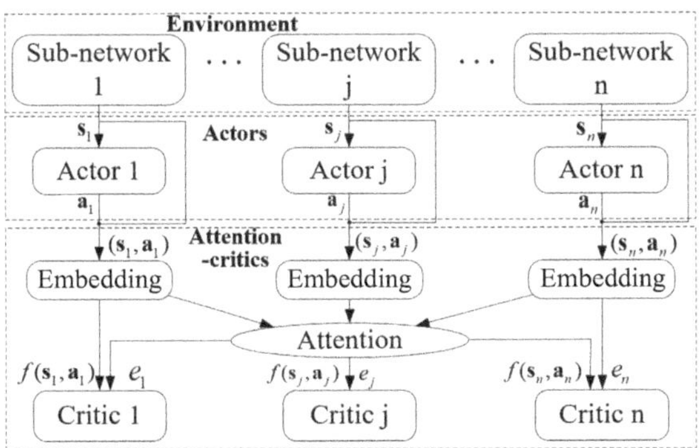

Figure 5.2 Architecture of the proposed attention-based MADRL method [14]

activation function; $\mathbf{o}_{i-1}$ is the output of the $(i-1)$th layer. The critic function of agent j is parameterized by the parameters of DNN $Q^{c_j} = \{\mathbf{W}_1^c, \mathbf{b}_1^c, ..., \mathbf{W}_l^c, \mathbf{b}_l^c\}$.

For MADRL, the complexity of the problem increases exponentially with the number of agents. Therefore, it is challenging for the agents to learn good policies when the population is large. To this end, the attention mechanism-based critic is developed in this paper. The input of critic function $(\mathbf{s}, \mathbf{a})$ is replaced with $f_j(\mathbf{s}_j, \mathbf{a}_j)$ and e_j, where $f_j(\cdot)$ represents the embedding function of agent j and e_j is the output processed by the attention, representing the weighted sum of other agents' value [15]:

$$e_j = \sum_{i \neq j} \alpha_i \cdot u_i = \sum_{i \neq j} \alpha_i \cdot \mathrm{ReLU}(\mathbf{T} \cdot f_i(\mathbf{s}_i, \mathbf{a}_i)) \tag{5.15}$$

where ReLU represents the activation function; $\mathbf{T}$ is the linear transformation matrix; α_i represents the attention weight obtained by comparing the similarity between the embedding of agent i, $f_i(\mathbf{s}_i, \mathbf{a}_i)$ and that of agent j, $f_j(\mathbf{s}_j, \mathbf{a}_j)$ using the query-key system [15]:

$$\alpha_i \propto \exp((f_i(\mathbf{s}_i, \mathbf{a}_i))^T \mathbf{W}_k^T \mathbf{W}_q f_j(\mathbf{s}_j, \mathbf{a}_j)) \tag{5.16}$$

where $\mathbf{W}_k$ and $\mathbf{W}_q$ are the transformation matrices. The calculated similarity value between the two embeddings is then passed to a softmax to obtain the attention weight α_t^i. The parameters of the attention model $Q^a = \{\mathbf{W}_k, \mathbf{W}_q, \mathbf{T}\}$ give rise to a weighted sum of contributions from all other agents for agent j. The parameters of attention-critic Q^{Q_j} include the parameters of the critic function Q^{c_j} and those of the attention model Q^a, which are optimized by minimizing the following loss function [15]:

$$L(Q^{Q_j}) = (Q_j(f_j(\mathbf{s}_j, \mathbf{a}_j), e_j) - y)^2 \tag{5.17}$$

$$y = r + \gamma Q_j(f_j(\mathbf{s}'_j, \mathbf{a}'_j), e_j)|_{\mathbf{a}'_j = p_j(\mathbf{s}'_j)} \tag{5.18}$$

where y is the target. The critic function is optimized by minimizing the "distance" between $Q_j(\cdot)$ and the target. However, the training process may be unstable since the critic being updated is also used for calculating the target y. To solve this problem, the target functions $Q'_j(\cdot)$ and p'_j are introduced. A pair of critics $(Q_{j,1}, Q_{j,2})$ is used for the calculation of target y to address the overestimation problem caused by the function approximation error in actor-critic-based methods. Then, (5.17) is rewritten as [16]:

$$y = r_t^j + \gamma \min_{n=1,2} Q'_{j,n}(f_j(\mathbf{s}'_j, \mathbf{a}'_j), e_j)|_{\mathbf{a}'_j = p_j(\mathbf{s}'_j)} \tag{5.19}$$

where $Q'_{j,n}(\cdot)$ represents the nth target critic of agent j.

Actor functions

The actor is the policy function that aims at maximizing the output of the critic by making the decision $\mathbf{a}_j$ under the state $\mathbf{s}_j$. DNN is advocated to deal with the nonlinearities when solving the Optimal Power Flow (OPF) problem, yielding:

$$\mathbf{a}_j = p_j(\mathbf{s}_j) = p_{j,l}[...p_{j,1}(\mathbf{s}_j)] \tag{5.20}$$

$$p_{j,i} = \sigma(\mathbf{W}_i^a * \mathbf{o}_{i-1} + \mathbf{b}_i^a), i = 2, 3, ..., l \tag{5.21}$$

where $p_j(\cdot)$ represents the parameterized policy function of agent j approximated by DNN; $p_{j,i}$ represents the function map of the ith layer NN; $\mathbf{W}_i^a$ and $\mathbf{b}_i^a$ represent the weight matrix and bias vector of the ith layer NN, respectively; σ is the activation function; $\mathbf{o}_{i-1}$ is the output of the (i-1)th layer NN. Then, the policy function of agent j is parameterized by $\mathcal{Q}^{\mu_j} = \{\mathbf{W}_1^a, \mathbf{b}_1^a, ..., \mathbf{W}_l^a, \mathbf{b}_l^a\}$. They are optimized via the policy gradient [17]:

$$\nabla_{\mathcal{Q}^{\mu_j}} J(\mathcal{Q}^{\mu_j}) = E_{S_t, A_t \sim D}[\nabla_{\mathcal{Q}^{\mu_j}} p_j(\mathbf{a}_j|\mathbf{s}_j) \nabla_{\mathbf{a}_j} Q_j(f_j(\mathbf{s}_j, \mathbf{a}_j), e_j)|_{\mathbf{a}_j = p_j(\mathbf{s}_j)}] \tag{5.22}$$

Replay buffer mechanism

Since DNN is utilized to fit the actor and critic functions, the input data for training should be independent and identically distributed. However, the data are highly correlated with the DRL algorithm. To this end, a replay buffer is utilized, where each agent employs memory to store the transitions $(\mathbf{s}_j, \mathbf{a}_j, r, \mathbf{s}'_j)$. The mini-batch experiences are sampled at each time step to calculate the gradient and optimize the parameters of networks. This mechanism helps to break the correlation between data and improves the stability of the training process.

5.2.2.4 Centralized training and decentralized execution

The implementation of the proposed approach can be divided into two main steps: centralized training for the formulation of coordinated strategies and decentralized execution for voltage regulation with only local information. They are explained below.

Centralized training

In the MGs with M agents, the parameter set to be optimized is $\mathcal{Q} = \{\mathcal{Q}_1, ..., \mathcal{Q}_M\}$. For agent j, the parameter set is denoted as $\mathcal{Q}_j = \left\{\mathcal{Q}^{\mu_j}, \mathcal{Q}^{\mu'_j}, \mathcal{Q}^{Q_j}, \mathcal{Q}^{Q'_j}\right\}$, where $\mathcal{Q}^{\mu_j}$ and $\mathcal{Q}^{\mu'_j}$ are parameters of Actor and Target actor network of agent j; $\mathcal{Q}^{Q_j}$ and $\mathcal{Q}^{Q'_j}$ are parameters of attention-critic and target attention-critic network. The training process of the proposed approach is shown in Algorithm 5.1.

The parameters of NN start to update when the replay buffer is full. At each time step, each agent samples a mini-batch of experiences $(\mathbf{s}_j, \mathbf{a}_j, r, \mathbf{s}'_j)_k, k = 1, 2, ..., B$ from its memory. Each actor NN takes the local state $\mathbf{s}_j$ as input and adjusts its parameters to output an action that maximizes the action value. The gradient is calculated according to

$$\nabla_{\mathcal{Q}^{\mu_j}} J(\mathcal{Q}^{\mu_j}) = \frac{1}{B}\sum_{k=1}^{B} \nabla_{\mathcal{Q}^{\mu_j}} p_j(\mathbf{a}_j|\mathbf{s}_j) \nabla_{\mathbf{a}_j} Q_j(f_j(\mathbf{s}_j, \mathbf{a}_j), e_j)|_{\mathbf{a}_j = p_j(\mathbf{s}_j)} \tag{5.23}$$

Then, the parameters of actor network are updated through

$$\mathcal{Q}^{\mu_j} \leftarrow \mathcal{Q}^{\mu_j} + \eta_\mu \nabla_{\mathcal{Q}^{\mu_j}} J(\mathcal{Q}^{\mu_j}) \tag{5.24}$$

where η_μ is the learning rate for the actor function. Each critic NN takes the global state as input, which includes the states and actions of all agents, and predicts an action value to minimize the following loss function:

$$L(Q^{Q_j}) = \frac{1}{B}\sum_{k=1}^{B}(Q_{j,n}(f_j(\mathbf{s}_j, \mathbf{a}_j), e_j) - y)^2, n = 1, 2 \tag{5.25}$$

The parameters are updated according to

$$Q^{Q_j} \leftarrow Q^{Q_j} + \eta_Q \nabla_{Q^{Q_j}} L(Q^{Q_j}) \tag{5.26}$$

where η_Q is the learning rate for the critic function. Then, the parameters of the target networks are optimized by slowly tracking the online ones:

$$Q^{Q'_j} \leftarrow \tau Q^{Q_j} + (1-\tau)Q^{Q'_j}, Q^{\mu'_j} \leftarrow \tau Q^{\mu_j} + (1-\tau)Q^{\mu'_j} \tag{5.27}$$

where $\tau \ll 1$ is the soft update coefficient.

Algorithm 5.1: Training process of the proposed algorithm

Algorithm Training the proposed algorithm
Input: Nodes active and reactive power in the distribution network, the PV output, and the reward.
Output: DNN's parameters Q
1: Randomly initialize parameters of critic networks Q^{Q_j} and actor network Q^{μ_j} for each agent j
2: Initialize target networks $Q^{Q'_j} \leftarrow Q^{Q_j}, Q^{\mu'_j} \leftarrow Q^{\mu_j}$ for each agent j
3: for episode = 1, 2, ..., H
  receive initial observation $\mathbf{s}_j$ for each agent j for t = 1, 2, ... T
4: choose action according to $\mathbf{a}_j = p_j(\mathbf{s}_j)$ for each AgentExecute actions $\mathbf{a} = (\mathbf{a}_1, ..., \mathbf{a}_M)$ and obtain reward r and new observation $\mathbf{s}'_j$for each agent j
5: store transition $(\mathbf{s}_j, \mathbf{a}_j, r, \mathbf{s}'_j)$ in the replay buffer for agent $j = 1, \ldots, M$
7: sample a random mini-batch B of transitions from the replay buffer
8: calculate target y according to (5.24)
9: update critic networks according to (5.26) and (5.27) if t mod d
  update actor network according to (5.24) and (5.25)
10: update target networks according to (5.28)
  end if
  end for
11: end for
12: end for

During the training process, the critic of each agent requires information from other agents. This can be achieved since the training of DRL is implemented off-line. The centralized critics that are augmented with information about other

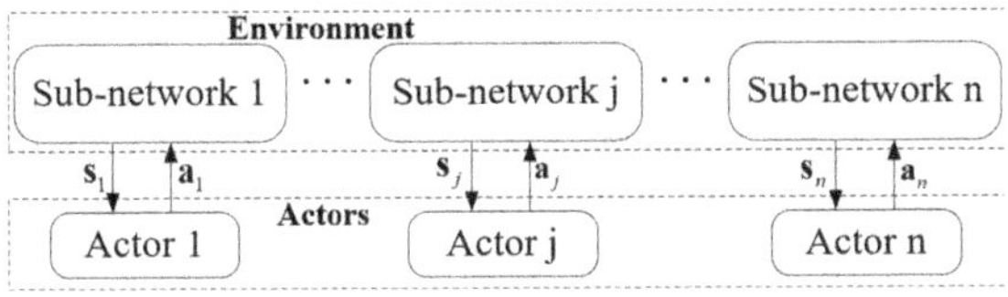

Figure 5.3 Workflow of the decentralized execution [14]

agents' policies during training help identify coordinated strategies. The explicit modeling of other agents' decision-making processes allows each agent to provide decisions with better robustness to the system dynamics based on local information only. This differentiates the existing work and allows us to deal with scalability issues in the presence of large-scale systems.

Decentralized execution

When the training process is completed, the parameters of DNN are fixed and only the actor network of each agent is kept for real-time voltage regulation. The workflow of decentralized execution is shown in Figure 5.3. Each agent is in charge of a sub-network. Since the actors only need local information, the proposed approach can be executed in a decentralized manner. The real-time reactive power control scheme of the proposed approach is shown in Algorithm 5.2.

Algorithm 5.2: Real-time reactive power control of the proposed approach

---

Algorithm Real-time reactive power control
Input: Nodes active and reactive power in the distribution network, the PV output.
Output: Reactive power schedules $\mathbf{a}$.
1: Read the parameters of actor network of each agent Q^{μ_j}
2: For time step $t = 1, 2, \ldots T$
3: for agent $j = 1, \ldots, M$
4: obtain the local observation $\mathbf{s}_j$
5: calculate action $\mathbf{a}_j$ according to $\mathbf{a}_j = p_j(\mathbf{s}_j)$
6: end for
7: concatenate actions of all agents $\mathbf{a} = (\mathbf{a}_1, \ldots, \mathbf{a}_M)$
8: end for

5.2.3 Numerical results on a distribution system

In this section, simulation results are provided to evaluate the performance of the proposed approach on IEEE 33-bus and 123-bus systems [18], whose parameters can be found in [5]. The network partition results are first illustrated, followed by the control performance comparison results with other methods.

5.2.3.1 Simulation setup

To simulate more realistic scenarios, real-world PV data are used, i.e., 1-year PV generation data of Xiaojin, a county in the Sichuan province of China. These data are divided into a training set and a test set, which contains 300- and 10-day data, respectively. The sampling frequency of the data is 1 h. Note that the strategy learned by the proposed approach can easily be extended to scenarios with different sampling times in practice. The parameters of the control devices are shown in Table 5.2. The maximum allowed voltage deviation is set to $\pm 5\%$. For the proposed method, each sub-region is modeled as an agent, which is composed of actor and critic networks. All the networks have two shallow layers, the number of neurons in each of which is 100 and 100, respectively. The hyperparameters are shown in Table 5.3. The power flow analysis is carried out using MATLAB® and the training of the proposed method is implemented in Python with TensorFlow. A workstation with an NVIDIA GeForce 1080Ti GPU and an Intel Xeon E5-2630 CPU is used for the simulation.

Sensitivity of Volt-VAR control with network partition

The spectral clustering algorithm is applied to partition the ADN into several sub-regions. After that, the proposed approach is trained for 25,000 epochs on the training data to learn the coordinated control strategy for voltage regulation. Each epoch corresponds to a day, which is randomly selected from the training set at the beginning of each epoch. The convergence curve of the cumulative reward is plotted in Figure 5.4. It can be seen that the proposed approach cannot make balanced decisions at the beginning of the training procedure and therefore achieves a low reward. With the training process going on, the reward increases significantly and finally converges to around -0.96 with small fluctuations. This illustrates that the proposed method can learn the coordinated control strategy from training data.

Table 5.2 Parameters of control devices for IEEE 33-bus system

Type	Capacity of Each Device	Location
SVC	0.3 MVar	5, 10, 30
PV	0.8 MW/0.8 MVA	15, 18, 22, 24, 27, 33

Table 5.3 Parameter settings of the proposed method

Parameters	Values
Batch size for updating NN	32
Replay buffer size	48,000
Discount factor	0
Soft update coefficient	0.001
Policy update frequency	2
Target policy smoothing coefficient	0.2
Learning rate for actor network	0.001
Learning rate for critic network	0.002

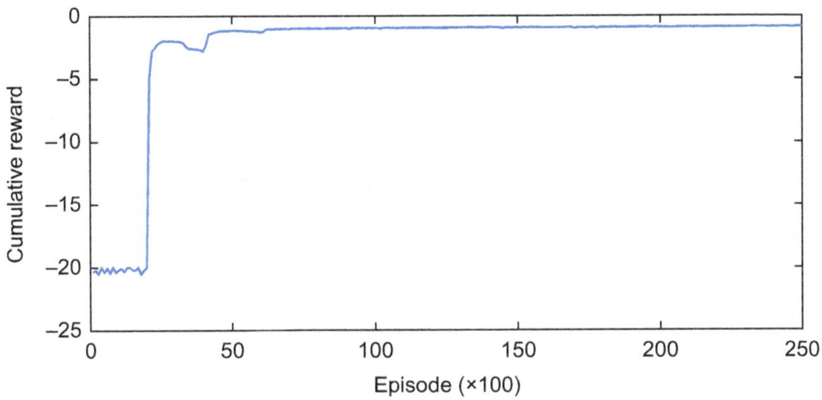

Figure 5.4 The evolution of the reward during the training procedure of the IEEE 33-bus system [14]

Table 5.4 Voltage deviation of various methods on 33-bus system

Cluster number	**3**	**4**	**5**	**6**
Ave. vol. dev. using MADDPG [15]	0.13%	0.13%	0.14%	0.17%
Ave. vol. dev. of proposed method	0.13%	0.13%	0.13%	0.13%

After training, the average voltage deviations on test data for multi-agent deep deterministic policy gradient (MADDPG) algorithm [15] and the proposed approach under different partition results are obtained and shown in Table 5.4. The MADDPG method employs centralized training and a decentralized execution framework and models each sub-network as a DDPG agent. The network structure and hyperparameter settings are the same as in the proposed method. *It is worth noting that the MADDPG method has been applied to the transmission network in* [15], *and we customized it for the voltage regulation of the distribution system for a fair comparison.* It can be seen from Table 5.4 that the control performance of the MADDPG method decreases when the number of sub-regions is gradually increased. This is because the complexity of the environment increases with the growing number of agents. This makes it difficult for the MADDPG method to search for a good policy. However, the proposed method can maintain control performance with an increased number of agents thanks to the developed attention model. The latter helps each agent attend to specific information that is mostly related to the reward. The partition results may be different according to the actual requirements of the operator, and our proposed method can easily adapt to them.

In subsequent tests, the number of sub-networks is selected as six in the 33-bus system for illustrations, and the clustering result is shown in Figure 5.5.

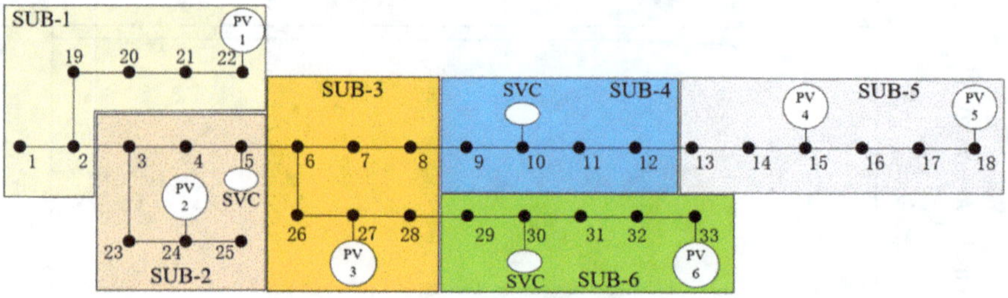

Figure 5.5 Partition results (SUB1-SUB6) of an IEEE 33-bus system [14]

Table 5.5 Voltage deviation of various methods

Method	Average	Max rise	Max drop	Com type
No control	1.46%	5.22%	7.11%	–
TD3-D	0.77%	3.88%	2.03%	D
MADDPG	0.17%	1.32%	1.25%	D
SP	0.13%	1.69%	1.24%	C
Proposed	0.13%	0.73%	1.25%	D
Centralized	0.08%	0.55%	1.25%	C

Comparison with other alternative control method

To demonstrate the benefits of the proposed method, comparative tests are carried out against other existing methods. They include (1) **No control method**; (2) **TD3-D method**, where each sub-network is controlled by a TD3 agent based on local observation. The TD3 agents are trained separately, and they sequentially minimize the voltage deviation of their sub-network. Note that there is no information exchange between TD3 agents during the training process; (3) **Stochastic programming-based (SP) approach**, where all the sub-networks are optimized separately, and the objective of each sub-network is to minimize the voltage deviation based on local information. Three hundred scenarios are randomly generated to represent the uncertainty, and scenario reduction is used to obtain 20 representative scenarios [16]; (4) **Model-based centralized control method**, where load demand and PV generation are assumed to be known beforehand and the commercial second-order cone programming (SOCP) solver MOSEK is applied to solve the deterministic cases based on the global information. It results in a *perfect model*, which is considered a benchmark.

The average, maximum rise, and maximum drop of voltage deviations, as well as the computing times for all methods, are shown in Table 5.5. It can be observed that when reactive power compensation is not applied (no control), the voltage will exceed the upper and lower limits. The TD3-D, MADDPG, and the proposed approach can ensure that the voltage is within the limited ranges. Note that these three methods all make decisions based on local information. However, the MADDPG method achieves better performance due to the coordinated control strategies learned during the training process. The proposed approach further enhances control performance thanks to the use of the attention mechanism. The proposed method achieves similar performance to that obtained by the SP method, which adopts a centralized control framework and informs decisions based on global information. To evaluate the control accuracy of the proposed method, the average optimization error is defined as:

$$ERR = \left|\frac{\Delta V_{pro} - \Delta V_{cen}}{\Delta V_{cen} - \Delta V_{ori}}\right| \times 100\% \tag{5.28}$$

where *ERR* represents the average deviation of the proposed approach to the global optimal solution; ΔV_{cen}, ΔV_{pro}, and ΔV_{ori} represent the average voltage deviations of the centralized approach, the proposed approach, and the original value on test data, respectively. The *ERR* is 3.6% for the proposed approach, which means that it can reach 96.4% optimality based on local information. However, it assumes that the uncertain variables are to be known beforehand, which is not possible in practice. Also, this method depends on complete two-way communication links. The infrastructure for this type of control framework is rarely available in ADNs. Note that the proposed approach is based on a decentralized control framework, where each sub-region is controlled using only local information without communications between agents. Thus, the communication requirements are reduced and the privacy concerns are mitigated.

To further demonstrate the capability of the proposed method in dealing with fluctuations of PV and load outputs, a sunny day is selected as a case study. The PV output and load demand for a sunny day are shown in Figures 5.6 and 5.7, respectively. The voltage profile of each node obtained by various control strategies is shown in Figure 5.8 when $t = 13$ in Figures 5.6 and 5.7. It can be observed that the voltages at Nodes 17–18 go beyond the upper voltage limit when there is no reactive compensation. Using the TD3-D method, the voltages can be adjusted to a

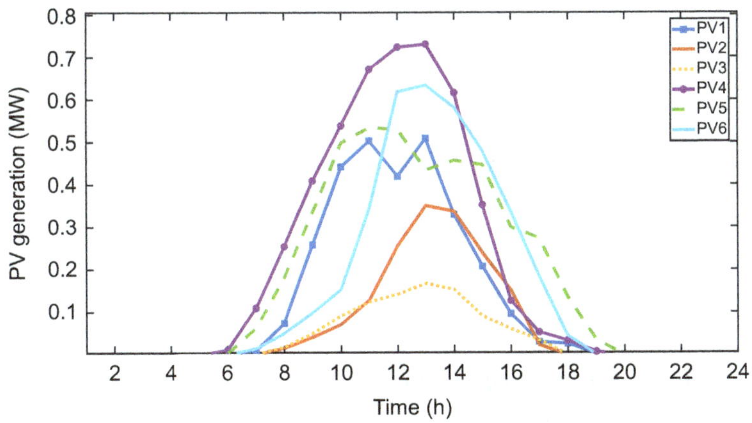

Figure 5.6 PV outputs for the selected sunny day [14]

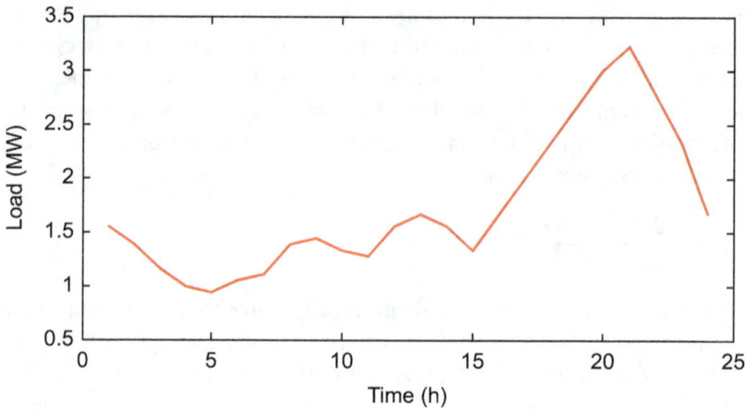

Figure 5.7 Load demand for the selected sunny day [14]

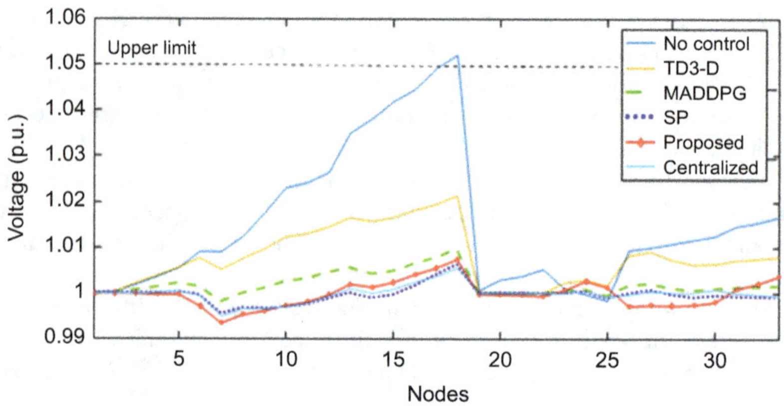

Figure 5.8 Voltages of each node before and after control when t = 13 [14]

limited range. However, due to the lack of coordination among multiple devices, the capabilities of voltage regulation devices are not fully utilized. Since the MADDPG method and the proposed approach explicitly model the decision-making process of other agents via centralized training, the agents can exhibit cooperative behaviors with only local information. The proposed approach further enhances the control performance of MADDPG using the attention model. The voltage profile of the proposed approach is very close to that of the SP and centralized method, demonstrating the effectiveness of the proposed method.

5.2.3.2 Robustness to large stochasticity

More simulations are carried out to demonstrate the advantages of the proposed approach for real-time controls. The PV output profile is shown in Figure 5.9. Due to cloud dynamics, the PV output changes fast within 1 min, i.e., its output rises

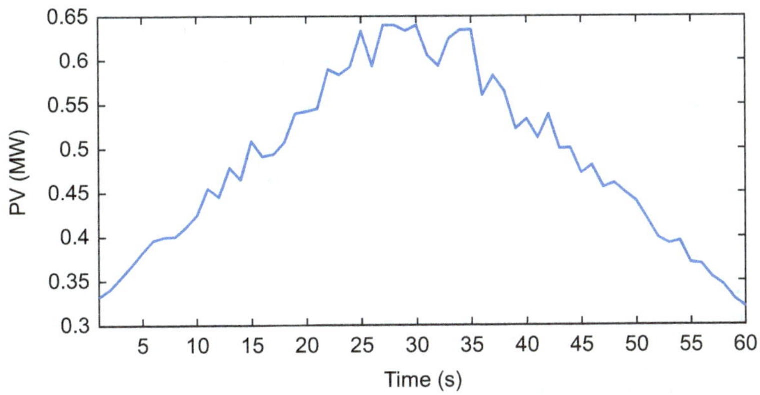

Figure 5.9 PV outputs with large variations [14]

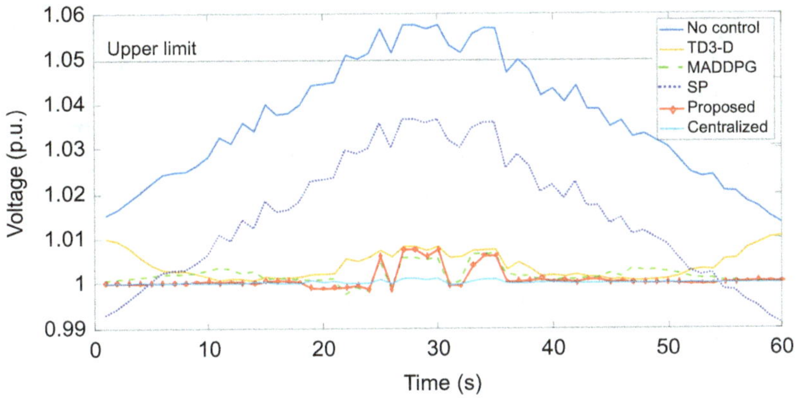

Figure 5.10 Voltage change of Node 18 with different control strategies when the PV outputs have large fluctuations [14]

from 0.33 MW at $t = 1$ s, reaches 0.65 MW at $t = 30$ s, and then returns to 0.33 MW in 30 s. The voltage profiles of Node 18 under various control strategies are shown in Figure 5.10. Since the SP method suffers from a heavy computational burden, it provides a predetermined control solution to deal with the rapid PV output variation during such a short period. The TD3-D, MADDPG, and proposed methods are DRL-based approaches, the strategy learned by which can generalize to new situations and inform decisions in milliseconds, according to the latest observation. In this test, decisions are made at each second. The centralized method ignores the communication delay and provides a theoretical limit for the problem (an ideal condition that could not be achieved in practice). It can be observed from Figure 5.9 that Node 18 suffers from an overvoltage problem if no controls of the SVC and PV inverter are applied, namely the original method. With the SP method,

Table 5.6 Control devices in IEEE 123-bus system

Type	Capacity	Location
SVC	0.3MVar	9,35,54,62,68,105
PV	0.8MW/0.84MVA	5,12,27,50,65,76,81,83,100,114,118

Table 5.7 Control performance on IEEE 123-bus system

Cluster number	6	7	8
Ave. vol. dev. of MADDPG [15]	0.60%	0.63%	0.80%
Ave. vol. dev. of proposed method	0.45%	0.45%	0.46%

the problem can be suppressed. However, since the control decisions provided by this approach are predetermined and cannot react dynamically to the fast-changing PV outputs, it has much larger voltage fluctuations than those of the TD3-D, MADDPG, and the proposed method. By contrast, the TD3-D, MADDPG, and the proposed approach can make decisions based on the latest states of the ADN and thus achieve a better response to the dynamic changes of the PV outputs. The proposed approach outperforms TD3-D and MADDPG due to the coordinated control strategy learned during the training process and the utilization of the attention model (Tables 5.6 and 5.7).

5.3 Surrogate model-based physical-model-free voltage control method

In this section, a physical-model-free voltage control method is proposed and the corresponding theory is specifically given.

5.3.1 Model-free framework for voltage control

The proposed model-free framework consists of four main components: (1) Surrogate Modeling via Sparse Pseudo-Gaussian Process (SPGP); (2) Network partition; (3) Formulation of multiple sub-regions voltage regulation as a Markov game; (4) Application of Muti-agent Soft Actor-Critic (MASAC) algorithm to solve the developed Markov game.

5.3.1.1 Gaussian process for surrogate modeling

It has been elaborated in the introduction that existing deep reinforcement learning (DRL) based voltage control algorithms have to use the original power flow model for reward calculations during the training process. To deal with that, a surrogate model is developed that yields the same input-output relationship as the power flow equations. In this section, we advocate the use of SPGP as it only needs a few numbers of measurements and can achieve good performance.

A Gaussian process (GP) is denoted as $f(x) \sim GP(m(x), k(x, x'))$, which is specified by a mean function $m(x) = E[f(x)]$ and covariance $k(x, x') = \text{cov}(f(x), f(x'))$ [19]. Typically, $m(x)$ is assumed to be 0 since no prior knowledge exists. For the regression problem $y = f(x) + \varepsilon$, where x is the input and y represents the response by $f(x)$ distributed by noise $\varepsilon \sim N(0, \sigma_n^2)$. The prior distribution of the observed value is

$$y \sim N(0, K(X, X) + \sigma_n^2 I_n) \tag{5.29}$$

where $K_{ij} = k(x_i, x_j)$ represents the kernel function to express the correlation between x_i and x_j; I_n is an $N \times N$ identity matrix. GP aims to forecast f_* given new input x_*. The joint prior distribution of y and the predicted value f_* is

$$\begin{bmatrix} y \\ f_* \end{bmatrix} \sim N\left(0, \begin{bmatrix} K(X, X) + \sigma_n^2 I_n K(X, x_*) \\ K(x_*, X) k(x_*, x_*) \end{bmatrix}\right) \tag{5.30}$$

where $K(X, x_*)$ denotes the N-dimensional covariance vector between the training points $X$and test points x_*; $k(x_*, x_*)$is the variance of x_*. Then the posterior distribution of f_* is

$$f_* | X, y, x_* \sim N(\bar{f}_*, \text{cov}(f_*)) \tag{5.31}$$

$$\bar{f}_* = K(x_*, X)[K(X, X) + \sigma_n^2 I_n]^{-1} y \tag{5.32}$$

$$\text{cov}(f_*) = k(x_*, x_*) - K(x_*, X) \cdot [K(X, X) + \sigma_n^2 I_n]^{-1} K(X, x_*) \tag{5.33}$$

where $\hat{\mu}_* = \bar{f}_*$ and $\hat{\sigma}_{f_*}^2 = \text{cov}(f_*)$ represent the mean and covariance of f_*. GP may suffer from a heavy computational burden when dealing with a large amount of data. To this end, SPGP is proposed. The pseudo set $\bar{D} = (\bar{X}, \bar{f})$ is used to substitute for the real dataset, where $\bar{X} = \{\bar{x}_i\}_{i=1}^m$ and $\bar{f} = \{\bar{f}_i\}_{i=1}^m$ represent the input and output of the pseudo points. The observed output distributed by noise is

$$y | x_*, \bar{X}, \bar{f} \sim N(\bar{k}_m(x_*)^T \bar{K}_{mm}^{-1} \bar{f}, \bar{k}(x_*, x_*) - \bar{k}_m(x_*)^T \bar{K}_{mm}^{-1} \bar{k}_m(x_*) + \sigma_n^2) \tag{5.34}$$

where $[\bar{k}_m(x_*)]_i = K(\bar{x}_i, x_*)$, $[\bar{K}_{mm}]_{ij} = K(\bar{x}_i, \bar{x}_j)$. Then the posterior distribution of f_*^{SP} is derived as:

$$\bar{f}_*^{SP} = \bar{k}_m(x_*)^T \bar{Q}_{mm}^{-1} \bar{K}_{mn} (\Lambda + \sigma_n^2 I)^{-1} y \tag{5.35}$$

$$\text{cov}(f_*^{SP}) = k(\bar{x}_*, \bar{x}_*) - \bar{k}_m(x_*)^T \times (\bar{K}_{mm}^{-1} - \bar{Q}_{mm}^{-1}) \bar{k}_m(x_*) + \sigma_n^2 \tag{5.36}$$

where

$$\begin{aligned} &[\bar{K}_{nm}]_{ij} = K(x_i, \bar{x}_j), \Lambda = diag(\lambda) \\ &\lambda_n = K_{nn} - \bar{K}_{nm}^T \bar{K}_{mm}^{-1} \bar{K}_{mn}, \bar{Q}_{mm} = \bar{K}_{mm} + \bar{K}_{mn} (\Lambda + \sigma_n^2 I)^{-1} \bar{K}_{nm} \end{aligned} \tag{5.37}$$

The square exponential covariance function is used as the kernel function, which can be expressed as:

$$k(x, x') = \sigma_f^2 \exp(-\frac{1}{2}(x - x')^T M^{-1}(x - x')) \tag{5.38}$$

where $M = diag(l^2)$ and l represents the variance scale; σ_f^2 is the signal variance. The hyperparameter set can be denoted as $\theta_s = \{M, \sigma_f^2, \sigma_n^2\}$. When the hyperparameter set is fixed and the optimal location of the pseudo set is determined, the mean and covariance of the predicted value can be obtained according to (5.35) and (5.36).

5.3.1.2 Network partition

A network partition aims to divide the whole network into several sub-regions such that the centralized voltage regulation problem is separated into several sub-problems that can be solved in a distributed manner. In this paper, the voltage sensitivity based on electrical distance is used to aggregate the nodes with a similar property while considering the regional voltage regulation ability. The electrical distance d_{ij} is defined based on the voltage-active power sensitivity and voltage-reactive power sensitivity matrix as follows [15]:

$$d_{ij} = d_{ij}^{VP} + d_{ij}^{VQ} \tag{5.39}$$

where d_{ij}^{VP} and d_{ij}^{VQ} represent the electrical distance defined based on voltage-active power sensitivity and voltage-reactive power sensitivity, respectively. They are expressed as

$$d_{ij}^{VP} = S_{ii}^{VP} + S_{jj}^{VP} - S_{ij}^{VP} - S_{ji}^{VP}, d_{ij}^{VQ} = S_{ii}^{VQ} + S_{jj}^{VQ} - S_{ij}^{VQ} - S_{ji}^{VQ} \tag{5.40}$$

where S_{ij}^{VP} and S_{ij}^{VQ} represent the sensitivity of the voltage of node i to the injected active and reactive power of node j, respectively. The smaller the electrical distance d_{ij} is, the tighter the electrical connection between nodes i and j will be. The voltage regulation capability is defined as:

$$c = \min\{c_1, c_2, ..., c_k\}, c_k = clip((\sum_{i \in k} Q_i S_{ij}^{VQ} + P_i S_{ij}^{VP})/\Delta V_j, 0, 1) \tag{5.41}$$

where c represents the voltage regulation capability of the whole network; $c_k$is the regional voltage capability of the cluster k; ΔV_j denotes the voltage deviation of the node with the maximum voltage violation value; Q_i and P_i represents the maximum reactive power and active power curtailment that can be used by node i for voltage regulation, respectively. A clip function restricts the range of c_k within [0,1]. The performance index of the clustering is defined based on the modularity index:

$$\rho = \frac{1}{a}\sum_i \sum_j \left[W_{ij} - \frac{l_i l_j}{a}\right] * \delta(i,j) * c, \delta(i,j)$$

$$= \begin{cases} 1, if\ i, j \in k \\ 0, else \end{cases}, W_{ij} = 1 - \frac{d_{ij}}{\max\limits_{m,n \in N} d_{mn}} \tag{5.42}$$

where $a = \sum_i \sum_j W_{ij}$ represents the sum of the weights for all nodes; $l_j = \sum_j W_{ij}$ represents the weight of node j. For the performance index, the larger the value is the closer the electrical connection inside the cluster and a looser connection among clusters. A tabu algorithm can be used to search for the optimal clusters while maximizing the defined performance [15].

5.3.1.3 Formulation of Markov games

After the network partition, the centralized voltage regulation problem is separated into several sub-problems. In this section, the control of multiple sub-networks is formulated as a Markov Game (MG), whose components are defined as follows:

Agents: In the MG, each agent represents a sub-network.

Environment: The surrogate model that calculates the voltage deviation according to states and actions by agents.

State: The global state set at t, S_t, includes all agents' states. For agent j, the state s_t^j includes the local observations of the j-th sub-network. For the tests when PVs and SVCs are utilized for voltage regulation, the state of agent j is defined as $s_t^j = (P_{Load}(i,t), Q_{Load}(i,t), P_{PV}(j,t))$. When PVs and BSSs are used, the state is $s_t^j = (P_{Load}(i,t), Q_{Load}(i,t), P_{PV}(j,t), E(i,t))$.

Action: The action set at t, A_t, includes all agents' actions. For agent j, the action a_t^j contains the control variables within the j-th sub-network. When PVs and SVCs are utilized for voltage regulation, the action of agent j is defined as $a_t^j = (\alpha_{PV}(j,t), \alpha_{SVC}(i,t), \alpha_{cur}(j,t))$. Then, the control variables can be obtained by:

$$\begin{aligned} &Q_{PV}(j,t) = \alpha_{PV}(j,t)\sqrt{(S_{PV}(j,t))^2 - (P_{PV}(j,t))^2}, -1 \le \alpha_{PV}(j,t) \le 1 \\ &Q_{SVC}(i,t) = \alpha_{SVC}(j,t)Q_{SVC.\max}, -1 \le \alpha_{SVC}(j,t) \le 1 \\ &P_{cur}(j,t) = \alpha_{cur}(j,t)P_{PV}(j,t), 0 \le \alpha_{SVC}(j,t) \le \delta \end{aligned} \tag{5.43}$$

By contrast, when PVs and BSSs are utilized, the action is defined as $a_t^j = (\alpha_{PV}(j,t), \alpha_{cur}(j,t), \alpha_{BSS}(i,t))$. Then, the control variable $P_{BSS}(i,t)$ can be obtained by $P_{BSS}(i,t) = \alpha_{BSS}(i,t)P_{BSS.\max}$. To avoid the violation of the constraint, the action can actually be performed as

$$P_{BSS}(i,t) = \begin{cases} \eta_{dis}(E_{\min} - E_t(i,t)), \mathit{if}\ P_{BSS}(i,t) < 0\ \mathit{and}\ E(i,t) + P_{BSS}(i,t)/\eta_{dis} < E_{\min} \\ (E_{\max} - E_t(i,t))/\eta_{ch}, \mathit{if}\ P_{BSS}(i,t) > 0\ \mathit{and}\ E(i,t) + \eta_{ch}P_{BSS}(i,t) > E_{\max} \\ P_{BSS}(i,t), \mathit{others} \end{cases} \tag{5.44}$$

where the charging/discharging actions of BSS are between $[\eta_{dis}(E_{\min} - E_t(i,t)), (E_{\max} - E_t(i,t))/\eta_{ch}]$. Then the inequality constraints can be satisfied. It is worth noting that since the proposed DRL is the off-policy algorithm, the bounded actions can also be used for training.

Reward: r_t represents the immediate reward the agent obtains when action A_t is executed under state S_t. All agents share the same reward, i.e.,

$$r_t = -\left(\sum_{i=1}^{N} |V(i,t) - V_0| + \beta \sum_{j=1}^{G} P_{cur}(j,t)\right) + \eta \tag{5.45}$$

where $V(i,t)$ is calculated by the surrogate model; $P_{cur}(j,t)$is calculated by (5.43) according to action $\alpha_{cur}(j,t)$ and state $P_{PV}(j,t)$;η is the penalty term if a voltage violation exists.

State transition: The state transition is denoted as $s_{t+1}^j = f(s_t^j, a_t^j, \omega_t)$. When PVs and SVCs are utilized for voltage regulation, the state transitions of $P_{Load}(i,t)$, $Q_{Load}(i,t)$, and $P_{PV}(j,t)$ are mainly affected by the randomness of the environment ω_t since the accurate values of load demand and PV generation are unknown. When PVs and BSSs are utilized, the state transition for the energy level of BSS is controlled by $P_{BSS}(i,t)$. It is difficult to find the accurate distribution of ω_t as it is affected by uncertain variables, such as load demand and PV generations. To this end, a model-free approach is proposed.

At each time step, each agent obtains a local observation of its corresponding sub-network s_t^j, based on which action a_t^j is made. Then all agents receive an immediate reward r_t according to the curtailment PV output and the voltage deviation calculated by the surrogate model.

5.3.2 *Solutions via MASAC algorithm*

The MASAC algorithm is used to solve the MG by modeling each sub-region as an SAC agent within the centralized training framework. Each SAC agent employs two functions for different purposes: the actor function takes the local observation of each sub-region s_t^j as input and outputs the action a_t^j; the critic function takes the global information (S_t, A_t) as input and outputs the judgment of the actor's decision. All agents are trained in a centralized manner to develop a coordinated control strategy [20].

Consider an MG with N agents and policy set $\pi = \{\pi_1, ..., \pi_N\}$, where policy π_i represents the actor function of agent i parameterized by μ_i. Different from the standard DRL algorithm that aims to maximize the cumulative reward, the objective of the actor function of SAC is to maximize the sum of the expected reward and an entropy term$\sum_{t=0}^{T} E[r_t + \alpha \mathrm{H}(\pi_i(\cdot|s_t^i))]$, where $\mathrm{H}(\pi^{\mu_i}(\cdot|s_t^i)$ is the entropy term; α represents the temperature parameter used to balance the two terms in the objective function. The introduced entropy term encourages the agents to explore more widely in the policy space by acquiring more diverse behaviors [16]. The parameters of the actor function are adjusted according to [21]:

$$\nabla_{\mu_i} J(\mu_i) = E_{S_t, A_t \sim D}[\nabla_{\mu_i} \log(\pi^{\mu_i}(a_t^i|s_t^i))\rho_i(S_t, a_t^1, ..., a_t^N)] \tag{5.46}$$

$$\rho_i(S_t, a_t^1, ..., a_t^N) = -\alpha \log(\pi^{\mu_i}(a_t^i|s_t^i)) + Q_i^{\pi}(S_t, a_t^1, ..., a_t^N) - b(S_t, a_t^{\backslash i}) \tag{5.47}$$

$$b(S_t, a_t^{\backslash i}) = E_{a_t^i \sim \pi^{\mu_i}(s_t^i)}[Q_i^{\pi}(S_t, (a_t^i, a_t^{\backslash i}))] \tag{5.48}$$

where $Q_i^{\pi}(S_t, a_t^1, ..., a_t^N)$ is the value of current action; $b(S_t, a_t^{\backslash i})$ is the baseline term, indicating the value of the average action for agent i; $Q_i^{\pi}(S_t, a_t^1, ..., a_t^N) - b(S_t, a_t^{\backslash i})$represents the advantage of current action as compared to the baseline term.

In (5.47), the action value of agent i $Q_i^{\pi}(S_t, a_t^1, ..., a_t^N)$ is calculated by the critic function $Q_i^{\pi}(\cdot)$, which optimizes its parameters following a Q-learning iteration. The parameters of the critic function are optimized by minimizing the following loss objective [21]:

$$L = (y_t - Q_i^{\pi}(S_t, a_t^1, ..., a_t^N))^2 \tag{5.49}$$

$$y_t = r_t^i + \gamma E_{a_{t+1}^i \sim \pi^{\mu_i'}}[-\alpha \log(\pi^{\mu_i'}(a_{t+1}^i | s_{t+1}^i)) + Q_i'\pi(S_{t+1}, a_{t+1}^1, ..., a_{t+1}^N)] \tag{5.50}$$

where y_t represents the target value; $\pi^{\mu_i'}$ and $Q_i'\pi'$are the target actor and critic introduced to stabilize the training process. The parameters of the target networks are typically updated by slowly tracking the online ones. The SAC algorithm follows the calculation of y_t in the double Q-learning method by adopting a pair of critics $(Q_{i,1}^{\pi}, Q_{i,2}^{\pi})$, yielding

$$y_t = r_t^i + \gamma E_{a_{t+1}^i \sim \pi^{\mu_i'}}[-\alpha \log(\pi^{\mu_i'}(a_{t+1}^i | s_{t+1}^i)) + \min_{n=1,2} Q_{i,n}'\pi(S_{t+1}, a_{t+1}'1, ..., a_{t+1}'N)] \tag{5.51}$$

where $Q_{i,n}'\pi(\cdot)$ represents the nth target critic function of agent i. $\pi^{\mu_j'}(\cdot)$ is the target actor function of agent j. By maintaining a pair of critics and taking the minimum value between them, the typical overestimation can be efficiently reduced.

The SAC algorithm also introduces the experience replay mechanism to promote training stability. The proposed MASAC algorithm consists of N agents, each with an M-sized replay buffer $D_i = \{m_1^i, ..., m_M^i\}$. At each time step, the transition experience $m_t^i = \{s_t^i, a_t^i, r_t^i, s_{t+1}^i\}$ is stored in the memory D_i. When the number of stored experiences reaches the upper limit, the old memory is replaced with the new ones. In the training process, a mini-batch of experience is sampled from the replay buffer for calculating gradients and performing optimization.

To cope with the complex nonlinearities in the voltage regulation problem, deep neural networks (DNNs) are used to approximate the actor and critic functions. Then, the parameters of functions are replaced by the weight matrix and bias vector. For agent i, the parameter set to be optimized can be represented as $\theta_i = \{\theta_i^{\mu}, \theta_i^{\mu'}, \theta_i^{Q_1}, \theta_i^{Q_1'}, \theta_i^{Q_2}, \theta_i^{Q_2'}\}$, where $\theta_i^{\mu}$and $\theta_i^{\mu'}$represent the parameters of the actor and target actor networks; $\theta_i^{Q_1}$ and $\theta_i^{Q_1'}$, $\theta_i^{Q_2}$ and $\theta_i^{Q_2'}$ are the parameters of the critic and target critic networks of $Q_{i,1}^{\pi}$ and $Q_{i,2}^{\pi}$.

5.3.2.1 Centralized training with a surrogate model

The proposed method contains two sets of parameters: the parameters of the surrogate model θ_s and the MADRL algorithm θ_c. The training procedure is shown in Algorithm 5.3.

Algorithm 5.3: Centralized training of model-free MADRL

Algorithm Training of proposed MADRL
1: Randomly initialize θ_s and obtain the prior model
2: Update θ_s and determine the location of the pseudo-set
3: Output the posterior model as the surrogate model
4: Randomly initialize parameters of NNs θ_c
5: for episode =1,2,..., M do
6: Receive initial observation s_0^i for each agent
7: for t = 1,2,... T do
8: Determine control action $a_t^i = \pi_i(s_t^i|\theta_i^u)$ for each agent,
Execute actions $A_t = (a_t^1, ..., a_t^N)$, calculate reward r_t by surrogate model
According to (5.35) and (5.45), and observe the new state s_{t+1}^i
9: For agent i = 1,..., N do
10: Store transition $m_t^i = \{s_t^i, a_t^i, r_t^i, s_{t+1}^i\}$ in memory D_i
11: if memory capacity is full, do
12: sample a random mini-batch B of transitions from memory D
13: update critic networks according to (5.52) and (5.53)
14: update actor networks according to (5.54) and (5.55)
15: update target networks according to (5.56)
16: end if
17: end for
18: end for
19: end for

The parameter set of the surrogate model is represented as $\theta_s = \{M, \sigma_f^2, \sigma_n^2\}$. The input of the algorithm is $P(i,t)$, $Q(i,t)$, and $V(i,t)$ and the output is θ_s. First, the hyperparameters are initialized and the prior model is obtained. Then, optimize the parameters through the maximum likelihood method and determine the optimal location of the pseudo set. After that, the posterior model is used as the surrogate model.

The parameter set of the control model is represented as $\theta_c = \{\theta_1, ..., \theta_N\}$, where $\theta_i = \{\theta_i^u, \theta_i^{u'}, \theta_i^{Q_1}, \theta_i^{Q_1'}, \theta_i^{Q_2}, \theta_i^{Q_2'}\}$ represents the parameter set of all the actor and critic NNs of agent i. The input includes $P_{Load}(i,t)$ and $Q_{Load}(i,t)$ of each node, $P_{PV}(j,t)$ of all the PVs, and the reward r_t calculated by the surrogate model. The output is θ_c. The training of DNN is based on the DRL algorithm and the centralized training framework.

In the beginning, θ_c is randomly initialized. The parameters of the target NNs are copied from online NNs. Then, the algorithm is trained for M episodes to update the parameters. An episode includes 24 time steps, each corresponding to an hour. At each time step, each agent makes a decision a_t^i based on its local observation s_t^i, yielding an immediate reward r_t calculated by the surrogate model via (5.35) and (5.45); then the system transfers to the next state s_{t+1}^i. Next, the transition experience $m_t^i = \{s_t^i, a_t^i, r_t^i, s_{t+1}^i\}$ is stored in its memory D_i. From $0 - M_1$ episodes, the actions are randomly selected to fully explore the environment. From $M_1 - M$ episodes, the actions of each agent are selected according to its policy function π^{μ_i} and the parameters are optimized utilizing the transition stored in the replay buffer. Specifically, a mini-batch of experiences is sampled from D to calculate the gradient of the NNs. The updated equations of the critic networks are:

$$L(\theta_i^{Q_n}) = \frac{1}{B}\sum_{k=1}^{B}(Q_{i,n}^{\pi}(S_k, a_k^1, \ldots, a_k^N) - y)^2, n = 1,2 \tag{5.52}$$

where B represents the size of the mini-batch. The loss function is minimized by optimizing the parameters of critic NNs based on the gradient rule:

$$\theta_i^{Q_n} \leftarrow \theta_i^{Q_n} + \eta_Q \nabla_{\theta_i^{Q_n}} L(\theta_i^{Q_n}), n = 1,2 \tag{5.53}$$

where η_Q represents the learning rate for critic networks. The parameters of the actor NNs are updated via

$$\nabla_{\mu_i} J(\mu_i) = E_{S_t, A_t \sim D}[\nabla_{\mu_i} \log(\pi^{\mu_i}(a_t^i|s_t^i))\rho_i(S_t, a_t^1, \ldots, a_t^N)] \tag{5.54}$$

Then, the gradient rule is applied:

$$\theta_i^{\mu} \leftarrow \theta_i^{\mu} + \eta_\mu \nabla_{\theta_i^{\mu}} J(\theta_i^{\mu}) \tag{5.55}$$

where η_μ represents the learning rate for actor networks. The parameters of target NNs are optimized via

$$\theta_i^{\mu'} \leftarrow \tau\theta_i^{\mu} + (1-\tau)\theta_i^{\mu'}, \theta_i^{Q_n'} \leftarrow \tau\theta_i^{Q_n} + (1-\tau)\theta_i^{Q_n'}, n = 1,2 \tag{5.56}$$

where $\tau \ll 1$ represents the tracking coefficient.

5.3.2.2 Real-time decentralized execution

The parameters of NNs will be fixed when the training procedure is completed, and only the actor NNs are kept for real-time control. Each actor NN takes the local information of its corresponding sub-network as input and output control decisions in real time. Since the critic NNs explicitly model the policies of other agents during training, the actor learns a coordinated strategy and can exhibit cooperative behavior using only local information. This allows a decentralized execution using only local measurements, achieving a significant reduction in communication requirements, and enabling its scalability to large-scale systems. Algorithm 5.4 shows the specific process of decentralized execution.

Algorithm 5.4: Decentralized execution

Algorithm Decentralized control by the proposed MADRL
1: Load the parameters of actor network of each agent θ_i^{μ}
2: for time step t = 1,2,... T do
3: for agent i = 1,..., N do
4: obtain the local observation s_t^i
5: determine action a_t^i according to $a_t^i = \pi_i(s_t^i|\theta_i^{\mu})$
6: end for
7: execute the concatenated actions $A_t = (a_t^1, \ldots, a_t^N)$
8: end for

5.3.3 Numerical results on a distribution system

In this section, simulations are carried out on the IEEE 123-bus system [21] to evaluate the performance of the proposed model-free MADRL method. Comparative tests with other regression algorithms are provided, followed by comparative results using other voltage control methods. For the modified IEEE 123-bus system, the parameters of the installed PVs and SVCs are listed in Table 5.8. In this study, the total number of nodes for each cluster is limited to 14–38. The optimal partition results are shown in Figure 5.11. The performance index is 0.09. For the PVs, actual data lasting 360 days from Xiaojin, a county in the Sichuan province of China, are used. The PV output data is separated into a training set (300 days) and a test set (10 days). The hyperparameters of the control model are listed in Table 5.9. Both the surrogate model and the control model are

Table 5.8 Parameters of controllable devices in the IEEE 123-bus system

Type	Capacity of each device	Location
SVC	0.3MVar	45, 58, 79, 114
PV	1.2MW/1.26MVA	11, 22, 35, 53, 62, 68, 90, 100, 105, 118

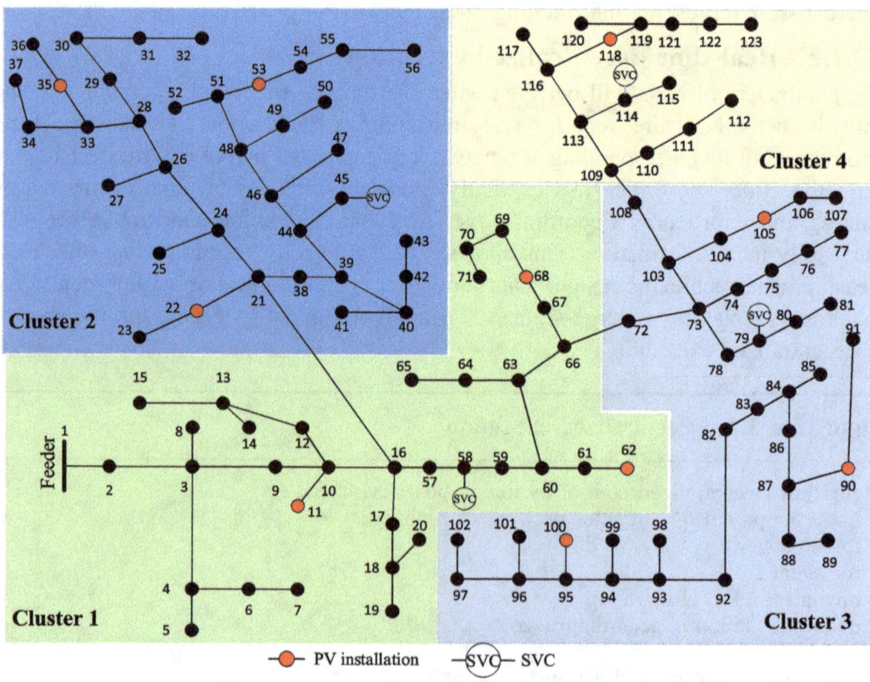

Figure 5.11 Partition of modified IEEE 123-bus system with PV and SVC included [22]

Table 5.9 Parameters of the control model

Parameter	Value
Batch size for updating NN	256
Replay buffer size	24,000
Temperature parameter	1.25e-3
Discount factor	0
Soft update coefficient	0.001
Learning rate for actor/critic-network	0.001/0.001
Neuron number of hidden layer	100/100/100

Table 5.10 MAE comparisons of IEEE 128-bus system under different training instances

Training instances	LR	DNN	SPGP
200	3.02e-3	1.29e-3	9.70e-4
500	2.23e-3	7.74e-4	4.88e-4
1500	1.96e-3	4.48e-4	2.58e-4

implemented in Python. A workstation with an Intel i9-7900X CPU and an NVIDIA GeForce 2080Ti GPU is used for the simulation.

5.3.3.1 Performance evaluation of the surrogate model

Comparative tests are carried out among the linear regression (LR), Deep Neural Networks (DNN), and Sparse Gaussian Process (SPGP) to evaluate the forecasting accuracy of the proposed surrogate model when there are few recorded samples. Three thousand instances of data with $\boldsymbol{P}$, $\boldsymbol{Q}$, and $\boldsymbol{V}$ are generated by the AC power flow analysis. Among these, 1,000 instances are used as a test set to evaluate the performance of the trained model. The mean absolute error (MAE) is used as a performance index.

Table 5.10 displays the MAE of each method under a different number of training instances. DNN has three hidden layers, the neuron numbers of which are 400, 200, and 200, respectively. The batch size and learning rate are selected as 32 and 0.001, respectively. The DNN is trained for 30,000, 17,500, and 17,500 epochs to learn the relationship between $\boldsymbol{P}$, $\boldsymbol{Q}$, and $\boldsymbol{V}$ when 200, 500, and 1500 instances of data are utilized for training. The proposed SPGP adopts the square exponential covariance function. It can be observed that when the training set consists of only 200 instances of data, the forecasting accuracy of the SPGP method has better performance than the LR and DNN. The DNN yields poor performance on the test data because it involves too many parameters to be optimized and therefore suffers from over-fitting issues. The performance of the LR method is the worst because it is unable to capture the complex nonlinear relationship between the power injection and the voltage magnitude of each node. The improvement in forecasting accuracy

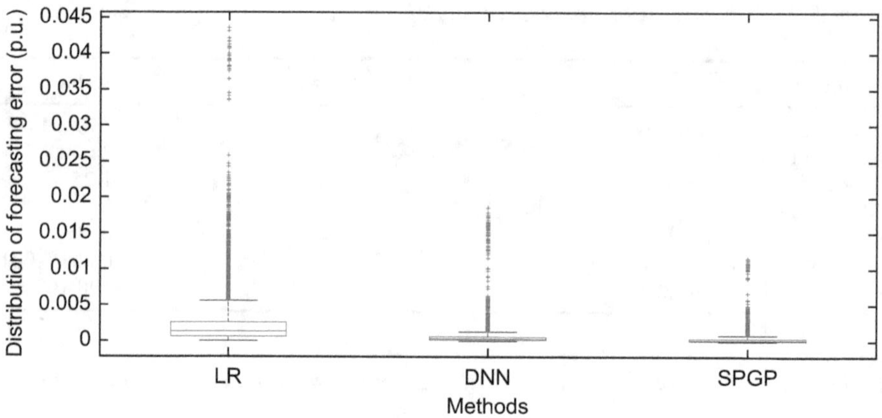

Figure 5.12 The distributions of the forecasting errors for three methods [22]

of the three methods can be observed when the number of training instances is gradually increased. Note that the proposed SPGP can always achieve the best performance. The distributions of the prediction error for various methods are shown in Figure 5.12 when the number of training instances is selected as 1,500. It can be observed that the prediction error of the SPGP method is distributed in a very small range, which is close to zero. The maximum prediction errors of the LR, DNN, and SPGP are 0.044, 0.019, and 0.012, respectively.

5.3.3.2 Performance evaluation of the control model

To demonstrate the benefits of the proposed model-free MADRL control method, comparative results with other ones are also carried out. First, the evaluation of cumulative reward during the training procedure of the proposed method is compared with other methods, including (1) **multi-agent twin delayed deep deterministic policy gradient (MATD3) algorithm** [16], where each sub-region is modeled as a TD3 agent and all agents are trained in a centralized manner according to the reward signal calculated by the surrogate model. The neuron number of each hidden layer, replay buffer size, soft update coefficient, and discount factor are set to the same values as those in the proposed method. The batch size, policy update frequency, target policy smoothing coefficient, and learning rate for actor/critic network are set as 128, 2, 0.2, and 0.001/0.002, respectively; (2) **the SAC method adopts a centralized control framework** (SAC-C) [15], where the whole distribution system is modeled as a SAC agent, which is trained by continuous interaction with the surrogate model. The learning rate for the actor/critic network is set to 3e-4/3e-4, while the other parameters are the same as the proposed method.

Centralized training evaluation

All agents are trained with 50,000 episodes to learn a coordinated voltage regulation strategy. The evolution of the cumulative reward during the training procedure

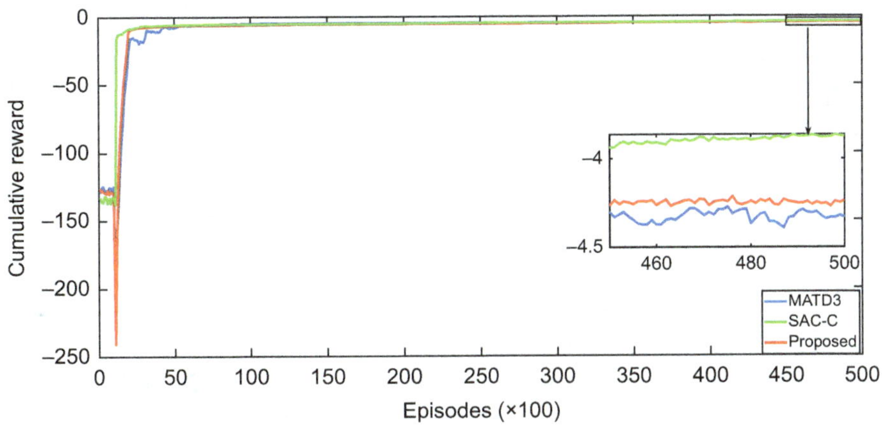

Figure 5.13 The evolution of the cumulative reward for different methods during the training procedure on the IEEE 123-bus system [22]

is shown in Figure 5.13 when β is set to 0.1. In the first 1,000 episodes, the parameters of NNs are fixed, and the actions of agents are randomly selected to explore the environment. From 1,000 episodes onward, the actions are selected by the NNs, whose parameters are optimized concurrently. It can be observed that the agents obtain a low reward value at the beginning. This demonstrates that the initialized policy is incapable of making good decisions to obtain a high reward. However, the cumulative reward continuously increases, and this indicates that the agents gradually evolve and learn a policy to obtain a high reward. Thanks to the entropy term that can improve the exploration capability of the SAC algorithm, the proposed method finally converges to a higher value than the MATD3 method. Since the SAC-C method takes global information as input, it can learn a better control strategy than the MADRL method based on only local information. The results validate the integration of the surrogate model with MADRL during the training stage.

Evaluation of the learned strategy on test data

When the training process is completed, comparative tests with other benchmark methods are carried out on 10-day test data to evaluate the generalization ability of the learned strategy, including (1) **the original method** without reactive power control and PV curtailment for voltage regulation; (2) **the SAC algorithm** in a decentralized training manner (SAC-D), where each sub-region is controlled by SAC agents and the agents are trained separately and sequentially based on local observations to minimize the regional voltage deviation. The parameter setting of the SAC-D method is the same as the proposed method; (3) **the stochastic programming method (SP)** is based on global information, where 300 scenarios are generated according to the assumed distribution of the predicted PV outputs and load demand. Note that the scenario reduction method is applied to select 20 representative ones; (4) **the MASAC method** is implemented in a centralized

Table 5.11 Comparison results for different strategies on test data

Method	Ave. Dev.	Ave. curt.	Max. rise	Max. drop	Para. dep.
Original	2.14%	–	6.02%	5.88%	–
SAC-D	3.20%	1.02	9.42%	5.81%	✓
MATD3	0.16%	0.24	0.70%	0.80%	×
SP	0.16%	1.61	0.91%	0.97%	✓
Proposed	0.15%	0.20	0.67%	0.84%	×
MASAC	0.14%	0.21	0.68%	0.78%	✓
SAC-C	0.12%	0.12	0.63%	0.78%	✓

training and decentralized execution manner. Note that the SAC-D, MASAC, and SAC-C methods use the Z-bus method for the calculation of the immediate reward during training, and therefore an accurate distribution system physical model is required.

The perfect physical model is assumed, while our proposed method is based on the surrogate model learned from the data. The key idea is that even without an accurate physical model, the proposed MADRL can achieve similar performances or outperform the known ones. Comparative results for all methods are shown in Table 5.11. For the parameter dependency index, "✓" represents that the corresponding approach relies on the exact physical model when taking control decisions, while "×" means model-free. There are two types of control frameworks, where SAC-D, MATD3, proposed, and MASAC methods make decisions according to the information collected from their sub-networks; by contrast, SP and SAC-C inform decisions using global information.

It can be observed from Table 5.11 that when SAC-D agents are trained separately, they do not provide appropriate decisions to regulate the voltage within an acceptable range if only local information is used. This is due to the lack of coordination between the agents. By contrast, the centralized SAC-C method fully leverages the coordination and yields much better performance than the SAC-D. The MASAC method can also effectively control the voltage with only local information. Note that during the centralized training and decentralized execution framework, the coordinated control strategy has been learned, which justifies its effectiveness. The performance of the proposed method is only slightly worse than that of the MASAC method but is better than that of the SP. It should be noted that the proposed method does not need any physical model, while MASAC needs an accurate physical power flow model, which is very difficult to achieve in practice. Both MATD3 and the proposed methods are model-free, decentralized approaches. Thanks to the enhanced exploration capability of the entropy term in the SAC algorithm, the proposed method achieves better performance than the MATD3 algorithm. The active power curtailments of PVs for different strategies are also shown in Table 5.11. Due to the lack of coordination, the SAC-D method curtails more PV generation to reduce voltage deviations. By contrast, the MATD3, the proposed method, and the MASAC method learn a coordinated strategy during the centralized training, and thus they can achieve better voltage regulation performance while curtailing much fewer PV generations. The SAC-C method can coordinate the reactive power and PV curtailments based on global information, yielding the best voltage regulation performance with the least PV curtailment. Since the SP method calculates a predetermined decision based on sampled scenarios, it tends to be conservative and curtails more PV generations. As a result, the effectiveness of the proposed model-free MADRL is validated.

A sunny day in the test set that suffers from voltage violation issues is also selected from the test set to demonstrate the effectiveness of the proposed method. The load demand and PV generation of the selected day are plotted in Figure 5.14. The voltages of all nodes using various control strategies when $t = 12{:}00$ are shown in Figure 5.15. It can be observed that the voltages at nodes 87–91 and 100 go beyond the upper bound if no reactive control and PV curtailment are applied. When the SAC-D method is used, the voltages return to the allowed range.

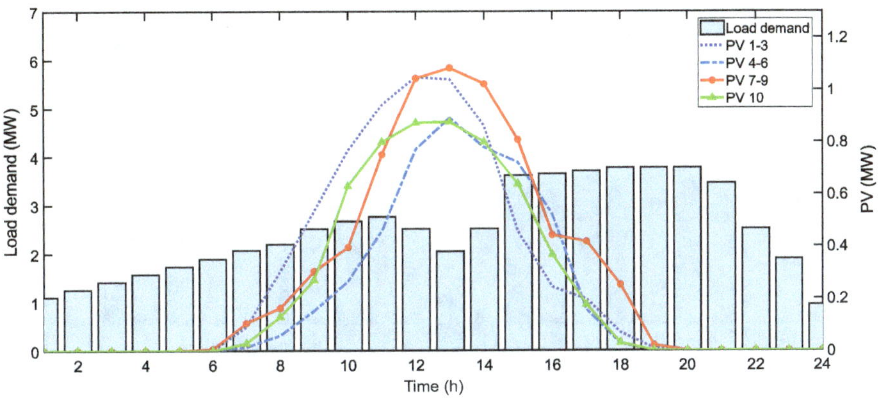

Figure 5.14 Load demand and PV generations of the test day [22]

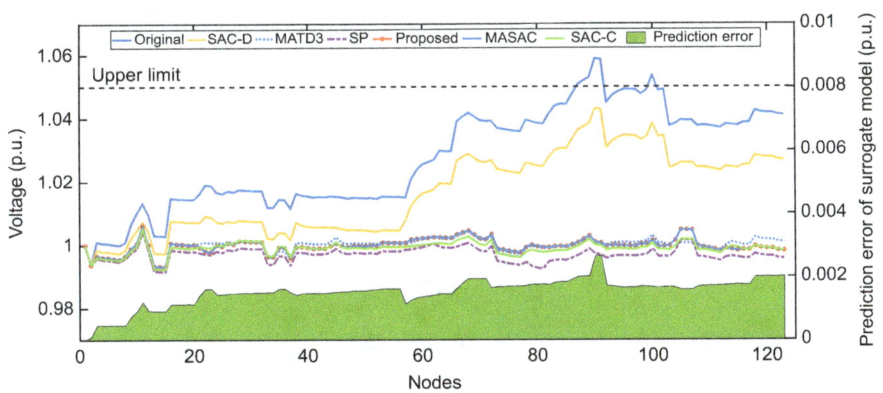

Figure 5.15 Voltage distributions for different strategies when t = 12 [22]

However, it suffers from large deviations due to the lack of coordination. By contrast, the proposed model-free MADRL, MATD3, and the model-based MASAC with only local measurements can achieve a control performance that is close to that of the centralized SAC-C. Most of the prediction errors for nodes at this moment are less than 0.002 p.u., demonstrating that the surrogate model can provide an accurate reward signal to guide the learning of the control strategies. The PV curtailments for different control strategies on this test day are shown in Figure 5.16. The proposed MADRL, the MATD3, the MASAC, and the SAC-C can achieve good voltage control performance with very few PV curtailments. This justifies their economic benefits.

Evaluation of the generalization ability of the learned strategy
Further tests are carried out under large PV outputs and fluctuations to evaluate the generalization ability of the learned strategy. In particular, a varying PV output profile in 90 s due to cloud dynamics is shown in Figure 5.17. In the 30th and 90th

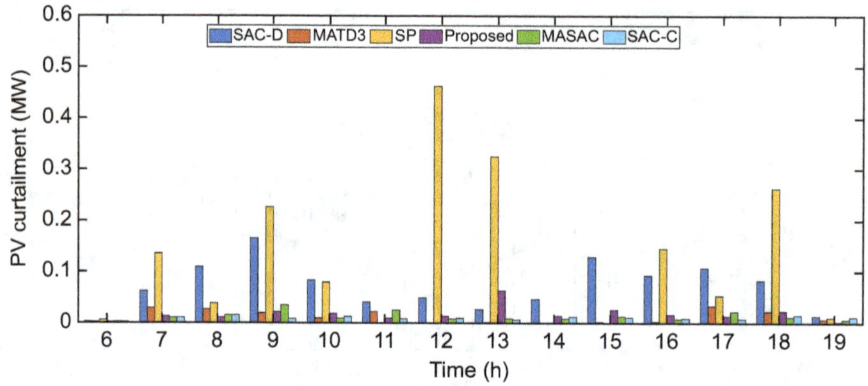

Figure 5.16 PV curtailment for different strategies from 6:00 to 19:00 on the day [22]

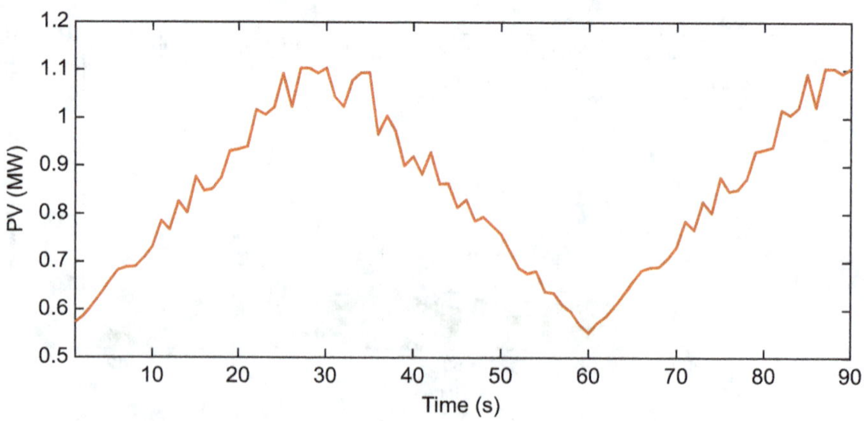

Figure 5.17 Varying PV output profiles in 90 s due to clouds [22]

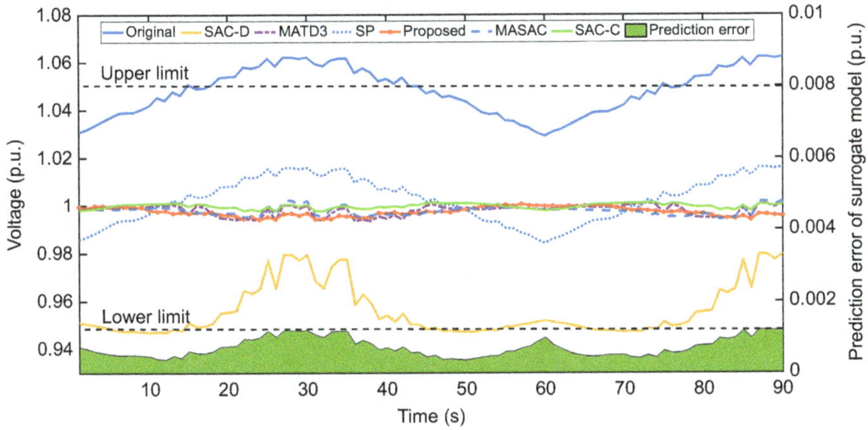

Figure 5.18 Voltage distributions of Node 90 for different control strategies [22]

second, the active power generations of all PVs reach 92% of their rated power, which is an extreme situation not seen by the agent during the training. The voltage distributions of node 90 using different voltage control methods are shown in Figure 5.18. The controllable devices are operated following pre-determined scheduling for the SP method since it is time-consuming to resolve the optimization problem. The SAC-D, MATD3, MASAC, SAC-C, and the proposed model-free MADRL methods can take decisions in milliseconds since the learned knowledge during training is scalable to newly encountered situations. It can be observed from Figure 5.13 that if no reactive power control and PV curtailments are applied, Node 90 has an over-voltage issue. The voltage comes down when the SAC-D method is applied, but its voltage falls out of the lower limit and suffers from overshoot issues due to the lack of coordination between controllable devices. Although the SP method can adjust the voltage to the allowed range, it suffers from large fluctuations. The proposed model-free MADRL, the MATD3, MASAC, and the SAC-C method can adjust the scheduling of various control devices based on real-time observations, therefore yielding much better performance than other methods. This demonstrates that the voltage control strategy learned by the proposed method can be generalized to extreme situations with large PV outputs and fluctuations. The SAC-C method achieves the best control performance since accurate global information is assumed, see t = 15–50 s in Figure 5.18 for example. The proposed approach can achieve a performance that is close to the MASAC and the SAC-C methods. Again, the proposed model-free method and accurate power flow models must be assumed for the SAC-C and MASAC methods. In summary, our model-free MADRL with only local measurements can achieve comparable performance with those methods that have accurate global information and physical models.

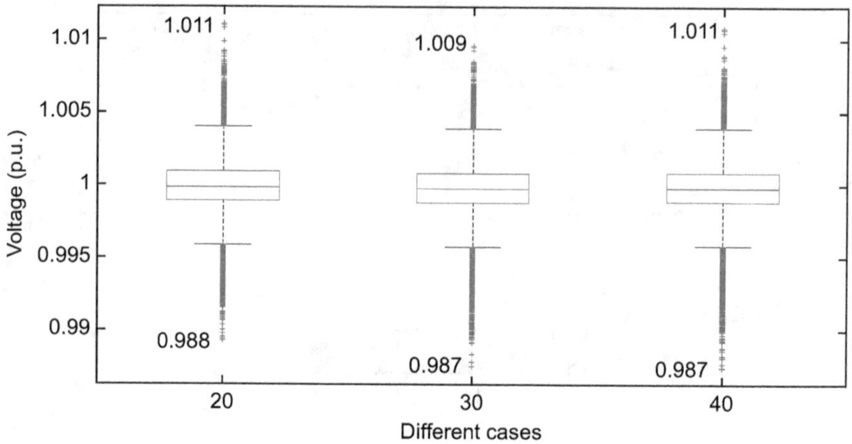

Figure 5.19 Voltage distributions of all nodes on test data by the proposed method under different cases [22]

Further tests are carried out to evaluate the generalization ability of the learned strategy when the actual load demand has large deviations from the forecasted values. In this test, the actual load demands of a certain number of nodes are assumed to have a 20% deviation from the forecasted ones. At each instant, the nodes that suffer from large deviations are randomly selected from all nodes. The voltage distributions of all nodes on 10 days of test data achieved by the proposed method are shown in Figure 5.19 when the numbers of the randomly selected nodes with large forecasting errors are set as 20, 30, and 40. The average voltage deviations in the three cases are 0.16%, 0.17%, and 0.17%, respectively. It can be observed from Figure 5.19 that the maximum voltage rise and drop under three cases are larger than those obtained by the proposed method under normal conditions, see Table 5.11. This may be because the extreme conditions are different from what the agent sees during training. Nevertheless, the suggested approach can regulate voltage within permissible ranges and attain average voltage deviations that closely resemble normal conditions, thereby showcasing the generalization capability of the acquired strategy.

With those comparative results, it can be concluded that (1) the proposed method can learn a strategy to reduce voltage deviation while minimizing the PV curtailment from the training data; (2) the learned strategy can generalize to test data that are unseen by the agent during training; (3) the learned strategy can also deal with extreme situations in the test stage; (4) the proposed model-free decentralized approach can achieve control performance that is close to that by the centralized method with accurate physical model.

Sequential tests considering BSSs

Further tests are also carried out on the IEEE 123-node systems to evaluate the performance of the proposed method when using Battery Storage System (BSS)

Table 5.12 Comparison results for different strategies on test data in the presence of BSSs

Method	Ave. Dev.	Ave curt. (MW)	Max. rise	Max. drop	Para. dep.
Original	2.42%	-	5.62%	5.87%	-
SAC-D	2.67%	2.34	6.99%	3.58%	✓
Proposed	0.20%	1.21	1.32%	0.86%	×
MASAC	0.18%	1.39	0.84%	0.84%	✓
SAC-C	0.31%	2.14	0.92%	0.95%	✓

and PV inverters for voltage regulation. The lower and upper bounds for the SOC of BSS are set as 0.2 and 0.8, respectively. The maximum charging/discharging power limit of BSS is 0.3 MW, with a charging/discharging coefficient η_{ch} and η_{dis} as 0.9. For the surrogate model, 1,500 instances of data with $\boldsymbol{P}$, $\boldsymbol{Q}$, and $\boldsymbol{V}$ are used for the SPGP algorithm. The setting of the surrogate model is the same as the single-shot control problem.

When the training process is complete, comparative tests are carried out on the test set to evaluate the performance of the proposed approach. Comparative results on 10 days' test data are shown in Table 5.12. It can be observed that there is a voltage violation issue when no control is applied. The SAC-D method also suffers from voltage security issues due to a lack of coordination. The centralized SAC-C method can reduce the voltage deviations and adjust the voltage to the allowable range. However, since the control of BSS needs to consider future uncertainties, it is difficult for a single agent to schedule multiple BSSs simultaneously. By contrast, the proposed MADRL-based decentralized control method can achieve better control performance. This demonstrates the superiority of the proposed decentralized method compared with the centralized ones in the presence of multiple BSSs. The performance of the proposed method is only slightly worse than that obtained by the MASAC method. However, the MASAC method depends on the perfect physical model of the DNs, which is difficult to obtain in practice.

A sunny day is selected from the test set to further evaluate the performance of the proposed method. The voltage profiles achieved by different methods at $t = 20{:}00$ are shown in Figure 5.20. It can be seen that when no control is applied, the voltage violates the lower limit. The SAC-D method suffers from over-adjustment due to a lack of coordination. The MADRL methods, with their centralized training and decentralized execution framework, can effectively reduce the voltage deviation and achieve better performance than the SAC-C method. The MASAC method obtains the best control performance. Again, it requires accurate information on the physical system. The proposed model-free method can achieve a similar performance as MASAC method without the accurate information. The voltage distributions of Node 90 for different control strategies are shown in Figure 5.21. The results shown here are consistent with those in Table 5.14.

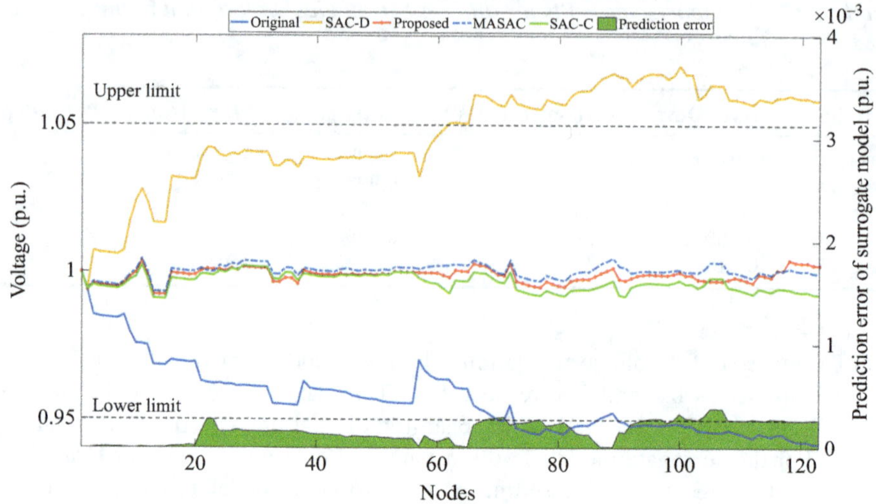

Figure 5.20 Voltage distributions for different strategies considering PVs and BSSs when t =20:00 [22]

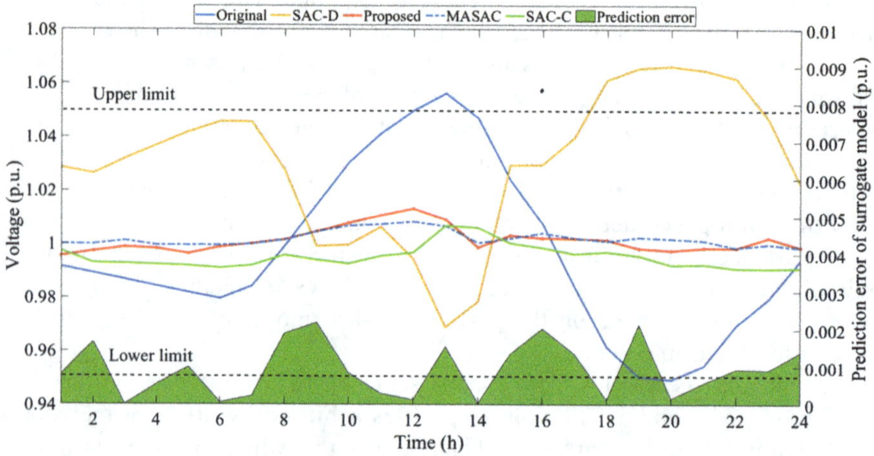

Figure 5.21 Voltage distributions of Node 90 for different control strategies considering PVs and BSSs [22]

5.4 Summary

In this chapter, two network partition-based decentralized voltage control methods are proposed. The conclusions are summarized as follows:

A distributed coordination control method is first proposed for distribution system Volt-VAR control, considering both PV inverters and SVCs. The spectral clustering algorithm allows for a partition of the large distribution system into several sub-networks from a voltage control perspective. Then, the control of each sub-network is formulated as the MGs and solved by an attention-based MATD3 algorithm. The proposed method is centralized-trained and distributed-implemented and can easily be used for real-time voltage regulation. Compared with centralized control, the proposed approach mitigates issues, such as communication bottlenecks and privacy concerns. Compared with other distributed control methods, only local information is needed without communications between agents. The proposed method can be adapted to the flexible network partition requirements of the operator rather than the typical MADRL algorithm. Comparative results with several other existing model-based and data-driven methods demonstrate that the proposed method can achieve 96.4% optimality based on local information while also considering uncertainties. However, the model-based approach could not achieve satisfactory outcomes in the case of rapid variations in PV outputs. A novel MADRL framework is introduced for voltage regulation in distribution systems with a high PV penetration, featuring centralized training and decentralized execution without physical models. The SPGP is first leveraged to build a surrogate model that learns the mapping relationship between the active and reactive power injections and the voltage magnitude of each node using few-shot recorded data. This surrogate model is further integrated with the MADRL to assist in the formulation of a coordinated control strategy. In particular, the voltage regulation problem is cast into the MADRL framework by partitioning the whole network into several sub-regions, considering both the regional voltage regulation ability and the electrical distance. Each sub-region is treated as an agent and solved by the DRL algorithm. All agents are trained in a centralized framework to learn the coordinated control strategy guided by the reward given by the surrogate model. The proposed method can achieve real-time scheduling using only local information. Comparative results demonstrate that: (1) the proposed decentralized control strategy can achieve similar performance as the centralized one; (2) the performance of the suggested model-free approach closely matches that of one relying on a flawless physical model; (3) the control strategy can be taken in real-time with the usage of the battery storage system to mitigate the influence of violent PV fluctuations.

The future works include the development of a new control method that can coordinate the inverters and utility-owned equipment, which is a two-timescale control problem. We will also propose a meta-learning-based MADRL algorithm to deal with topology changes in the distribution networks.

References

[1] C.-S. Wang and P. Li, "Development and Challenge of Distributed Generation, Micro Network and Intelligent Distribution Network," *Automation of Electric Power Systems*, vol. 34, no. 2, pp. 10–14, 2010.

[2] N. Mahmud and A. Zahedi, "Review of Control Strategies for Voltage Regulation of the Smart Distribution Network with High Penetration of

Renewable Distributed Generation," *Renewable & Sustainable Energy Reviews, Review,* vol. 64, pp. 582–595, 2016.

[3] Y. You, D. Liu, W. Yu, F. Chen, and F. Pan, "Active Distribution Network Technology and Its Progress," *Automation of Electric Power Systems*, vol. 36, no. 18, pp. 10–16, 2012.

[4] Wang C, Song B, Li P. "Current Status and Prospect of Active Distribution Network Optimization Technology," *Power Construction*, vol. 36, no. 1, pp. 8–15, 2015.

[5] Y. Xu, Z. Y. Dong, R. Zhang, and D. J. Hill, "Multi-Timescale Coordinated Voltage/VAR Control of High Renewable-Penetrated Distribution Systems," *IEEE Transactions on Power Systems*, vol. 32, no. 6, pp. 4398–4408, 2017.

[6] C. Zhang and Y. Xu, "Hierarchically-Coordinated Voltage/VAR Control of Distribution Networks Using PV Inverters," *IEEE Transactions on Smart Grid*, vol. 11, no. 4, pp. 2942–2953, 2020.

[7] M. Azarnia and M. Rahimiyan, "Robust Volt-Var Control of a Smart Distribution System Under Uncertain Voltage-Dependent Load and Renewable Production," *International Journal of Electrical Power & Energy Systems*, vol. 134, p. 107383, 2022

[8] T. Ding, C. Li, Y. Yang, J. Jiang, Z. Bie, and F. Blaabjerg, "A Two-Stage Robust Optimization for Centralized-Optimal Dispatch of Photovoltaic Inverters in Active Distribution Networks," *IEEE Transactions on Sustainable Energy*, vol. 8, no. 2, pp. 744–754, 2016.

[9] S. Kaisler, F. Armour, and J. A. Espinosa, "Introduction to Big Data: Challenges, Opportunities, and Realities Minitrack," in *2014 47th Hawaii International Conference on System Sciences*, 2014: IEEE, pp. 728–728.

[10] S. M. Amin, "Smart grid security, privacy, and resilient architectures: Opportunities and challenges," in *2012 IEEE Power and Energy Society General Meeting*, 2012: IEEE, pp. 1–2.

[11] W. Zheng, W. Wu, B. Zhang, H. Sun, and Y. Liu, "A Fully Distributed Reactive Power Optimization and Control Method for Active Distribution Networks," *IEEE Transactions on Smart Grid*, vol. 7, no. 2, pp. 1021–1033, 2015.

[12] A. Ng, M. Jordan, and Y. Weiss, "On Spectral Clustering: Analysis and an Algorithm," *Advances in Neural Information Processing Systems*, vol. 14, 2001.

[13] S. Fujimoto, H. Hoof, and D. Meger, "Addressing Function Approximation Error in Actor-Critic Methods," in *International Conference on Machine Learning*, 2018: PMLR, pp. 1587–1596.

[14] D. Cao, J. Zhao, W. Hu, F. Ding, Q. Huang, and Z. Chen, "Attention Enabled Multi-Agent DRL for Decentralized Volt-VAR Control of Active Distribution System Using PV Inverters and SVCs," *IEEE Transactions on Sustainable Energy*, vol. 12, no. 3, pp. 1582–1592, 2021.

[15] R. Lowe, Y. I. Wu, A. Tamar, J. Harb, O. Pieter Abbeel, and I. Mordatch, "Multi-Agent Actor-Critic for Mixed Cooperative-Competitive Environments," *Advances in Neural Information Processing Systems*, vol. 30, pp. 6379–6390, 2017.

[16] S. Iqbal and F. Sha, "Actor-Attention-Critic for Multi-Agent Reinforcement Learning," in *International Conference on Machine Learning*, 2019: PMLR, pp. 2961–2970.

[17] S. Wang, J. Duan, D. Shi, *et al.*, "A Data-Driven Multi-Agent Autonomous Voltage Control Framework Using Deep Reinforcement Learning," *IEEE Transactions on Power Systems*, vol. 35, no. 6, pp. 4644–4654, 2020.

[18] IEEE PES, Distribution Test Feeders, Sep. 2010. [Online]. Available: https://site.ieee.org/pes-testfeeders/resources/.

[19] Y. Chai, L. Guo, C. Wang, Z. Zhao, X. Du, and J. Pan, "Network Partition and Voltage Coordination Control for Distribution Networks with High Penetration of Distributed PV Units," *IEEE Transactions on Power Systems*, vol. 33, no. 3, pp. 3396–3407, 2018.

[20] T. Haarnoja, A. Zhou, P. Abbeel, and S. Levine, "Soft Actor-Critic: Off-Policy Maximum Entropy Deep Reinforcement Learning with a Stochastic Actor," in *International Conference on Machine Learning*, 2018: PMLR, pp. 1861–1870.

[21] H. Zhu and H. J. Liu, "Fast Local Voltage Control Under Limited Reactive Power: Optimality and Stability Analysis," *IEEE Transactions on Power Systems*, vol. 31, no. 5, pp. 3794–3803, 2015.

[22] D. Cao, J. Zhao, W. Hu, F. Ding, Q. Huang, Z. Chen, and F. Blaabjerg, "Data-Driven Multi-Agent Deep Reinforcement Learning for Distribution System Decentralized Voltage Control With High Penetration of PVs," *IEEE Transactions on Smart Grid*, vol. 12, no. 5, pp. 4137–4150, 2021.

Chapter 6

Energy management of hybrid systems using artificial intelligence

Weihao Hu[1], Guozhou Zhang[2], Zhenyuan Zhang[1], Sayed Abulanwar[3,4] and Frede Blaabjerg[5]

6.1 Introduction

Wind generation is currently considered one of the most rapidly increasing resources among other distributed generation technologies. By 2023, it is expected that the total wind power generation supply will be around 12% of the total world electricity demand. However, the integration of such wind resources into power system grids presents a major challenge to power system operators and planners. This is because of the high uncertainty and variability in the characteristic nature of wind power. Therefore, integrated energy systems (IESs) have great potential to enhance the flexibility of power systems to accommodate more renewable energy (RE) sources, such as wind and solar [1,2].

Conventional energy management (EM) of the IES is based on deterministic rules and abstract models, which suffer two main limitations: deterministic rules can only be optimally operated in a stationary system, and these are limited by fixed variables; building abstract models, the process of which is complex, usually relies on the establishment of real models, and its performance is related to the skills and experience of the modelers. Thus, the EM methods of IES can be further improved [3,4].

In recent years, with the rapid development of artificial intelligence (AI), an increasing number of studies have used machine learning methods to solve decision-making problems. Reinforcement learning (RL), which is inspired by behaviorist psychology, is an important machine learning method, and it is concerned with how an agent ought to take actions to maximize the cumulative reward

[1]School of Mechanical and Electrical Engineering, University of Electronic Science and Technology of China, China
[2]College of Engineering and Technology, Southwest University, China
[3]Electrical Engineering Department, Faculty of Engineering, Mansoura University, Egypt
[4]Faculty of Engineering, Horus University-Egypt, Egypt
[5]Department of Energy Technology, Aalborg University, Denmark

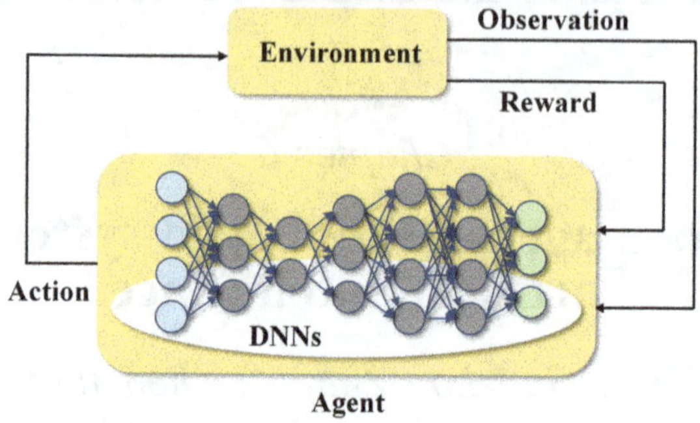

Figure 6.1 Reinforcement learning setup using DNNs [3]

derived from a stochastic environment [5]. As shown in Figure 6.1, the agent interacts with its environment to obtain state information and choose an action from a set of available actions in each time step. Then, the agent receives a reward for the current time steps, and the environment moves to a new state. The goal of the agent is to learn how to choose actions in the corresponding state to achieve as much reward as possible, through iterations [6]. The approach based on approximate dynamic programming (ADP) among RL methods proved to be superior compared to existing deterministic algorithms when the forecast information is inaccurate or even missing [7,8].

Because the state and the action space are discrete in conventional RL algorithms, the value function of the state can be represented by a lookup table corresponding to the state. However, more complex and actual tasks often have a large state space and a continuous action space. RL will encounter "the curse of dimensionality" when dealing with high-dimensional input data, which greatly limits the practical application of the RL algorithm. DRL combines the strong nonlinear perceptual capability of deep neural networks (DNNs) with the robust decision-making ability of RL. In recent years, the learning-based approach has achieved great breakthroughs in image processing [9], parameter estimation [10,11], and complex decision-making applications. In 2019, OpenAI defeated the world champion with a DRL-trained agent in a computer game, which demonstrates the tremendous potential of DRL. In the Internet of Things and smart city scenarios, the work in [12] extends DRL to a semi-supervised paradigm, which consumes both labeled and unlabeled data to improve the performance and accuracy of the learning agent. Various DRL-based methods have been applied in energy management systems, and a number of papers have demonstrated their superiority. In [13,14], DRL-based approaches were compared to dynamic programming algorithms and Q learning. Considering the randomness of the commuting behavior, simulation results demonstrated that DRL-based approaches have better performance. Thus, in cases of continuously changing factors, e.g., dynamic electricity prices, energy generation, and consumption, the DRL approach is one of the most promising methods for obtaining the optimal operation situation of energy management systems.

This chapter is organized as follows: Section 6.2 provides a proximal policy optimization (PPO)-based renewable energy conversion strategy in an integrated electrical and heating system. In Section 6.3, soft actor critic (SAC)-enabled multi-objective optimization of the electricity–gas–heat hybrid energy system is introduced. Section 6.4 gives the conclusion on energy management of hybrid systems using artificial intelligence.

6.2 PPO-based energy management strategy in an integrated electrical and heating system

This section considers RE as the energy source for the IES. In Figure 6.2, the system operator (SO) formulates the dynamic energy conversion strategy according to the customers' (CUs) heat and electricity demand profiles and the electricity price of the upper-level grid at each time slot. To be specific, the SO (agent) decides the wind power conversion (action), and the action value is sent to the IES (environment) at each time slot. Then, the SO calculates the operating cost of the system as a reward, and the states are denoted by wind power generation, CUs' energy demands, and the electricity price of the upper-level grid. This study uses the DRL approach to analyze how the SO can learn and ultimately obtain dynamic wind power conversion strategies through constant interaction with the IES to minimize the operating cost of the whole system.

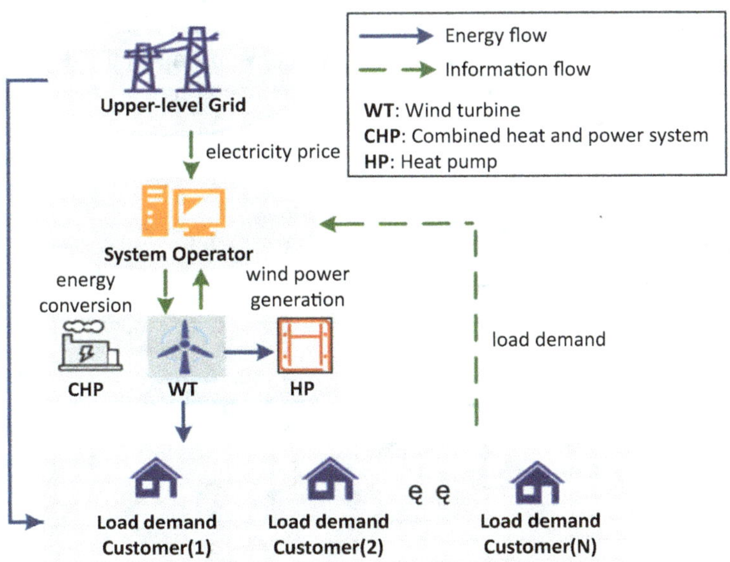

Figure 6.2 Integrated energy management model including the energy flow and information flow [3]

6.2.1 Description of an integrated electrical and heating system

The structure of a modern hybrid energy system is depicted in Figure 6.3. It consists of five parts: wind turbine (WT), heat pump (HP), combined heat and power plant (CHP), customers (CUs), and the upper-level grid. These five parts are connected by two networks: the electricity subnetwork and the heat subnetwork. In the IES operation mode, the SO can purchase from or sell electricity to the upper-level grid according to the electricity price. The electricity subnetwork is used to meet the CUs' electricity demand, whereas the heat subnetwork is used to meet the thermal demand. The CHP can generate heat and electricity at the same time. It acts as the source for both the heat and electricity networks. The WT is connected "directly" to the electricity system and "indirectly" to the heat system through the HP. Hence, part of the CUs' heating demand can also be supplied by the HP by converting the excess wind power into heat. Finally, the heat sources, i.e., CHP and HP, are connected to the distributed heat network, which is used to transport the generated heat from the sources to the CUs.

It can be observed from the above energy conversion process that the wind power curtailment can be reduced in this IES. In addition, if the heat demand is supplied by wind power, the operating time of the CHP will be decreased, resulting in a reduction in fuel consumption and carbon emissions. However, determining the optimal amount of energy conversion that can reduce the operating cost is a major challenge. If too much wind power is converted into heat energy, it will lead to a power deficit, which requires importing power from the upper-level grid, thereby increasing the operating cost. Conversely, when not enough energy is converted to heat, it will increase the operating cost of the CHP because it has to accommodate more heat demand. Therefore, this study will develop an optimal wind power conversion strategy through DRL to minimize the operating cost.

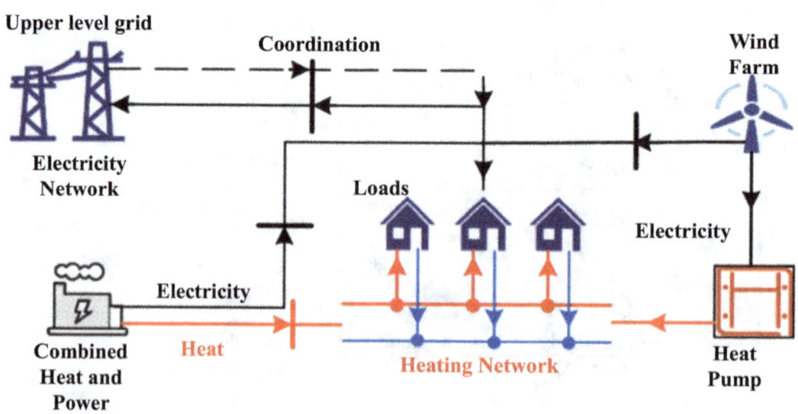

Figure 6.3 Integrated electrical and heating system [3]

6.2.1.1 Mathematic model of integrated electrical and heating system

A common and simple way of modeling the output of the HP is provided by Petrović and Karlsson [15]:

$$\phi_{th}(t) = COP_{ave}\Delta P_{HP}(t) \tag{6.1}$$

where $\phi_{th}(t)$ stands for the thermal power produced and $\Delta P_{HP}(t)$ is the power input required for the HP heat generation at time slot t. COP_{ave} represents the conversion ratio between the input power and output heat. As for the CHP, it generates both heat and electricity (in the principle of pricing the electricity based on the amount of heat, it is used as a heat source, and the electricity can be considered as a byproduct). In this study, a gas-fed CHP is applied in the IES. A heat-to-power ratio can be expressed as follows [16]:

$$\alpha = \frac{Q_{CHP}}{P_{CHP}} \tag{6.2}$$

where Q_{CHP} and P_{CHP} stand for the output heat and electricity of the CHP, respectively.

District heating plays an increasingly important role in heat networks, as illustrated in Figure 6.4. It connects the heat sources to the heat loads. The heat generation unit, a network of insulated pipes, and heating loads constitute a district heating network. Water is selected as the heat transfer medium because it has a larger specific heat capacity and can store more heat when raising the same temperature. The water is heated by the heat generation unit and then transported to the heating loads through a network of pipes. An exchanger extracts the heat from the hot water and sends it to the building's pipe network to satisfy the thermal demand. Finally, a return pipe network brings the cold water back to the heating unit, and the process restarts.

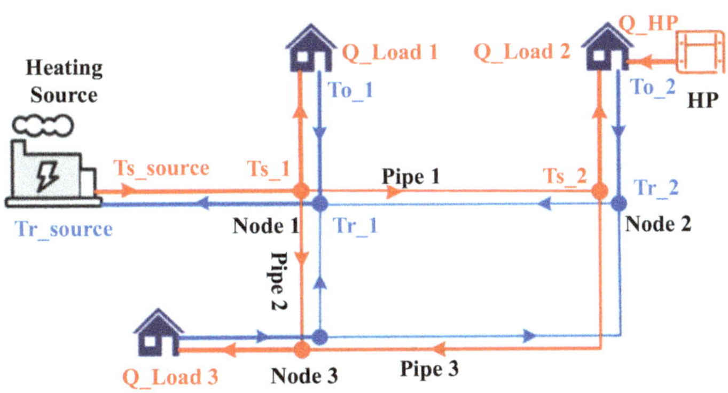

Figure 6.4 District heating network in the integrated energy system from Figure 6.3 [3]

In the district heating system, the necessary variables are the thermal power Q_{load} required by the CUs, supply temperature T_{s_i} of node i in the supply pipe network, and the return temperature T_{r_i} of node i in the return pipe network. It is assumed that the following three variables, namely the thermal power load demand, supply temperature T_{r_source} of the heat generation unit, and return temperature T_{o_i} of the water when it comes out of the CUs' buildings before entering the return network, are specific. The remaining required node temperatures (T_{s_i} and T_{r_i} of the node i) can be calculated by the hydraulic principle of the pipeline system and some other parameters, such as the topology and characteristics of the pipe and the mass and pressure of the water flowing through each pipe.

The district heating modeling process can be divided into two sequential steps: hydraulic and thermal. The hydraulic sub-model calculates information about the nodal and pipe mass flow in the network. Then, the obtained flow parameters derived from the hydraulic model are substituted into the thermal model to estimate the temperature of the water (T_{s_i} and T_{r_i} of the node i) and the heat (Q_{CHP}) generated by the CHP.

The relationships between the variables of the hydraulic and power networks have certain similarities, as presented in Table 6.1 [17].

Here, $\dot{m}_k$ represents the mass flow rate in pipe j, and m_q is the mass flow rate in the pipe directly connected to the load. The structure and internal values of the matrix A are related to the topology of the network. The number of its rows is equal to the number of nodes, and the number of its columns is the number of pipes in the entire network. Hence, in this case, A is a $3 * 3$ matrix. Each element can be "+1" if the flow in the pipe arrives at a node, "−1" if the flow leaves a node, or "0" if there is no connection between the pipe and the node. Hence, as shown in the Figure 6.4, the matrix A is defined as:

$$A = \begin{pmatrix} -1 & -1 & 0 \\ 1 & 0 & -1 \\ 0 & 1 & 1 \end{pmatrix} \tag{6.3}$$

Equations in Table 6.1 can be combined to obtain an additional formula, which is expressed as:

$$\sum_{j}^{n\_pipe} B_{ij} R_j \times \dot{m}_j \times |\dot{m}_j| = 0 \tag{6.4}$$

Table 6.1 Hydraulic model and electrical model

Electrical network	Hydraulic network	Hydraulic equation
Kirchhoff's current law	Continuity of flow	$A \times \dot{m}_k = \dot{m}_q$
Kirchhoff's voltage law	Loop pressure equation	$\sum \dot{P}_f = 0.B \times \dot{P}_f = 0$
Ohm's law	Head loss equation	$P_f = R \times \dot{m}_j \times \lvert \dot{m}_j \rvert$

Matrix B is also related to the topology of the network. The number of its rows is the number of loops of the network, which is made up of pipes, and the number of its columns is the number of pipes in the corresponding loop. Moreover, when the flow direction of the liquid in one of the tubes of the loop is counterclockwise, the value of the corresponding position in matrix B is "1" and the opposite is set as "−1". "0" means that no liquid flows inside the pipe.

The purpose of the thermal model is to estimate the temperature of each node using the liquid mass in each pipe calculated from the previous hydraulic model. In the model shown in Figure 6.4, Node 1 and Node 2 are non-mixing nodes, the temperature of which is determined by the water temperature of only one pipe. Thus, the temperatures of Nodes 1 and 2 (T_{s_1} and T_{s_2}) in the supply heating network are obtained by Equations (6.5) and (6.6), respectively, because the heating source supply temperature (T_{s_source}) is already specified. ϕ_i represents the temperature drop coefficient of the liquid flowing through pipe i, and T_{a_i} is the ambient temperature of pipe i. The temperature of Node 3 (T_{s_3}) in Figure 6.4, a mixing node, can be calculated by Equation (6.8) [18,19].

$$T_{s\_2} = \phi_i \times T_{s\_i} \tag{6.5}$$

$$T_{s\_1} = T_{s\_source} \tag{6.6}$$

$$\phi_i = e^{-\frac{\gamma \cdot L_i}{C_p \cdot m_i}} \tag{6.7}$$

$$T_{s\_3} \cdot (m_1 + m_2) = m_2 \cdot \phi_2 \cdot T_{s\_2} + m_1 \cdot \phi_1 \cdot T_{s\_1} \tag{6.8}$$

In the above equations, C_n and γ represent the specific heat of water [J/(kg K)] and the thermal conductivity [W/(m K)].

Conversely, in the return heating network, Node 3 is a non-mixing node. In Equation (9), the return temperature of Node 3 (T_{r_3}) is equal to the outlet temperature (T_{o_3}), which is specified. On the other hand, Node 1 and Node 2 are mixing nodes. Hence, their return temperatures (T_{r_2} and T_{r_3}) depend on the mass and temperature of the water reaching the nodes as indicated in Equations (6.10) and (6.11). Finally, (6.12) is used to estimate the heat (QCHP) produced by the CHP.

$$T_{r\_3} = T_{o\_3} \tag{6.9}$$

$$T_{r\_2} \cdot m_2 = m_{q2} \cdot T_{o\_2} + m_3 \cdot \phi_3 \cdot T_{r\_3} \tag{6.10}$$

$$T_{r\_source} \cdot m_q = m_{q1} \cdot T_{o\_1} + m_2 \cdot \phi_2 \cdot T_{r\_2} + m_3 \cdot \phi_3 \cdot T_{r\_3} \tag{6.11}$$

$$Q_{CHP} = C_P \times \dot{m}_q \times (T_{s\_source} - T_{r\_source}) \tag{6.12}$$

After deriving the CHP's heating output, its electrical output can be calculated by the heat-to-power ratio. Hence, the line flow, voltages, and losses in the system can be obtained by using the Newton–Raphson (NR) method.

The whole sequential process is depicted in Figure 6.5. In summary, the district heating system can actually be divided into two submodels. First, the mass and pressure of the flow are calculated by the hydraulic model, and then flow

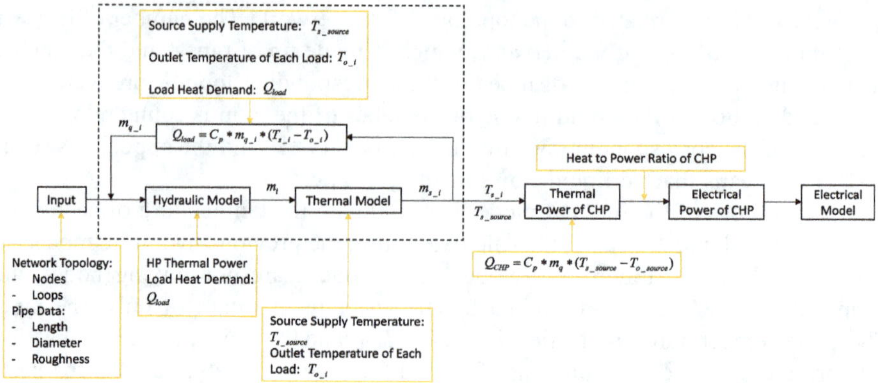

Figure 6.5 Sequential modeling process of the integrated energy system in Figure 6.2 [3]

parameters derived from the hydraulic model are substituted into the thermal model to estimate the temperature of the water and Q_{CHP}. Finally, P_{CHP} and the remaining amount of wind power are substituted into the power grid to meet the electricity needs of CUs. The sub-electricity network can be solved by using a power flow method (such as the NR method). Hence, the IES modeling process can be divided into three sequential steps: hydraulic → thermal → electrical.

6.2.1.2 Formulation of the optimal operation problem

The objective of the optimization procedure is to find how much power should be converted into heat (through the HP) in order to minimize the total operating cost (cost of operation for both the electrical and heating systems). The cost functions associated with all the components operating in the IES are presented in (6.13) to (6.15). For the CHP, the cost function (EUR/MW) can be represented by a quadratic function of both power and heat [20]:

$$C_{CHP}(t) = a + b \cdot P_{CHP}(t) + c \cdot P_{CHP}(t)^2 + d \cdot H_{CHP}(t) + e \cdot H_{CHP}(t)^2 + f \cdot P_{CHP}(t) \cdot H_{CHP}(t) \tag{6.13}$$

where $C_{CHP}(t)$ stands for the hourly cost function of the CHP unit at time slot t, $P_{CHP}(t)$ and $H_{CHP}(t)$ signify the power and heat generated by the CHP, respectively, and letters a to f are constant coefficients.

The operating cost of wind power production is given by the cost function (EUR/MW) in (6.14):

$$C_{WT}(t) = g \cdot P_{WT}(t) + h \tag{6.14}$$

where $P_{CHP}(t)$ represents the power generation of the wind turbine at a time slot t, and letters g and h are constants [21].

Regarding the cost of HP utilization, no cost function is considered. The main idea behind this is that when there is a surplus of electrical power, some of the excess energy is converted into heat. Conversely, when the electricity demand is higher than the production, extra power is purchased from an external grid at a specific cost (the spot price). Some of the purchased power can be fed to the HP to provide heat. On both occasions, though, the efficiency of the HP is taken into consideration.

The mathematical formulation of the optimization problem's objective function is

$$C = C^e + C^{th} = \sum_{t=1}^{24} C_{CHP}(t) + C_{WT}(t) + C_{grid}(t) \tag{6.15}$$

where C stands for the operating cost in [EUR] (indices e and th are for the electrical and thermal subnetworks, respectively) and t for each hour of the day. $C_{grid}(t)$ is associated with the cost of importing electricity to meet the electricity demand (this cost can be negative indicating that the SO needs to purchase electricity from the external grid).

Table 6.2 presents the power level of units. The constraints of the optimization procedure are listed (on an hourly basis) as follows:

$$0 \leq P_{CHP}(t) \leq P_{CHP}^{\max}(t) \tag{6.16}$$

$$0 \leq P_{HP}(t) \leq P_{HP}^{\max}(t) \tag{6.17}$$

$$0.95 \leq V_i(t) \leq 1.05 \tag{6.18}$$

Equation (6.16) indicates the range of power and heat output of the CHP. For the selected CHP, the maximum power is 30 MWe and the maximum heat output is 45 MWth. Similarly, in (6.17), the HP's operation ranges from 0 MW to the maximum input and $P_{HP}^{\max}(t)$ is the maximum input of the HP at the time slot t. This study sets 8 MW as the upper boundary for the HP (the maximum wind power production is 8 MW; therefore, the HP has the potential to convert all the wind power into heat). A total of 153 heat pump units are selected (Vitocal 350 GPro [22]) to achieve that value. Finally, (6.18) states that the voltage at any bus must remain within 0.95 pu to 1.05 pu, where $V_i(t)$ is the voltage of bus i at the time slot t. Specifically, it is assumed that the pipes/lines in the IES have no problems in accommodating the water/electricity flow. In addition, it should be noted that because the optimization procedure considers both the heat and power flow, no energy imbalance will occur; thus, no extra constraints are required.

Table 6.2 Power level of generation units

Unit	CHP	HP	Wind turbine
Power level	30 MW	8 MW	8 MW

6.2.2 Hybrid energy system operating cost optimization using PPO algorithm

As shown in Figure 6.6, DRL is a method that can effectively solve a hierarchical decision-making framework, where the agent is the SO, the action is the wind power distribution rate α at each time slot, the load demand information of the CUs and electricity price serves as the state, and the SO's operating cost represents the reward. In this study, a dynamic integrated energy distribution problem is formulated as a Markov Decision Process (MDP). Then, an efficient dynamic energy distribution algorithm based on the PPO without full knowledge of the dynamics and uncertainties of the system is developed.

6.2.2.1 Formulating the operating cost optimization task to the MDP

The MDP is an effective way of building a model of RL, which depicts that the next state of the system is related to not only the current state but also the current action taken. In this study, because the dynamic energy conversion problem is a decision-making framework in a stochastic environment, it can be modeled as a discrete finite horizon MDP. Instead of depending on historical data, the reward of the MDP model depends only on the wind power conversion ratio at time slot t.

The pivotal elements, which are to be modeled in the MDP, include discrete time slot t, action a_t, state s_t: $\{WP(t), \pi(t), \mu(t), \sigma(t)\}$, and reward r_t: $r(\alpha_t|s_t)$.

1. t is a finite discrete time slot at which wind power conversion ratio action α_t is executed and acted on the environment.
2. α_t is the wind power distribution ratio that the SO chooses at a time slot t to access RE in the network.

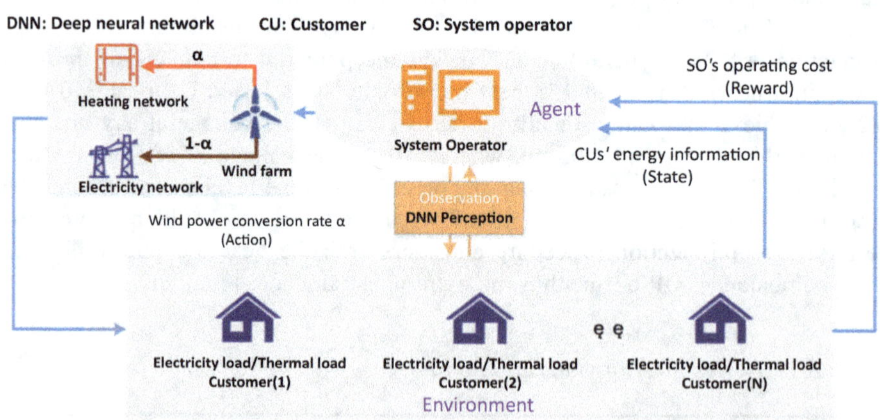

Figure 6.6 Framework of the IES model with deep reinforcement learning. SO: system operator, CU: customer, DNN: deep neural network [3].

3. $WP(t)$ stands for the power generation of WT at a time slot t, $\pi(t)$ represents the wholesale electricity price, and $\mu(t)$ and $\sigma(t)$ indicate the CUs' electricity and heat demand, respectively.
4. r_t: $r(a_t|s_t)$ is the opposite of the operating cost of the system. r_t specifies the expected immediate reward.

The MDP consists of a number of concrete episodes, each of which consists of a finite sequence of time slots, states, actions, and rewards. Considering one episode of the MDP, the total reward for one run can easily be calculated:

$$R = r(a_1|s_1) + r(a_2|s_t) + \ldots + r(a_T|s_T) \tag{6.19}$$

However, due to the randomness of the environment, if the agent performs the same action the next time, it can never be guaranteed that the same reward will be obtained. Hence, the cumulative discount future reward indicates that as the process continues, it becomes easier to converge:

$$R = r(a_t|s_t) + \gamma \cdot r(a_{t+1}|s_{t+1}) + \ldots + \gamma^{T-t} \cdot r(a_T|s_T) \tag{6.20}$$

where $\gamma \in [0, 1]$ is a discount factor that represents the influence on the future, which is decreasing due to the progress after performing an action a_t at a time slot t.

6.2.2.2 Adopting PPO to a dynamic wind power conversion problem in the IES

DRL is an approach that efficiently combines the decision-making ability of RL with the perception of deep learning. It uses a DNN to represent the value function to provide target values and uses the RL to make the reward the estimated value. The parameters of the DNN are updated constantly until the difference between the target value and the estimated value converges.

PPO is a type of model-free RL technique. This study uses the PPO to acquire the optimal policy (a sequence of wind power conversion ratio in this study). The basic principle behind PPO is to assign a policy π: $a_t = P_\theta(WP(t), \pi(t), \mu(t), \sigma(t))$ to each state-action set at time slot t, and to use two similar neural networks to update the policy at each iteration, in a way that reinforces good behavior. The optimal policy π:$a_t = P_\theta(WP(t), \pi(t), \mu(t), \sigma(t))$ denotes that when taking action a_t at state s_t at each iteration, the maximum discounted future reward can be achieved and updated continuously.

PPO is a DRL algorithm based on an actor–critic structure. The flowchart in Figure 6.7 shows the operation process of the PPO algorithm. The PPO algorithm consists of critic network (CN), actor network (AN) parameterized by θ' and another AN parameterized by θ. The purpose of AN is to continuously adjust the parameters of the policy ($s_t \rightarrow a_t$) through the reward expectation to increase the probability of a good reward. The role of CN is to obtain the potential rewards of the current state by learning the relationship between the environment and rewards. Therefore, PPO uses CN to guide AN so that AN is updated every step of the way.

Traditional policy updates are based on policy gradient (PG) [23]. However, the PG will slow down the learning rate of the AN. This is because the PG is on-

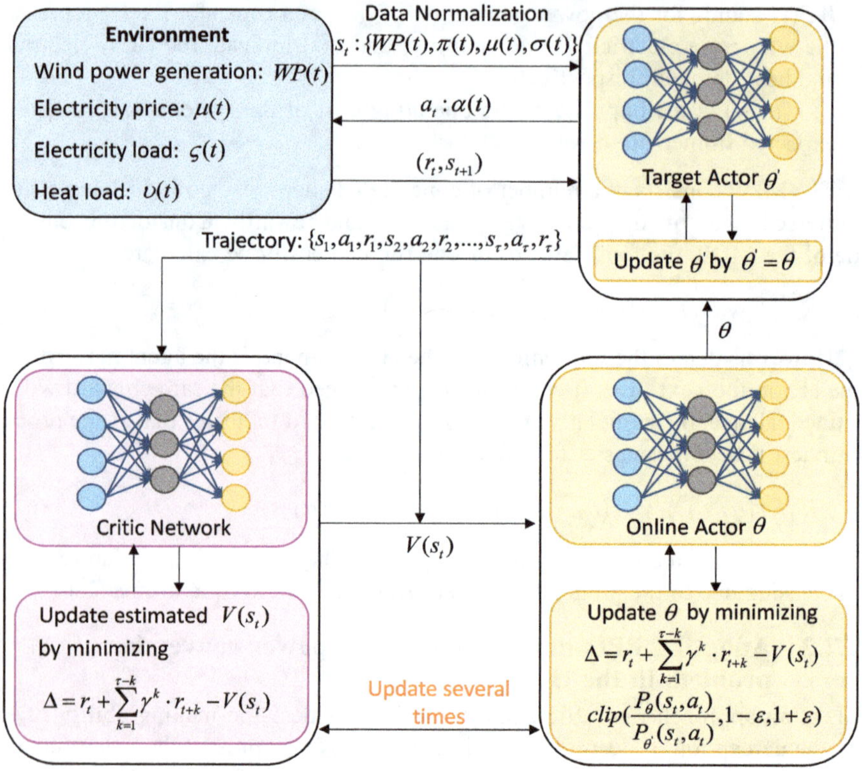

Figure 6.7 Flowchart for implementing the PPO algorithm for the hybrid energy system in Figure 6.2 [3]

policy, which takes much time to re-interact with the environment after each episode update of the AN. Hence, another actor, parameterized by θ, is introduced to interact with the environment to extract the data. Then, the previous actor uses the data to perform multiple learning updates, so that the sampled data are reused many times, and the learning rate is improved, which turns to an off-policy strategy. When the same state is input, the probability distribution of the actions obtained by the two actor networks cannot be too far. Thus, ε is used to limit the probability difference within a certain range. This idea is the core of the PPO algorithm.

The update mechanism of the CN and the AN parameterized by θ is based on the loss function and backpropagation to update the neural network parameters. After the number of the AN parameterized by updates has reached UPDATE_STEPS, the parameter of the AN parameterized by θ is assigned to the AN parameterized by θ updating the parameters.

As shown in Figure 6.7, the PPO-RECA is implemented to obtain the maximum future reward expectation V_θ' (optimal wind power conversion ratio) in the IES. In Figure 6.7, the PPO runs at the beginning of the day. The actor θ' interacts with the environment to obtain the wind power generation, the CUs' heat and electricity demand, and the wholesale electricity price for the succeeding time slots. Upon receiving these variables, the SO computes the optimal wind power conversion ratio in an iterative way. At each iteration i, the SO observes the CUs' energy demand, electricity price information, and wind power generation (state) at each time slot t. Then, the actor θ' makes a decision based on these data and calculates the reward value of the decision to obtain a sequence of trajectories at an iteration. Afterward, the CN and actor θ use the trajectory to repeatedly update the parameters of the network to improve the efficiency of the policy. Actor θ provides feedback on the parameters to update actor θ'. The termination condition of the iteration is set as $|R^i - R^{i-1}| \leq \delta$. This termination condition means that when the gap between the current future reward expectation and the previous future reward expectation is less than δ, the reward expectation converges to the maximum value. The value δ depends on the system design. Finally, the SO will achieve the optimal wind power conversion ratio for the succeeding time slots.

6.2.3 Numerical simulations

Figure 6.8 presents the details of the simulation process. The performance of the proposed PPO algorithm is evaluated through numerical experiments on an IES. First, the training process is presented in detail, and it is shown that the algorithm

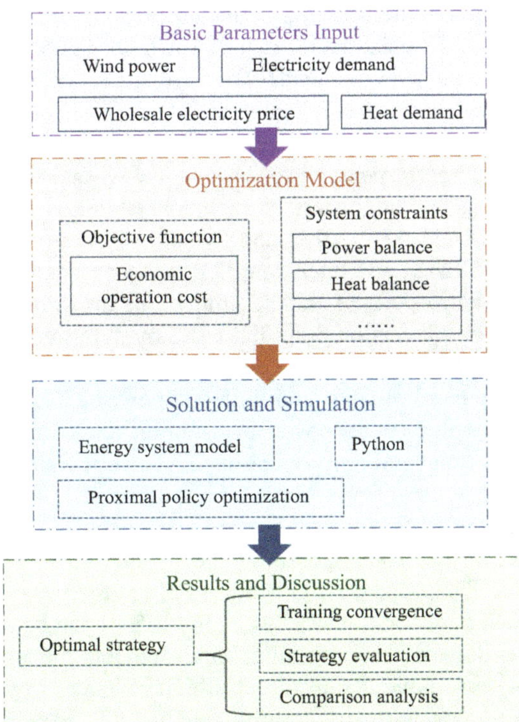

Figure 6.8 Simulation process of the proposed energy management strategy [3]

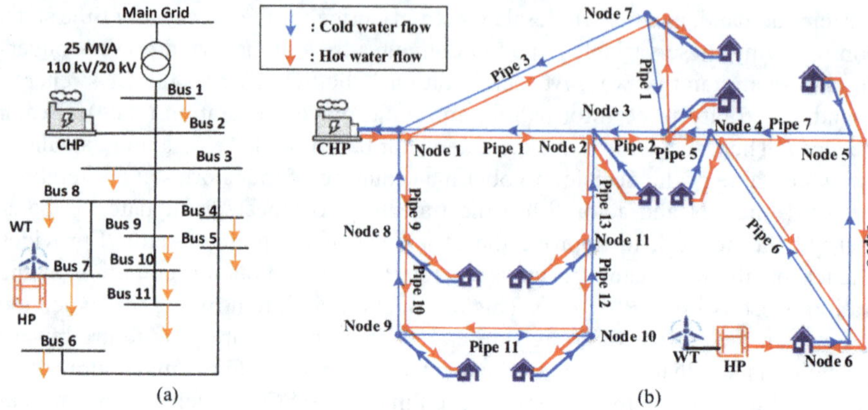

Figure 6.9 Structure of IES in Figure 6.2: (a) electrical sub-network, (b) thermal sub-network [3]

can learn a satisfactory policy in an uncertain situation. Second, the training results are tested on different days, and the simulation results show that the algorithm has good adaptability to uncertainties. Finally, the superiority of the algorithm is demonstrated by comparing the simulation results with those of other methods.

For ease of illustration, the networks of the electrical and thermal submodels are displayed in Figure 6.9. The simulation is performed based on one CHP, one load-connected WT and HP, and ten CUs' loads. The CHP connected at Bus 2/Node 1 is the source of electricity and heat. The WT at Bus 7/Node 6 is for accommodating the power of the HP and the electrical load of Node 6. This simulation defines 24 time slots as the entire time cycle corresponding to 24 h in 1 day, and the value of T is 24. Specifications regarding the parameters of each subnetwork can be found. All case studies are run using MATLAB® 2018b and Python 3.6 on a 64-bit Windows-based computer with 4 GB of RAM and Intel Core i5 processor clocking at 2.7 GHz.

Based on the scenario defined above, simulations are conducted to determine the optimal wind power conversion ratio in the IES by iteration. At the beginning of the day, the SO receives the wholesale electricity price from the upper grid, the electricity and heat demand of CUs, and the wind power generation. Then, the SO calculates the reward value (wind power distribution rate), until it finally obtains the maximum reward (the minimum operating cost) for the succeeding time slots. The 120-day historical data from May 1 to August 28, 2018, obtained from SDG&E [24] and ComEd [25], is used as the training set.

Figure 6.10 shows the convergence of the reward value during the training process by iterations. The proposed algorithm is trained for 100,000 epochs to learn an optimal wind power conversion strategy in the IES. It can be observed that, due to the initial perception of the unfamiliar environment, the agent (SO) has no experience choosing a reliable action to get a high reward. However, because the agent constantly interacts with the environment to gain experience, the reward increases and eventually converges to its maximum value. This illustrates that the

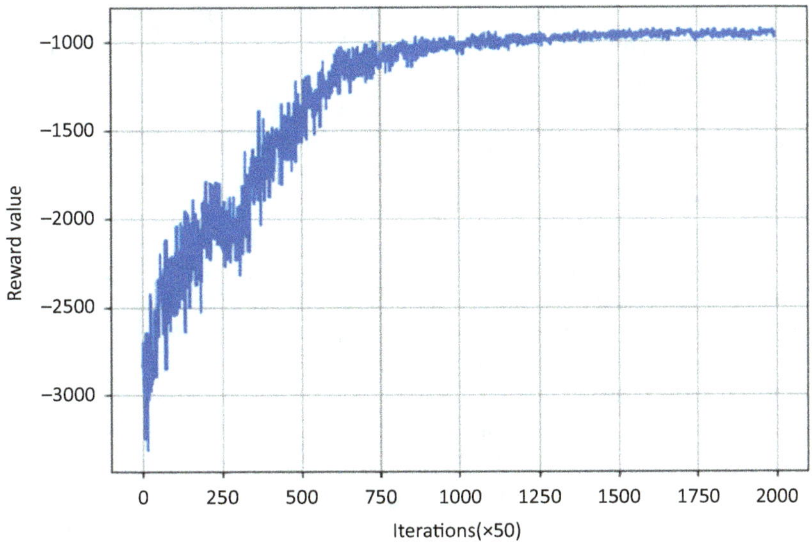

Figure 6.10 Cumulative reward change with an episode during the training process [3]

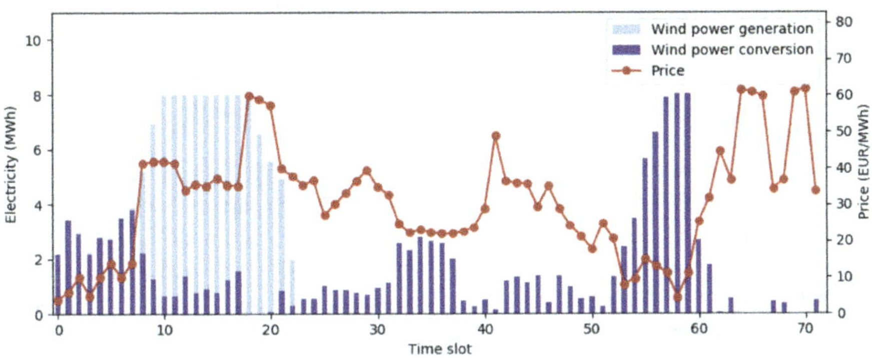

Figure 6.11 Cumulative reward change with an episode during the training process for the system in Figure 6.2 [3]

agent has learned the optimum policy to minimize the operating cost of the IES. Because the 24 h of data per day of training are randomly selected within the 120 days, and the daily input data are different, the reward value is increasing and oscillating in the training process.

After the offline training with the 120-day historical data for 100,000 epochs, in the following cases, the dynamic wind power distribution model will be tested to demonstrate the real-time optimization ability of the PPO algorithm in the IES.

To evaluate the system's performance, the test simulations are based on 3 days of data. Figure 6.11 shows the optimal wind power distribution for the IES along with the electricity price of the upper-level grid within these 3 days. The simulation results of the first day are discussed in detail. First of all, the wind power conversion matches the wind power generation from 00:00 to 07:00, meaning that during this period, all the wind power is transferred to act as the source of the HP, resulting in enough energy conversion. Because of the low electricity price of the upper-level grid, it is more economical to buy surplus electricity from the external grid. Similarly, interacting wind power with the electricity load would cause a chain of events: a portion of the produced wind power would be used for electricity purposes, resulting in less wind power for the thermal subnetwork. At the same time, the CHP's output would be increased because a portion of the heat demand cannot be accommodated by the HP units, resulting again in more power for the electricity subnetwork. Both scenarios result in less energy being purchased from the external grid (at a low price), which leads to a more expensive solution overall. Then, in the other hours of the day, the conversion of the wind power decreases as the electricity increases. In particular, from 18:00 to 20:00, the wind power conversion is approximately 0 due to the high electricity price. In the time range from 8:00 to 10:00, although the electricity price remains unchanged, the wind power conversion is decreasing. This is because the CUs' heat demand is large during this period, and the SO needs more wind power to supply heat to reduce the operating cost of the CHP. Thus, more electricity needs to be purchased from the external grid to meet the electricity demand. The results of the other 2 days are the same as on the first day.

Test simulations on different days further verify the previous training results analysis, indicating that the proposed PPO method has the ability to solve the real-time operating cost optimization problems in the IES.

To illustrate the effectiveness of the PPO method in the IES when dealing with uncertainty, the DQN (deep Q network)- [26] and PSO (particle swarm optimization)-based stochastic optimization methods are used for comparison. A comparison is carried out on test data composed of 30 days. The optimal solution of the PPO is the output of the actor policy, the parameters of which are fixed after the training. The optimal solution of the PSO-based stochastic optimization method is given by averaging the optimal actions of 120 samples. The DQN is a DRL method that uses neural networks to predict the value of action and learns the optimal path of action by continuously updating the neural networks. By this assumption, the wind power conversion rate is set as an action variable A comprising six options: convert 20%, 40%, 60%, 80%, or 100% of wind power to supply electricity, or convert all wind power to supply heat. During the DQN training process, the data and parameters are set up in the same way as in the previous PPO training process.

$$A = \{0, 0.2, 0.4, 0.6, 0.8, 1\} \tag{6.21}$$

The simulation is performed on the 30-day test set. The operating costs of the corresponding methodology for 30 days are illustrated in Figure 6.12. The optimal value is obtained by using PSO on every deterministic day for 30 days. It can be observed that, among the three approaches, the proposed one achieves the lowest

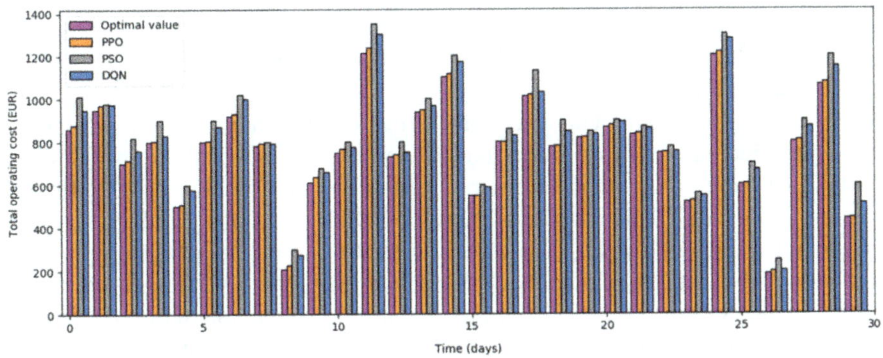

Figure 6.12 Comparison of test data using the proposed approach. DQN: deep Q network and PSO: particle swarm optimization [3].

Table 6.3 Comparison of PSO: particle swarm optimization, DQN: deep Q network, and PPO: proximal policy optimization regarding average cost reduction

Methods	Average cost (EUR/day)	Improvement
PSO	852.69	0
DQN	818.75	3.98%
PPO	781.02	8.41%

operating cost. The quantitative results of the different approaches to the test data are presented in Table 6.3. Compared with the DQN- and PSO-based stochastic optimization methods, the PPO can effectively reduce the operating cost. Because the DQN uses the action value function mapping state to action, it makes real-time decisions based on current conditions, resulting in better performance than the PSO-based stochastic optimization method. However, because the IES has a high-dimensional and continuous state space, it is more difficult for a DQN-based value function learning to obtain a behavior with the largest value. Thus, the DQN will oscillate around the optimal value function in the late stage of training without convergence.

Obviously, the simulation of the two DRL-based strategies is completed in much less time than that of the PSO, showing their real-time ability in a real application. Because the PPO and DQN require a large amount of model training in the early stages of the test, the optimal mapping relationship from state to action is obtained; thus, the efficiency can be improved in the test. In contrast, because the PSO does not have a memory function, feedback cannot be performed online in real-time. Therefore, compared with DRL, PSO, as a global optimization algorithm, performs unsatisfactorily in terms of computational efficiency.

6.3 SAC-enabled multi-objective optimization of electricity–gas–heat hybrid energy system

As an essential development direction of the energy internet, IES with interdisciplinary techniques is of great significance to promoting multi-energy cooperation, realizing low-carbon economic operation, and improving the flexible scheduling potential. This chapter presents the study of an integrated power, heat, and natural gas energy (IPHNGE) system consisting of energy coupling units and wind power generation interconnected via a power grid. A DRL-based energy scheduling strategy is proposed to optimize multiple targets, including minimizing operational costs and ensuring power supply reliability. By scheduling the output of energy units, the economy and reliability of the considered system are improved. Taking diversified uncertainties into account, like intermittent wind power and flexibility of load demand, the stochastic dynamic optimization problem is modeled as MDP and a SAC algorithm is introduced to solve the complex scheduling problem. The optimized decision-making action can be identified by the SAC algorithm through empirical learning without prediction information and prior knowledge. The simulation results verify the effectiveness of the SAC-based energy scheduling strategy. Moreover, compared with conventional methods, the proposed scheduling method has a better optimization result.

6.3.1 Electricity–gas–heat hybrid energy system description

In Figure 6.13, the IPHNGE system is composed of a power network, a natural gas network, a heating network, and some coupling components, which are interconnected via transmission lines. The arrow indicates the flow of energy. For example, the combined heating and power (CHP) plant consumes natural gas and

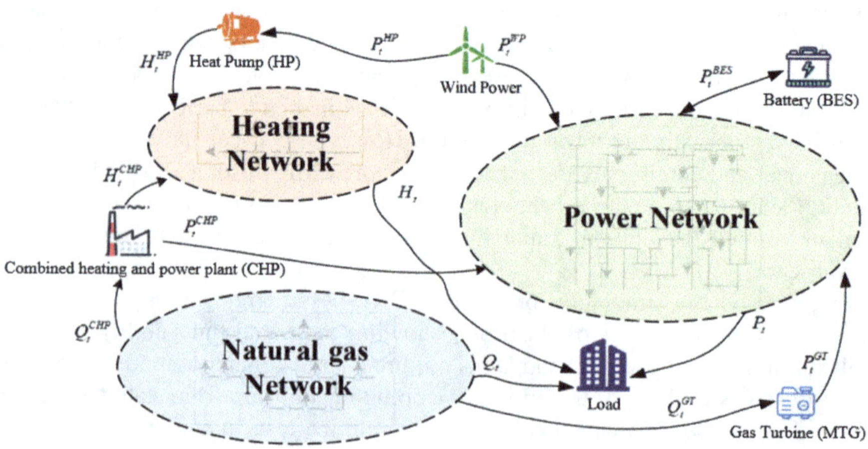

Figure 6.13 A framework of an integrated power, heat, and natural gas energy system [4]

acts as both heat and power sources. In the considered IPHNGE system, the system operator (SO) is utilized for standardized instant monitoring and control of the system. More concretely, the SO is capable of exchanging control information with the energy information fed back by each coupling unit, and then the coupling unit executes the command delivered from the SO. For simplicity, the centralized energy management model assumes the following components, including a heat pump (HP), wind turbine generator (WTG), battery energy storage (BES), micro-turbine generator (MTG), gas-fired combined heating and power plant (CHP), and loads.

6.3.1.1 Modeling of HP and CHP

HP is a device that transfers heat from a low-temperature heat source to a high-temperature heat source, which provides heat by consuming a small amount of electric energy. The relationship between the output Q_t^{HP} and the input P_t^{HP} can be expressed by [27]:

$$Q_t^{HP} = \eta^{HP} P_t^{HP} \tag{6.22}$$

where the coefficient of performance is denoted as η and is further analyzed in [28].

For a gas-fired CHP, the relationship between the generated heat and power can be represented by a Heat to Power ratio "α" as explained in [29]:

$$\alpha = \frac{Q_t^{CHP}}{P_t^{CHP}} \tag{6.23}$$

where Q_t^{CHP} and P_t^{CHP} stand for the heat and power generated by the gas-fired CHP respectively.

6.3.1.2 Modeling of BES

With similar functionalities as distributed generators, BES can effectively regulate the balance between energy generation and consumption in a power grid. A notable advantage of BESs is that, when the distributed energy generation, e.g., wind power, photovoltaic, far exceeds the amount of power demand, it is capable of storing the excess energy for later utilization. In this process, the performance of BESs would be damaged by excessive power throughput. Therefore, it is necessary to restrict the maximum amount of charging/discharging capacity. Meanwhile, for practical applications, there would be a reasonable task to limit the state of charge (SOC) of BESs within a certain boundary. Here, the control signal for BESs is denoted as u_t^{BES}, and the amount of the charge/discharge power at the time slot t is defined as P_t^{BES}. The restrictions for P_t^{BES} and SOC_t are expressed by

$$-P_{\max}^{BES} \leq P_t^{BES} \leq P_{\max}^{BES} \tag{6.24}$$

$$SOC_{\min} \leq SOC_t \leq SOC_{\max} \tag{6.25}$$

Furthermore, the dynamics of P_t^{BES} and SOC_t in the adjacent time can be described as

$$P_{t+1}^{BES} = P_t^{BES} + B_{BES} u_t^{BES} \tag{6.26}$$

$$SoC_{t+1} = SoC_t + \eta P_t^{BES}/Q \tag{6.27}$$

where B_{BES} represents the maximum allowed change of charge/discharge capacity between t and $t+1$; Q is the BESs capacity, and the charge/discharge coefficient for BESs is denoted as η, which is expressed in (6.28).

$$\eta = \begin{cases} \eta_0 - \eta_1 |P_t^{BES}|/P_{BES}^{\max}, P_{BES}^{\max} \geq 0 \\ 1/(\eta_0 - \eta_1 |P_t^{BES}|/P_{BES}^{\max}), P_{BES}^{\max} \leq 0 \end{cases} \tag{6.28}$$

where parameters η_0, η_1 are related to the charge/discharge efficiency of BESs, see, e.g., [30].

6.3.1.3 Modeling of power system

The AC power flow is used as the analysis method of the power system, and the power balance is shown as follows [31]:

$$\begin{cases} P_i = U_i \sum_{j \in i} U_j (G_{ij} \cos \theta_{ij} + B_{ij} \sin \theta_{ij}) \\ Q_i = U_i \sum_{j \in i} U_j (G_{ij} \sin \theta_{ij} - B_{ij} \cos \theta_{ij}) \end{cases} \tag{6.29}$$

where P_i and Q_i are the injected active power and reactive power of node i, respectively; i and j are node number; U_i is node voltage amplitude; G_{ij} and B_{ij} represent conductance and admittance of branch ij, respectively; θ_{ij} is voltage phase angle difference of node i and j.

In the optimal operation of the IPHNGE system, the power system is considered to operate under the constraints of power balance, generator output, and node voltage.

6.3.1.4 Modeling of thermal system

The thermal system is mainly composed of a heat source, supply network, and heat load. The heat supply network transmits the heat power generated by the source to the heat load [32]. The whole model of the thermal system including the water supply and return pipes is established by:

$$\begin{cases} A_h^w M_h = 0 \\ B_h^f (\Delta\pi_w - \pi_p) = 0 \end{cases} \tag{6.30}$$

where A_h^w is the complete correlation matrix between the node and the pipeline; M_h indicates the flow vector, and B_h^f is the basic loop matrix of the water supply and return network; note that 0 is the zero vector; $\Delta\pi_w$ and π_p are the loss vector of pipeline head and circulating pump or pressure pump lifting, respectively.

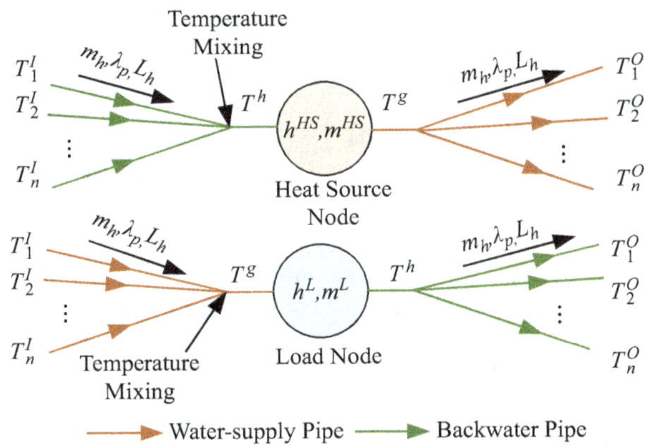

Figure 6.14 Heat source and load node model in thermal system [4]

For any pipeline numbered i in the thermal system, the head loss includes two parts: the frictional resistance loss and the local resistance loss:

$$\Delta\pi_w^i = \zeta\frac{L_h^i}{D_h^i}\frac{\rho v^2}{2} + \xi\frac{\rho v^2}{2} \tag{6.31}$$

where $\Delta\pi_w^i$ is the head loss of the pipeline i; L_h^i, D_h^i, and ρ are the length, diameter, and working fluid density of the pipeline i, respectively; v is the average flow rate of the pipeline i; ζ and ξ represent resistance coefficients.

As shown in Figure 6.14, the thermal model mainly involves the heat source, heat load, pipeline, and connection point among pipelines. Besides, the thermal model equations consist of the heat and energy balance equations of the node, and the pipeline temperature loss equation [33,34].

$$\begin{cases} h_{i,t}^{HS} = C_p m_{i,t}^{HS}(T_{i,t}^g - T_{i,t}^h) \\ (\sum m_{out})T_{out} = \sum(m_{in}T_{in}) \\ T_{end} = (T_{start} - T_o)\exp\left(-\dfrac{\lambda_p L_h^i}{C_p m_h}\right) \end{cases} \tag{6.32}$$

where $h_{i,t}^{HS}$, $T_{i,t}^g$, and $T_{i,t}^h$ are respectively the heat power, water supply temperature, and return water temperature provided at the heat source or the heat load; T_{start}, T_{end}, and T_o are the starting temperature, the terminal temperature, and the environmental temperature of the pipeline, respectively; λ_p and C_p represent heat transfer factor per unit length of the pipeline and the specific heat capacity of the thermal mass, respectively; m_{out} and T_{out} are the output flow and temperature; similarly, m_{in} and T_{in} are the flow mass and temperature at the joint of the mixing pipe; m_h is the mass flow rate of the pipeline.

On the basis of the above model, a matrix calculation model of the thermal system can be established, which is expressed by the matrix equations as follows:

$$\begin{cases} q_{in} = C_p M_h T_{in} \\ q_{out} = C_p M_h T_{out} \\ q_o = A_q q_{in} - A_z q_{out} \end{cases} \tag{6.33}$$

where q_{in} and q_{out} are the inflow and outflow mass vectors respectively; T_{in} and T_{out} represent the inlet and outlet temperature vectors at the connection point between the pipelines; A_q and A_z are the correlation matrices of the node and flow at the beginning and the end of the pipelines; q_o is the heat load vector at the node.

6.3.1.5 Modeling of natural gas system

A typical gas system includes a gas source, pipeline, compressor, gas storage point, load, etc. The natural gas pipeline with an electric compressor is shown in Figure 6.15. The relationship between electric power consumption P_{com} and natural gas flow m_g^{kn} is calculated as [35]

$$\begin{cases} P_{com} = K_{com} \cdot m_g^{kn} \cdot T_k[(\Pi_k/\Pi_m)^z - 1] \\ K_{com}^{\min} \leq K_{com} = \Pi_k/\Pi_m \leq K_{com}^{\max} \\ \Pi_k^{\min} \leq \Pi_k \leq \Pi_k^{\max} \\ m_g^{mk} = m_g^{kn} \\ z = (\alpha - 1)/\alpha \end{cases} \tag{6.34}$$

where K_{com}, $K_{com}^{\min}$, and $K_{com}^{\max}$ are compression ratio, lower and upper limits; $\{m_g^{mk}, m_g^{kn}\}$ and $\{\Pi_k, \Pi_m\}$ represent the natural gas flow and pressure at the inlet and outlet of the compressor, respectively; $\Pi_k^{\min}$ and $\Pi_k^{\max}$ are the lower and upper limits of outlet pressure; the gas temperature is denoted as T_k; α is compressor polytropic index.

The time period for operating and managing the integrated power, heat, and natural gas energy system (IPHNGE) is defined as $[0, T]$. Assuming that the time period $[0, T]$ is divided into M time steps, the length of two adjacent time steps is $\Delta t = T/M$. The objective functions in different scenarios are discussed in the following section.

6.3.1.6 Modeling of objective function

The main objective of the operation economy is to minimize the operating cost. The generation cost F_{oe} of the IPHNGE system in isolated grid mode considers the fuel

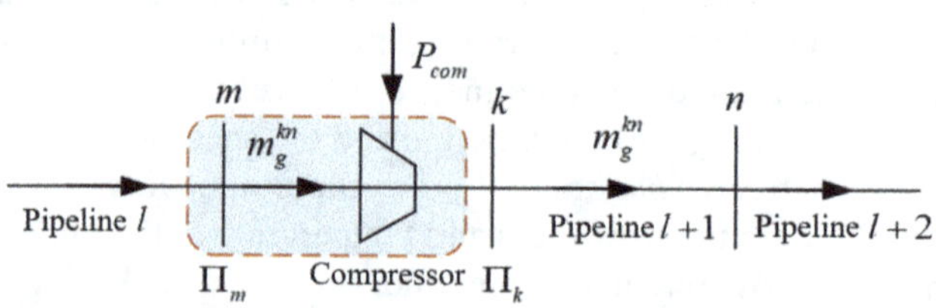

Figure 6.15 Model of a pipeline with an electric compressor [4]

cost C_f, investment depreciation cost C_{dp}, operation and maintenance cost C_{om}, environmental cost C_e and heating income C_s of the CHP system.

$$\min F_{oe} = C_f(t) + C_{dp}(t) + C_{om}(t) + C_e(t) - C_s(t) \tag{6.35}$$

$$C_f(t) = \sum_{i=1}^{N} C_f^{G_i}(P_t^{G_i}) \tag{6.36}$$

$$C_{dp}(t) = \sum_{i=1}^{N} \left[\frac{C_{az}^{G_i}}{8760k_{G_i}} \times \frac{r(1+r)^{n_{G_i}}}{(1+r)^{n_{G_i}} - 1} \times P_t^{G_i} \right] \tag{6.37}$$

$$C_{om}(t) = \sum_{i=1}^{N} C_{om}^{G_i} = \sum_{i=1}^{N} K_{om}^{G_i} \times P_t^{G_i} \tag{6.38}$$

$$C_e(t) = \sum_{k=1}^{M} 10^{-3}\beta^k \left(\sum_{i=1}^{N} \alpha_k^{G_i} P_t^{G_i} \right) \tag{6.39}$$

$$C_s(t) = Q_{h,t} \cdot K_{ph} \tag{6.40}$$

where the current time step during $[0, T]$ is denoted as t, and N is the number of power sources; $P_t^{G_i}$ is the output of the power source G_i, and $C_f^{G_i}$ is the corresponding fuel cost; $C_{az}^{G_i}$ and k_{G_i} represent installation cost per unit and a capacity factor of the power source G_i, respectively; r is the annual interest rate, and n_{G_i} is the investment repayment period of the power source G_i; $K_{om}^{G_i}$ is the operation and maintenance cost coefficient of the power supply G_i; k is the type number of pollutants (including CO_2, SO_2, NO_x, etc.); M is the total number of pollutants; $\alpha_k^{G_i}$ is the emission coefficient of pollutant k from the power supply G_i; β^k is the treatment cost of pollutant k; K_{ph} is the unit heat price.

Minimizing the loss of power supply probability (LPSP) [36] is the objective of power supply reliability. Here, the LPSP in (6.78) is defined as the load demand that cannot be met by the system divided by the total load demand during the dispatch period. Moreover, the load shedding rate is taken as a measure index of power supply reliability. The smaller the LPSP, the higher the power supply reliability of IPNGHS.

$$\min F_{psr} = LPSP(t) = \frac{P_t^d - \sum_{i=1}^{N} P_t^{G_i}}{P_t^d} \tag{6.41}$$

where F_{psr} is the LPSP, and P_t^d represent the load demand at time slot t.

Considering the economy and reliability of the IPHNGE system in isolated operation, the load shedding [37] amount P_t^l is unified as the cost compensated

$C_l(t)$ by the SO, which is included in the total generation cost.

$$\min F_{cb} = C_f(t) + C_{dp}(t) + C_{om}(t) + C_e(t) + C_l(t) - C_s(t) \tag{6.42}$$

$$C_l(t) = K_{bu} \times P_t^l \tag{6.43}$$

where the comprehensive generation cost is denoted as F_{cb}; K_{bu} is the compensation cost of unit outage paid by the SO.

6.3.2 Multi-objective optimization of IPHNGE system using SAC algorithm

6.3.2.1 Formulating the multi-objective optimization task to the MDP

The MDP [38] is introduced into the considered IPHNGE system, which signifies that the state at $t+1$ would only be related to the action and state information at the time step t, and it is independent of the previous state at time $t-1$, $t-2$, ..., which is described in (6.44).

$$P(s^{'}; s, a) = P\{s(t) = s^{'} | s(t-1) = s, a(t-1) = a\}, s, s^{'} \in S, a \in A \tag{6.44}$$

Furthermore, a DRL algorithm named SAC [39] is employed to find optimal solutions to the dispatch problem.

Here, it is necessary to convert the optimal control task into an appropriate form that can be solved by RL. The RL framework is composed of an agent and an environment. The IPHNGE system is regarded as the environment for agents, and the agent is the SO. The agent can control the power of each energy unit discussed.

Based on the state information $s_t \in S$ provided by the environment and the strategy π learned by itself, the agent interacts with the environment. Specifically, at the current time t, the agent gets a reward r_t by executing an action a_t. By exploring action space A, the agent learns the optimal strategy π^* that maximizes the cumulative reward.

Suppose that the simulation starts at a time slot t in one episode, the cumulative reward is given as

$$R(s_t, t) = \sum_{k=0}^{M-t} \gamma^k r_{t+k} \tag{6.45}$$

where $\gamma \in [0, 1]$ is the discount coefficient and reflects the relative importance of current and future rewards. Further, with the direction of a policy π, the value function for state s at time slot t can be described in (6.46).

$$V^{\pi}(s_t) = E[R(s_t, t) | s_t = s] \tag{6.46}$$

In this sense, the optimal policy π^* for the agent is to deliver solutions to maximize $V^{\pi}(s_0, 0)$.

Generally, it is a reasonable trick that the cost function described can be regarded as the negative form of a reward. Besides, the constraint violations for the system can be regarded as a penalty function, such that they are required to be involved in the reward item for optimization.

Here, for the constraints for BESs, the corresponding penalty function ϕ_t^{SoC} at the time step t is set as

$$\phi_t^{SoC} = \begin{cases} \varphi_1 e^{(SoC_{\min} - SoC_t)}, SoC_t < SoC_{\min} \\ 0, SoC_{\min} \le SoC_t \le SoC_{\max} \\ \varphi_2 e^{(SoC_t - SoC_{\max})}, SoC_t > SoC_{\max} \end{cases} \tag{6.47}$$

where φ_1 and φ_2 are penalty coefficients, usually larger and related to the model.

Additionally, considering the constraints for the energy balance $\{\Delta P_t, \Delta Q_t\}$ in both the heat network and power network, the penalty functions are defined as follows:

$$\phi_t^{power} = \begin{cases} \varphi_3 \Delta P_t^2, & \Delta P_t > 7 \\ e^{\Delta P_t}, & 0 < \Delta P_t \le 7 \end{cases} \tag{6.48}$$

$$\phi_t^{heat} = \begin{cases} \varphi_4 \Delta Q_t^2, & \Delta Q_t > 7 \\ e^{\Delta Q_t}, & 0 < \Delta Q_t \le 7 \\ \varphi_5 \Delta Q_t, & \Delta Q_t \le 0 \end{cases} \tag{6.49}$$

$$\Delta P_t = P_t^d - \sum_{i=1}^{N} P_t^i \tag{6.50}$$

$$\Delta Q_t = Q_t^d - Q_t^i \tag{6.51}$$

$$i = \{MTG, HP, CHP\} \tag{6.52}$$

where $\{\varphi_1, \varphi_2, \varphi_3\}$ are penalty coefficients; Q_t^d and Q_t^i are heating load and supply at time step t. The constraints include electricity balance and heat balance. ΔP_t is the amount of electricity mismatch, which is defined in (6.50). ΔQ_t is the amount of heat mismatch, which is defined in (6.51).

Then, the total penalty function of the IPHNGE system for the considered time period can be written as

$$\Phi = \sum_{t=1}^{M} (\phi_t^{SoC} + \phi_t^{power} + \phi_t^{heat}) \Delta t \tag{6.53}$$

On the basis of the above processes of formulating a problem and defining penalty functions, the immediate reward at the time slot t can be given as follows:

$$r_t = (F_i + \phi_t^{SoC} + \phi_t^{power} + \phi_t^{heat}) \Delta t, i = 1, 2, 3 \tag{6.54}$$

Thus, the state-value function V^π for (6.55) can be established.

$$V^\pi(s_0, 0) = E(F_i + \Phi), i = 1, 2, 3 \tag{6.55}$$

where E is denoted as expectation. Note that when the mentioned coefficients, i.e., $\{\varphi_1, \varphi_2, \varphi_3, \varphi_4, \varphi_5\}$, are properly fine-tuned, the optimal strategy π^* that maximizes $V^{\pi}(s(t_0))$ would be obtained by executing the optimal action a^*.

Thereby, a finite MDP (S, A, P, r, γ) is formulated to describe the optimization problem of the discrete system, and the RL approach [40] is applied to deal with the problem.

6.3.2.2 SAC algorithm description

Traditional model-free DRL algorithms are caught in two dilemmas: high sampling complexity and weak convergence, which rely critically on parameter tuning. In particular, as commonly used on-policy gradient algorithms, PPO [41] and asynchronous advantage actor-critic (A3C) [42] will encounter the contradiction of exploration and utilization, and an epsilon-greedy mechanism is a compromise under this condition. To improve the sample efficiency, off-policy algorithms such as DDPG [43] were introduced, but their performance is often brittle with hyperparameters. Thus, the state-of-the-art off-policy DRL algorithm based on the maximum-entropy theorem, SAC, is employed to provide learning robustness and sampling efficiency. Three key tricks are incorporated into the SAC algorithm: an actor-critic architecture with policy and value DNNs (named Actor and Critic, respectively), a replay buffer that reuses the past experience for an off-policy formulation, and maximum entropy to juggle exploration and learning stability.

The overall system architecture diagram for the SAC algorithm is illustrated in Figure 6.16, which comprises three parts, i.e., the Actor, the Critic, and the experience replay buffer. The Actor is used to map state to action, the Critic is responsible for estimating the value of state and state-action, and the replay buffer takes charge of storing experiences.

The Actor is composed of a parameterized DNN, i.e., dense layers and a Rectified Linear Unit (ReLU). The Critic includes a soft Q network, a state-value network, and a target state-value network. In addition, Temporal Difference (T.D.) error and stochastic gradient decent methods are used to update the parameters of the DNNs.

Before the approximating policy $\pi_{\varphi}(\cdot|s)$, a reparameterization trick, including noise sampled from the standard normal distribution $N(0, 1)$, is employed to solve sampling problems. Given the current environment state s_t, the Actor selects and executes an action a_{t+1} based on policy $\pi_{\varphi}(\cdot|s)$. When the reward r_t and the next state s_{t+1} are returned, the information tuple (s_t, a_t, r_t, s_{t+1}) is stored in the experience replay buffer M. To break up the correlation with input data, the Critic randomly samples batches from M to train the state-value $V_{\vartheta}(s)$ and Q-value $Q_{\theta}(s, a)$ networks. In the training process, the loss function w.r.t. ϑ, θ produced by T.D. error will be back-propagated to update the parameters of the corresponding DNNs, while $Q_{\theta}(s, a)$ and $V_{\vartheta}(s)$ are transferred to renew the parameters of the Actor. After several updating times, the parameters of the target state-value network are refreshed via a soft updating mechanism. By interacting with the environment to collect experience and updating parameters of the DNNs based on batches sampled from the replay buffer, the learning process continues to move forward until reaching the timeout and final state.

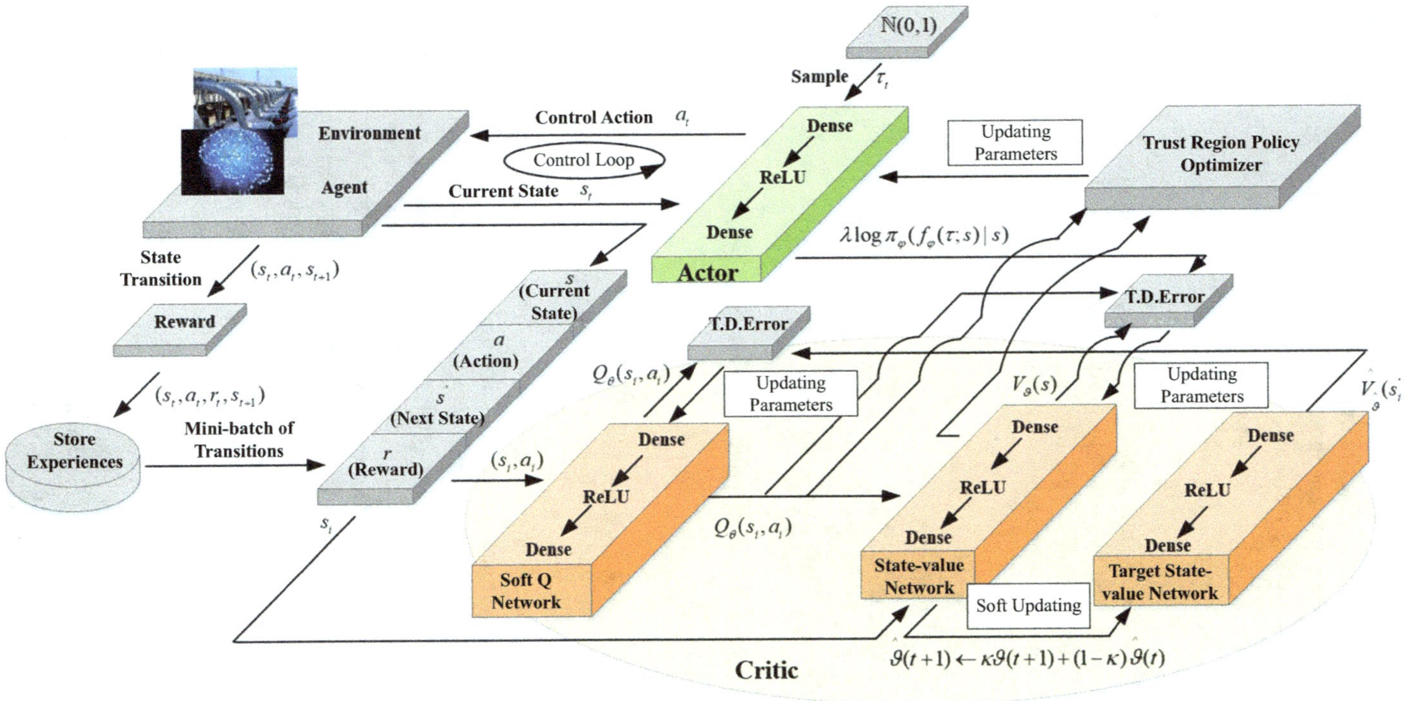

Figure 6.16 Architecture of the soft actor-critic algorithm [4]

The general objective of an RL agent is to only follow with interest the policy $\pi(a|s)$ that maximizes the expected reward for the future. However, distinct from the existing RL algorithms, a maximum entropy term $H(\pi(a|s)) = -\log \pi(a|s)$ is introduced into the reward to encourage exploration in this study. The redefined reward formulation can be written as:

$$J(\pi) = E\left\{\sum_{t=0}^{\infty} \gamma^t [r_t - \lambda \log \pi(a_t|s_t)] | \pi\right\} \tag{6.56}$$

where $\gamma \in [0, 1]$ is the discount factor and measures the importance of current and future rewards, and λ regulates the stochastic degree of the policy, which is named after the temperature parameter. In view of a specific initial state s and action a, the above entropy objective is rewritten into a Q-value function as shown in (6.57).

$$Q^{\pi}(s, a) = E\left\{\sum_{t=0}^{\infty} \gamma^t [r_t - \lambda \log \pi(a_t|s_t)] | s_0 = s, a_0 = a, \pi\right\} \tag{6.57}$$

Theorem 1. The state-value $V^{\pi}(s)$ based on Q-value function is written as

$$V^{\pi}(s) = \lambda \log \int_A \exp\left(\frac{1}{\lambda} Q^{\pi}(s, a)\right) da \tag{6.58}$$

and the optimal solution of policy is given by

$$\pi^*(\cdot|s) = \exp\left(\frac{1}{\lambda}(Q^{\pi}(s, a) - V^{\pi}(s))\right) \tag{6.59}$$

Since the operation *LogSumExp* is called Softmax, the value function is termed as soft value function, and the Q-value function is also named as soft Q-value function. Besides, note that the optimal policy $\pi^*(\cdot|s)$ is no longer the Gaussian distribution form, but an energy-based policy (EBP) model associated with the value function so that it has better generalization ability.

In the actor-critic framework, the two essential roles (the Critic and Actor) are responsible for *policy evaluation* and *policy improvement*, respectively. The learning process advances in the direction of achieving maximum return in the long run by the iteration of these two processes.

The method of policy evaluation is introduced in this section. Because of its good convergence and stability [44], a deep neural network (DNN) is employed to fulfill the task of approximating the value function.

The output value $Q_{\theta}(s, a)$ of a DNN parameterized by θ is used to estimate soft Q-value, i.e., $Q(s, a) \approx Q_{\theta}(s, a)$. To break the correlations between input data required by DNN, the experience replay mechanism is utilized, in which the experience of the agent interacting with the environment at each time step is stored in the replay buffer, i.e., $M \leftarrow (s, a, r, s') \cup M$. The DNN updates parameter θ based on b tuple information (s, a, r, s'), which is randomly batch-sampled from the replay buffer.

Similar to the DQN algorithm, the performance evaluation of the DNN θ depends on the mean square error (MSE) of the difference between the Q target value $Q(s,a)$ and Q estimated value $Q_\theta(s,a)$.

$$L(\theta) = E\left[\frac{1}{2}\left(Q_\theta(s_i,a_i) - \widehat{Q}(s_i,a_i)\right)^2\right] \tag{6.60}$$

$$\widehat{Q}(s_i,a_i) = r + \gamma \widehat{V}_{\widehat{\vartheta}}(s_i') \tag{6.61}$$

where $\widehat{V}_{\widehat{\vartheta}}(s')$ is the target state-value and will be discussed later. To ensure the accuracy of $Q(s,a)$ and minimize the loss function L, it is common to update the parameter θ by gradient descent, i.e.,

$$\theta_{t+1} \leftarrow \theta_t - \alpha_c \nabla_\theta L(\theta) \tag{6.62}$$

where $\alpha_c > 0$ is the learning rate of the Critic and satisfies a small value, and $\nabla_\theta L(\theta)$ denotes the gradient of $L(\theta)$ and calculated by

$$\nabla_\theta L(\theta) = \nabla_\theta Q_\theta(s_i,a_i)[Q_\theta(s_i,a_i) - r - \gamma \widehat{V}_{\widehat{\vartheta}}(s_i')] \tag{6.63}$$

To improve the training stability, another DNN, parameterized by ϑ, is introduced to serve as a state-value estimation function, i.e., $V(s) \approx V_\vartheta$. Equally, the performance of the DNN is evaluated by using MSE

$$L(\vartheta) = E\left[\frac{1}{2}\left(V_\vartheta(s) - E[Q_\theta(s_i,a_i) - \lambda \log_{\pi_\varphi}(a_i|s_i)]\right)^2\right] \tag{6.64}$$

Correspondingly, the updating process of the DNN w.r.t. ϑ is calculated by

$$\nabla_\vartheta L(\vartheta) = \nabla_\vartheta V_\vartheta(s)[V_\vartheta(s) - Q_\theta(s,a) + \lambda \log \pi_\varphi(a|s)] \tag{6.65}$$

$$\vartheta_{t+1} \leftarrow \vartheta_t - \alpha_c \nabla_\vartheta L(\vartheta) \tag{6.66}$$

Additionally, to stabilize the updating of the Critic, a DNN with parameters $\widehat{\vartheta}$ is utilized to estimate the target state-value function $\widehat{V}_{\widehat{\theta}}(s)$. The parameter $\widehat{\vartheta}$ is obtained by soft update based on the soft state-value network weights

$$\widehat{\vartheta}_{t+1} \leftarrow \kappa \vartheta_{t+1} + (1-\kappa)\widehat{\vartheta}_t \tag{6.67}$$

where $\kappa \in (0,1)$ denotes a smoothing factor.

After obtaining the estimated values, the sole objective of the Actor is to seek the direction of policy improvement. Note that the state space is continuous, it is essential to leverage an approximator function that can formulate a policy for each state. Another DNN, parameterized by φ, is employed to represent the policy $\pi_\varphi(\cdot|s)$, which will be trained following (6.68). It is intractable to sample the model directly like the general Gaussian distribution. Thus, this study employs a state-conditioned stochastic network $\pi_\varphi(\cdot|s)$ to sample action instead of EBP, and then KL divergence is used to measure the difference of $\pi^*(\cdot|s)$ and $\pi_\varphi(\cdot|s)$.

$$L(\varphi) = E[D_{KL}(\pi_\varphi(\cdot|s)||\pi^*(\cdot|s))] \tag{6.68}$$

Furthermore, a reparameterization trick $f_\varphi(\tau_t; s_t)$ is introduced, where τ is action noise signal sampled from a standard normal distribution. Specifically, the policy $\pi_\varphi(\cdot|s)$ outputs the mean $\pi_\varphi^\mu(s_t)$ and standard deviation $\pi_\varphi^\sigma(s_t)$ of a Gaussian distribution, and then the corresponding Gaussian distribution is sampled as the decision-making action of the Actor.

$$a_t = f_\varphi(\tau_t; s_t) = \pi_\varphi^\mu(s_t) + \tau_t \cdot \pi_\varphi^\sigma(s_t) \tag{6.69}$$

$$L(\varphi) = E\left[\lambda \log \pi_\varphi(f_\varphi(\tau; s)|s) - Q^\pi(s, f_\varphi(\tau; s)) + V^\pi(s)\right] \tag{6.70}$$

The gradient of $L(\varphi)$ w.r.t. φ.

$$\nabla_\varphi L(\varphi) = \nabla_\varphi \lambda \log \pi_\varphi(a|s) + (\nabla_a \lambda \log \pi_\varphi(a|s) - \nabla_a Q_\theta(s, a)) \nabla_\varphi f_\varphi(\tau; s) \tag{6.71}$$

The Actor with parameter φ is updated by

$$\varphi_t \leftarrow \varphi_t - \alpha_a \nabla_\pi L(\varphi) \tag{6.72}$$

where α_a is the learning rate of the Actor and satisfies a small positive value.

6.3.3 Simulation results

In this section, simulation for different scenarios is used to verify the superiority of the proposed SAC-based control strategy, which consists of four steps: parameter setting, model establishment, simulation, and discussion. The implementation details of the case study in this chapter are depicted in Figure 6.17.

Step 1. Input basic parameters of the discrete IPHNGE system, including customers' demand profiles, wind power generation, and energy equipment parameters.

Step 2. Under the IPHNGE system constraints, integrate the above input data into the coupling model; and the optimization in the economy, power supply, and comprehensive benefits is developed.

Step 3. According to the model set above, three scenarios are used for simulation through the SAC algorithm. The time period of one episode and each time step are set as a typical day (24 h) and 1 h, respectively.

Step 4. The scheduling optimization results based on the simulation in Step 3 are evaluated from the perspective of each component operation. Besides, a comparison is further analyzed to illustrate the effectiveness of the proposed control strategy.

To assess the performance of the proposed SAC-based dispatch strategy, a test case system is studied in this work, which is shown in Figure 6.18.

Different scenarios are first done to verify whether the learning curves of the SAC algorithm can converge and a satisfactory dispatch strategy can be learned; then the testing results and the sensitivity of the model are analyzed in detail, which further validates the rationality of the proposed SAC-based dispatch strategy; Next, two benchmark optimization-based algorithms for reliable economic dispatch problems are also implemented to illustrate the superiority of the proposed control strategy. All case studies were finished using Python 3.7 on a 64-bit Windows-based computer with 4 GB of RAM and Intel Core i5 processor clocking at 2.7 GHz.

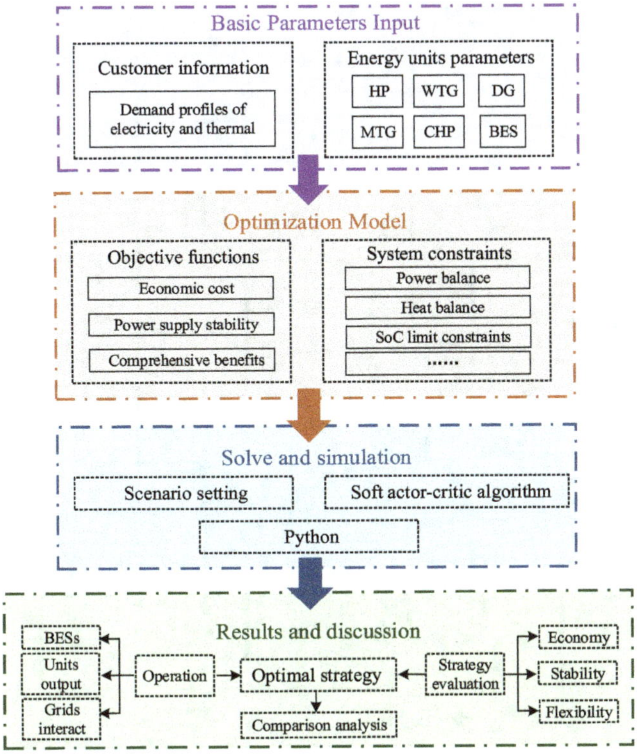

Figure 6.17 Optimization scheme of the proposed energy management strategy for the IPHNGE system [4]

The test system operations under islanded mode are built on the combination of IEEE 39-bus power system, a 6-node heating system, and a 20-node gas system [45]. The branches of the IPHNGE system are 29, 19, and 5, respectively. There are three generation sites in the power system, in which G3 and G7 are gas turbines connected at Node 6 and 19, respectively, and G1 is powered by a CHP unit and wind farm. The BES is placed on bus 4. In the natural gas system, there are three source nodes. Besides, Node 3, Node 6, and 16 supply natural gas to CHP units and gas turbines, respectively. Three compressors, including COM#1, COM#2, and COM#3, are responsible for boosting gas pressure. In the heating sub-system, the CHP units, which provide thermal energy, are installed at N1. The HP, powered by wind power, is set at N1 as well. Tables 6.4–6.8 present the data of the system described in Figure 6.18.

To completely demonstrate the effectiveness of the SAC-based dispatch strategy in minimizing the operating cost and maximizing the reliability of the power supply, three scenarios are analyzed:

Scenario 1: The objective function is the operating cost, which indicates that the SAC-based agent only focuses on the economy of the system.

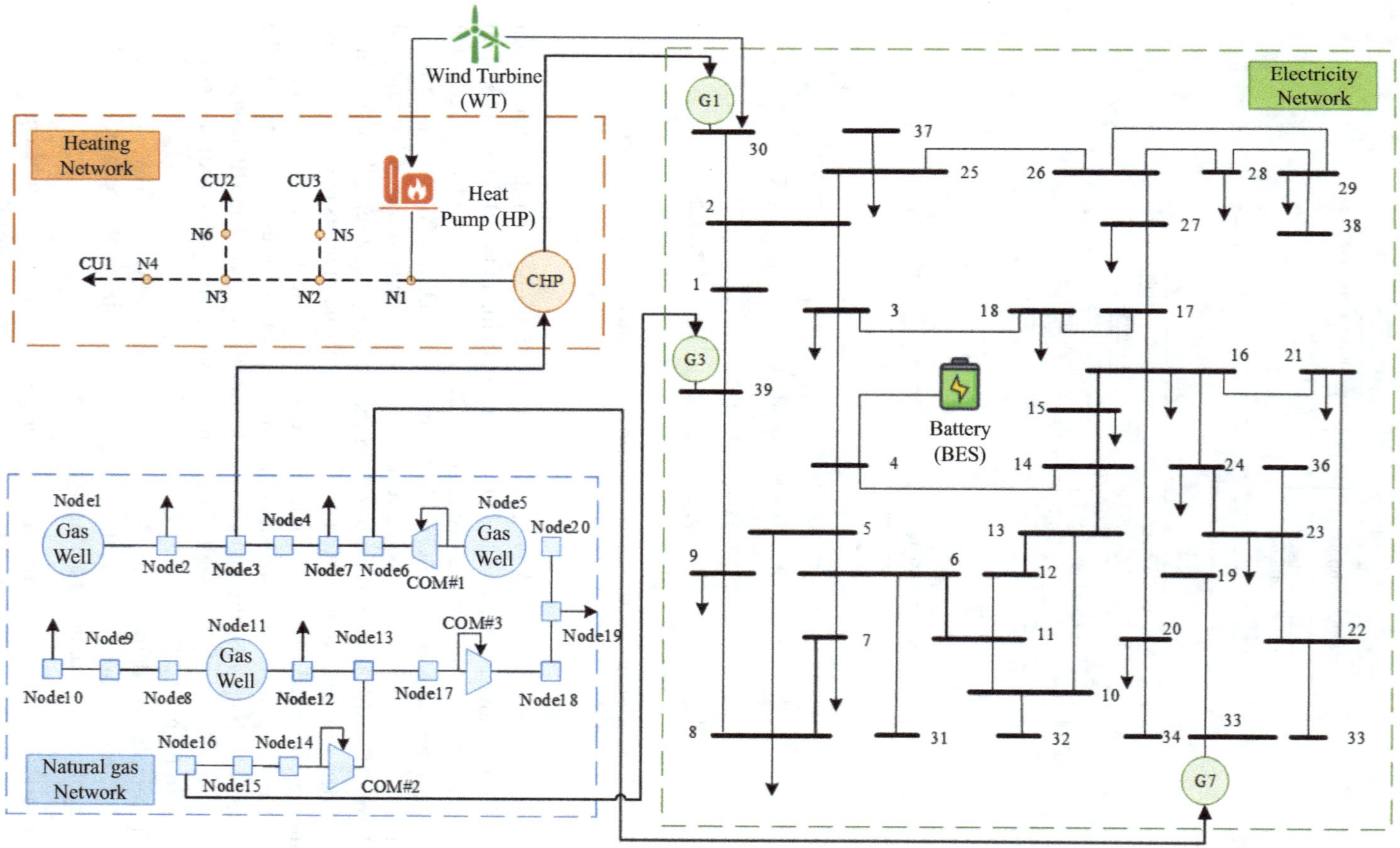

Figure 6.18 Studied integrated power, heat, and natural gas energy systems [4]

Table 6.4 Information of natural gas system

Pipeline number	Head node	Terminal node	σ	Compressor
1	1	2	115.0	/
2	2	3	112.5	/
3	3	4	65.1	/
4	4	7	63.4	/
5	7	6	66.1	/
6	6	5	60.0	COM#1
7	3	11	61.3	/
8	10	9	65.0	/
9	9	8	61.5	/
10	8	11	60.0	/
11	11	12	62.5	/
12	12	13	67.5	/
13	13	17	61.0	/
14	17	18	59.1	COM#3
15	18	19	69.0	/
16	19	20	67.0	/
17	13	14	66.1	COM#2
18	14	15	70.0	/
19	15	16	60.0	/

Table 6.5 Parameters of GT

Gas-fired generator	Power system node	Natural gas system node	Upper active power limit/ kWh	Lower active power limit /kWh	Conversion efficiency/%
G1	30	3	30	0	43%
G2	33	6	20	0	43%
G3	39	16	30	0	43%

Table 6.6 Parameters of CHP

Power system node	Natural gas system node	Maximum thermal generation/kWh	Maximum power generation/kWh	Power ramping constraints/kWh
30	3	200	80	60

Table 6.7 Parameters of BES

Power system node	Maximum capacity/kW	Minimum capacity/kW	Upper injection power limit/kWh	Lower injection power limit/kWh
4	100	20	50	0

Table 6.8 Parameters of HP

Maximum power input/kWh	**Maximum thermal output/kWh**	**Power ramping constraints/kWh**
50	75	50

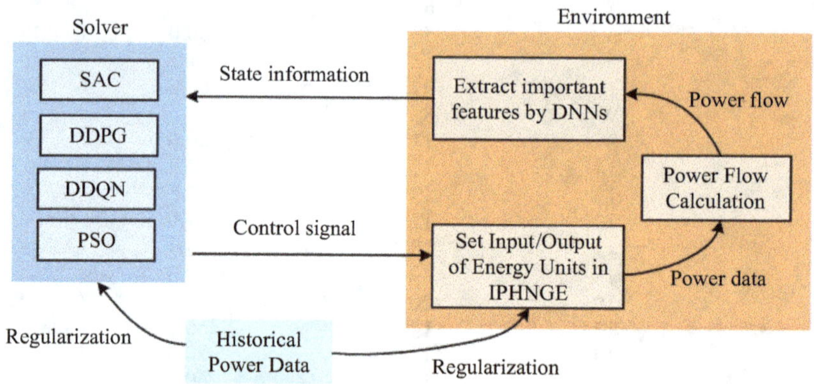

Figure 6.19 Simulation process for the system in Figure 6.18 [4]

Scenario 2: To guarantee the reliability of the power supply, the load-shedding rate is chosen as the optimization objective.

Scenario 3: The comprehensive profits, where both operating economy and reliability of power supply are considered.

As it is shown in Figure 6.19, the simulation process is established on three kinds of optimization methods, simultaneously.

The first one is the proposed SAC algorithm, described. At each time t, the current important features, i.e., $P^{wind}, SOC, P^{ele}, P^{heat}$ are fed to the SAC solver; then, the output of the DNNs in the SAC algorithm is treated as the control signal, provided to the environment; finally, referring to the above simulation results, the reward is calculated and the parameters of the actor-critic network are properly fine-tuned. The algorithm parameters refer to [23]. Table 6.9 gives the algorithm data for the optimization. For updating the weights of the actor and critic networks, the learning rates are $\alpha_c = 2 \times 10^{-3}$ and $\alpha_a = 2 \times 10^{-4}$, respectively. Besides, mini-batch data, the size of which is $m = 32$, serves as the updating samples of the networks. A discount factor γ discussed is set as 0.9, and the value of the soft update coefficient κ used for the target network updating rate 0.01. The entropy regularization coefficient $\lambda = 0.2$ aims to balance the stochastic degree of the policy. The input data consists of 1-year historical generation data and load data.

The double DQN (DDQN) [46] algorithm, which is an improved extension of the DQN algorithm, is employed to be the first benchmark RL algorithm. With

Table 6.9 Core hyperparameters setting for SAC algorithm

Name	Value
Replay size (M)	60,000
Discount factor (γ)	0.9
Mini-batch size (m)	32
Training episodes (N)	8,000
Step in each episode (t)	24
Soft update coefficient (κ)	0.01
Learning rate for value network (α_c)	2×10^{-3}
Learning rate for actor network (α_a)	2×10^{-4}
Entropy regularization coefficient (λ)	0.2

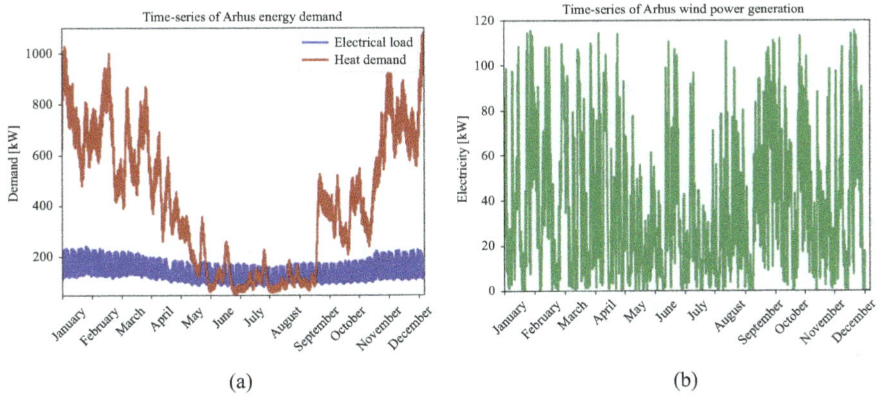

Figure 6.20 Simulation input data based on 1 year from the city of Aarhus: (a) electricity and heat load profiles, (b) wind power generation [4]

regards to some parameter settings, e.g., $\{M, \gamma, m, N, t\}$, DDQN shares the same values with the proposed SAC algorithm. In DDQN, there are three DNNs, namely predict network, target network, and evaluation network, each of which owns two hidden layers with 200 and 300 neurons, respectively. The updating rates for the above networks are set as 0.005. For further comparison, the random optimization method based on particle swarm optimization (PSO) will be used as the second benchmark algorithm, which is a classic heuristic algorithm.

When continuous iterations based on the above scenarios are carried out, the corresponding optimal energy management and dispatch strategy for the discrete IPHNGE system will be obtained. In Figure 6.20, the historical data of 1 year for Aarhus [47] is selected as the training set.

Figure 6.21 shows the cumulative reward convergence performance of Scenario 3, which is the most complex scenario. Moreover, the change in penalty values, w.r.t. constraints for SOC, heating balance, and power balance is presented in Figure 6.22. Note that in Figure 6.22, there are the dark-colored curves and

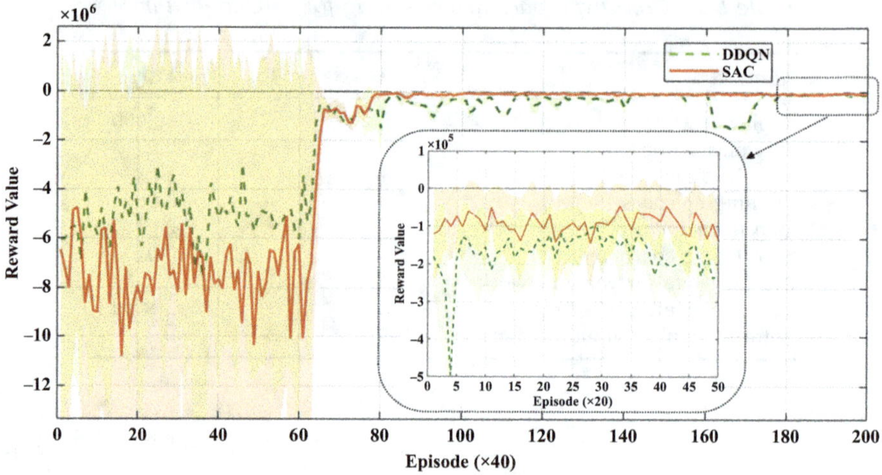

Figure 6.21 Cumulative reward change based on the episode during the training process for the system in Figure 6.18 [4]

the light-colored regions, which represent the average value of five random experiments and the error bounds, respectively. The above settings are to make the changes in the training curves easier to read.

It is observed that the optimal SAC-based scheduling strategy reaches convergence after the entire training process of 8,000 episodes. Specifically, the training curves of the SAC algorithm could be split into three sequential stages: policy exploration (before 2,500 epochs), policy learning (from 2,500 epochs to 5,600 epochs), and policy convergence (after 5,600 epochs). During the process of policy exploration, the proposed algorithm aims to store enough experience in the buffer replay by randomly selecting actions such that the training curve significantly oscillates. When the experience buffer is filled, the agent starts training. The weights of the corresponding neural network are properly modified following the refreshing principles. After continuous adjustment, the penalty value and cumulative reward tend to converge, which implies that the weights of DNNs are fixed and the optimal scheduling strategy is found. As shown in Figure 6.21, there are still few oscillations in the last 1,000 epochs. The reason for this is that the input data in every episode is selected from different historical days, and the corresponding reward gains are obviously distinct from each other.

To illustrate the effectiveness of the proposed method, a comparison between algorithms, e.g., the DDQN and the proposed SAC algorithms, is also done.

From the aspect of algorithms shown in Figure 6.21, to achieve the same level of performance, the proposed SAC algorithm requires fewer training episodes and achieves better episodic average profits. Furthermore, Table 6.10 presents comparative results among PSO-based stochastic optimization methods, DDQN, and

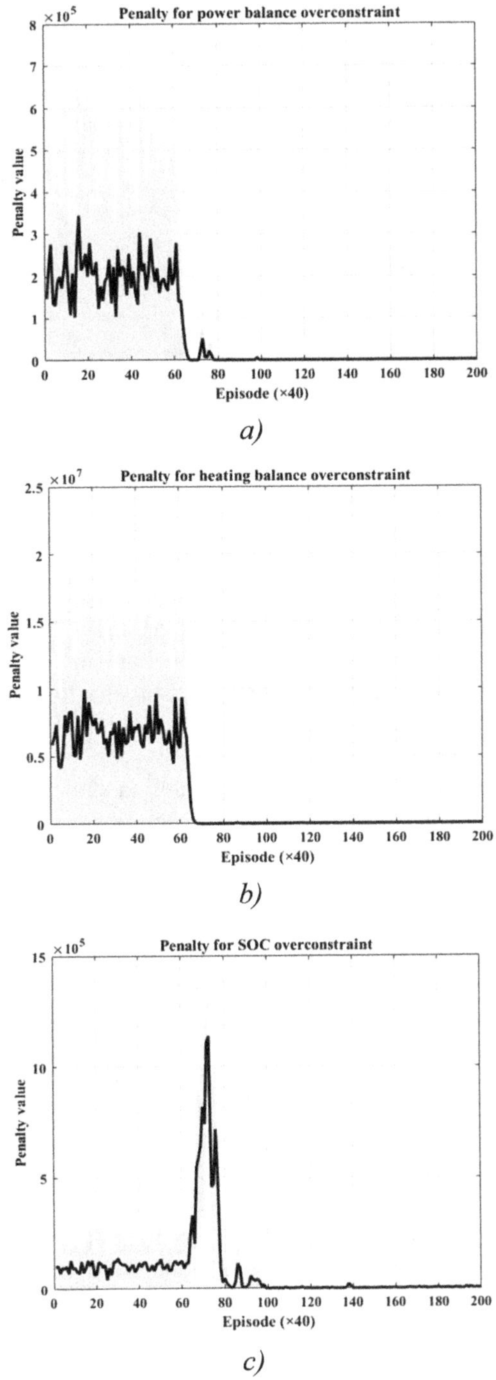

Figure 6.22 Penalty values change for constraint violations discussed in Section 6.3.2.1. (a) Penalty value curve for power balance overconstraint and (b) penalty value curve [4].

Table 6.10 Comparison results of different approaches regarding average costs and calculation time

Method	Average cost ($/day)	Improvement	Calculation time (s)
PSO-based stochastic optimization method	295,830.32	0	10,236.5
DDQN	192,552.96	34.91%	865.3
SAC	94,573.12	68.03%	3,203.2

the proposed SAC method. It can be observed that the RL-based methods, such as DDQN and SAC, are 34.91% and 68.03% higher than the PSO-based stochastic optimization method in minimizing economic cost, respectively. The IPNGHS being considered possesses a high-dimensional and continuous action space, whereas the DDQN algorithm is tailored for discrete action spaces. This incongruence stems from the tendency for actions with high rewards to be overlooked. The performance of the PSO-based stochastic method is based on past experience roughly, and it performs unsatisfactorily.

In terms of computational efficiency, since the DDQN algorithm has a relatively simple architecture and updating mechanism for DNNs, it takes the shortest time. To reach the final convergence, it takes a long time for the PSO to search all the action spaces among the samples, which results in low computational efficiency.

According to the scenarios set and the scheduling strategy based on the offline training, the operation and dispatch results of each energy coupling unit are analyzed further.

In this analysis, the demand profiles of a typical winter day [47] are selected as the dataset, and the curves of energy loads are shown in Figure 6.23. The CHP, WTG, HP, and MTG are the main generators. Besides, BES is capable of importing and exporting electricity, and the SOC curves of BES are displayed in Figure 6.24, as well.

In Figure 6.24(a), the operation of the IPHNGE system is optimized according to the economy, where it can be observed that in the time range from 00:00 to 06:00, the electric load demand is low and a high thermal generation is required to be met. To obtain the minimum operating cost, the WTG generation is fully utilized to act as "direct" and "indirect" sources for both the CUs' electricity demand and HP, respectively.

From 07:00, BES starts to export electricity to supply the increasing electric load. Since the WTG generation and CHP generation are limited by environmental factors and the principle of determining power generation by heat, respectively, the WTG and CHP cannot meet the demand of the electric load. Until 12:00, the SOC of BES has reached the lower limit (20%). To avoid the penalty term of SOC, the SAC-based scheduling strategy imports power to BES

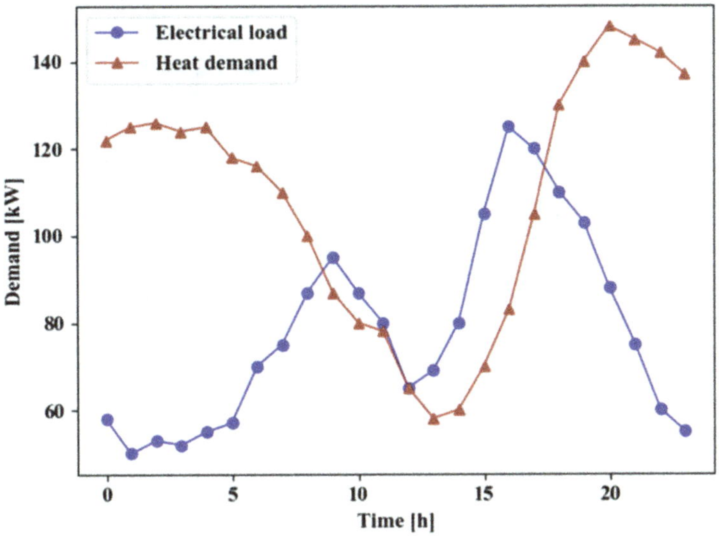

Figure 6.23 Energy loads curves on selected testing day in the system described in Figure 6.18 [4]

by increasing the MTG output. At hours 12–14, MTG has reached the output limit, and a small amount of load shortage is cut off. Then, the electrical load from 16:00 to 18:00 is mainly met by the more economical BES. During 18–22 hours, although the CHP provides a large amount of power support for the peak of electric and heat load, the MTG output and SOC have reached the limit such that load shedding still happens.

When the optimization is carried out according to Scenario 2, WT and CHP still supply electric energy during the low period of electric load at hours 1–6; the active power of the system is balanced by the BES from 07:00 to 09:00; and the BES is charged at hours 10–15 due to no residual power in the BES. To ensure the reliability of the power supply, the WT and MTG operate at the power limit. During the peak period of power load from 18:00 to 22:00, WT and MTG are all at full capacity. Since MT and the SOC of BES reach the output and lower limit point at hour 20, respectively, the IPNGHS appears to have a temporary power shortage, and the load demand is met in other periods.

When optimizing according to Scenario 3, it can be seen from Figure 6.24(c) that there is only a small amount of power shortage in hours 9–11 and 18, while the power supply can be maintained in other time periods; during hours 6–11 and 16–18, the BES operates at close to the limit power, and the MTG with relatively high generation cost produces less electricity and only operates at the limit power during the peak period of hours 12–14 and 19–22 to ensure the reliability of power supply.

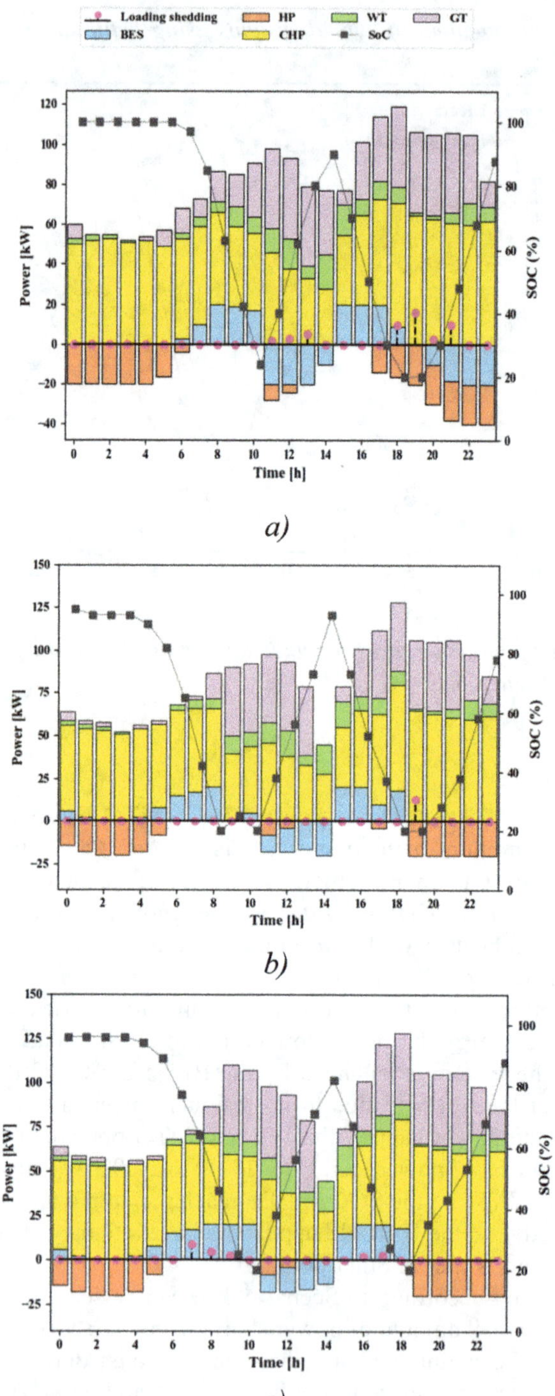

Figure 6.24 Electric dispatch and SOC of BES of three scenarios in the system described in Figure 6.18. (a) Electric dispatch and SOC of BES of Scenario 1, (b) electric dispatch and SOC of BES of Scenario 2, (c) electric dispatch and SOC of BES of Scenario 3 [4].

6.4 Conclusion

This study mainly introduces the energy management of a hybrid energy system using artificial intelligence, which is composed of PPO-based operating cost optimization and SAC-based multi-objective optimization.

In the PPO-based operating cost optimization, a dynamic power conversion algorithm between electricity and heat sub-models in the IES is investigated, wherein the SO can adaptively decide the wind power conversion ratio using the DRL methodology according to the electricity price of the upper-level grid, CUs' energy demand profiles, and wind power generation. This study first formulated the dynamic power conversion to a finite discrete MDP and employed PPO to solve this decision-making problem. Through the process of using the DRL methodology, the SO (agent) does not require a prespecified model of the IES (environment) on which an energy conversion ratio (action) needs to be selected. It is capable of responding to a dynamic and changing environment through constant online learning, which considers the flexibility of wind power generation and electricity prices, and the uncertainty in CUs' load demand profiles. Finally, the numerical simulation results showed that this proposed dynamic power conversion algorithm can effectively minimize the operating cost of the system to promote the SO's profit.

In the second study, a SAC-based energy management plan is put forth as a solution to scheduling issues that are both dependable and affordable. In addition, taking into account the various complex uncertainties present in the IPHNGE system—such as the variability of load demands and the intermittent nature of wind energy—the DRL-based agent can formulate an energy scheduling plan in real-time. Furthermore, this problem is converted into the constrained optimum control problem with several optimization targets in addition to a variety of restrictions. Subsequently, the optimization problem is formulated as a decision-making problem using a finite discrete MDP, making it amenable to solution by the SAC algorithm. Following the implementation of the current strategy, the state, the control signals (action), and the associated reward are recorded in the experience buffer replay. Based on the numerous interactions above, the strategy is significantly learned to improve the reward. Finally, numerical cases conducted on the IEEE 39-bus power system, a 6-node heating system, and a 20-node gas system demonstrate that the proposed SAC-based energy management strategy achieves better optimization results and constraint satisfaction than the benchmark RL algorithms and the conventional optimization algorithm.

References

[1] J. Fang, Q. Zeng, X. Ai, *et al.*, "Dynamic optimal energy flow in the integrated natural gas and electrical power system," *IEEE Transactions on Sustainable Energy*, vol. 9, no. 1, pp. 188–198, 2018.

[2] J. Li, J. Fang, Q. Zeng, *et al.*, "Optimal operation of the integrated electrical and heating systems to accommodate the intermittent renewable sources," *Applied Energy*, vol. 167, pp. 244–254, 2016.

[3] B. Zhang, W. Hu, D. Cao, *et al.*, "Deep reinforcement learning-based approach for optimizing energy conversion in integrated electrical and heating system with renewable energy," *Energy Conversion and Management*, vol. 202, no. 112199, 2019.
[4] B. Zhang, W. Hu, D. Cao, *et al.*, "Soft actor-critic -based multi-objective optimized energy conversion and management strategy for integrated energy systems with renewable energy," *Energy Conversion and Management*, vol. 243, no. 114381, 2021.
[5] R. El-Eman, I. Dincer, "Investigation and assessment of a novel solar-driven integrated energy system," *Energy Conversion and Management*, vol. 158, pp. 246–255, 2018.
[6] D. Cao, W. Hu, J. Zhao, *et al.*, "Reinforcement learning and its applications in modern power and energy systems: A review," *Journal of Modern Power System and Clean Energy*, vol. 8, no. 6, pp. 1029–1042, 2020.
[7] S. Hang, J. Fang, X. Ai, *et al.*, "Stochastic optimization of economic dispatch for microgrid based on approximate dynamic programming," *IEEE Transactions on Smart Grid*, vol. 10, no. 3, pp. 2440–2452, 2019.
[8] S. Hang, J. Fang, X. Ai, *et al.*, "Optimal real-time operation strategy for microgrid: an ADP based stochastic nonlinear optimization approach," *IEEE Transactions on Sustainable Energy*, vol. 10, no. 2, pp. 931–42, 2019.
[9] Q. Junfei, G. Wang, W. Li, *et al.*, "An adaptive deep Q-learning strategy for handwritten digit recognition," *Neural Networks*, vol. 107, pp. 61–71, 2018.
[10] W. Zhiqiang, H. Haibo, T. Bo, "A generative model for sparse hyperparameter determination," *IEEE Transactions on Big Data*, vol. 4, no. 1, pp. 2–10, 2017.
[11] D. Laredo, Z. Chen, O. Schutze, *et al.*, "A neural network-evolutionary computational framework for remaining useful life estimation of mechanical systems," *Neural Networks*, vol. 116, pp. 178–87, 2019.
[12] M. Mohammadi, A. Al-Fuqaha, M. Guizani, *et al.*, "Semi-supervised deep reinforcement learning in support of Io-T and smart city services," *IEEE Internet Things Journal*, vol. 5, no. 2, pp. 624–635, 2018.
[13] W. Jingda, A. Fuqaha, M. Guizani, *et al.*, "Continuous reinforcement learning of energy management with deep Q network for a power split hybrid electric bus," *Applied Energy*, vol. 222, pp. 799–811, 2018.
[14] H. Haochen, Y. Qin, C. Hao, *et al.*, "Optimal energy management strategies for energy Internet via deep reinforcement learning approach," *Applied Energy*, 239:598–609, 2019.
[15] S. Petrović, K. Karlsson, "Residential heat pumps in the future Danish energy system," *Energy*, vol. 114, pp. 787–797, 2016.
[16] W. Cho, K. Lee, "A simple sizing method for combined heat and power units," *Energy*, vol. 65, pp. 123–133, 2014.
[17] I. Hassine, U. Eicker, "Impact of load structure variation and solar thermal energy integration on an existing district heating network," *Applied Thermal Engineering*, vol. 50, no. 2, pp. 1437–1446, 2013.

[18] J. Coulson, J. Richardson, J. Backhurst, *et al.*, "Fluid flow, heat transfer and mass transfer," *Filtration & Separation*, vol. 33, no. 2, pp. 102, 1996.

[19] Z. Guang, T. Chow, V. Maisotsenko, *et al.*, "Design optimization and performance appraisal of a combined cooling, heating and power system primed with Maisotsenko combustion turbine cycle," *Energy Conversion and Management*, vol. 177, pp. 91–106, 2018.

[20] W. Lingfeng, C. Singh, "Stochastic combined heat and power dispatch based on multi-objective particle swarm optimization," *International Journal of Electrical Power and Energy Systems*, 30(3):226–234, 2008.

[21] Wind Energy. (2022). *The cost of energy generated by wind power*. [Online]. Available: https://www.wind-energy-the-facts.org/the-cost-of-energy-generated-by-wind-power.html.

[22] G. Viessmann. (2022). *Heat pumps up to 2000 kW*. [Online]. Available: http://www.viessmann.co.uk/en/residential-buildings/heat-pumps/large-heat-pumps/vitocal-300g-pro.html.

[23] J. Peters, S. Schaal, "Reinforcement learning of motor skills with policy gradients," *Neural Networks*, vol. 21, no. 4, pp. 682–697, 2008.

[24] Sdge.com. (2022). "Home—San Diego Gas & Electric." [Online] Available: https://www.sdge.com.

[25] Comed.com. (2022). "Powering lives—ComEd – an Exelon company." Available at: https://www.comed.com/Pages/default.aspx.

[26] V. Mnih, K. Kavukcuoglu, D. Silver, *et al.*, "Human-level control through deep reinforcement learning," *Nature*, vol. 518, no. 7540, pp. 529, 2015.

[27] S. Petrović, K. Karlsson, "Residential heat pumps in the future Danish energy system," *Energy*, vol. 114, no. C, pp. 787–797, 2016.

[28] A. Schafer, P. Baumanns, A. Moser, "Modeling heat pumps as combined heat and power plants in energy generation planning," in *Proceedings of IEEE Energytech*, Cleveland, OH, USA, 2012, pp. 1–6.

[29] A. Zahedi, H. Timasi, A. Kasaeian, *et al.*, "Design and construction of a new dual CHP-type renewable energy power plant based on an improved parabolic trough solar collector and a biofuel generator," *Renewable Energy,* vol. 135, pp. 485–495, 2019.

[30] Q. Wei, D. Liu, Y. Liu, *et al.*, "Optimal constrained self-learning battery sequential management in microgrid via adaptive dynamic programming," *IEEE/CAA Journal of Automatic Sinica*, vol. 4, no. 2, pp. 168–176, 2017.

[31] L. Ni, W. Liu, F. Wen, *et al.*, "Optimal operation of electricity, natural gas and heat systems considering integrated demand responses and diversified storage devices," *Journal of Modern Power Systems and Clean Energy*, vol. 6, no. 3, pp. 423–437, 2018.

[32] Q. Xu, S. Dubljevic, "Modelling and control of solar thermal system with borehole seasonal storage," *Renewable Energy*, vol. 100, pp. 114–128, 2017.

[33] W. Li, M. Paul, M. Rolley, *et al.*, "A coupled optical-thermal-electrical model to predict the performance of hybrid PV/T-CCPC roof-top systems," *Renewable Energy*, vol. 112, pp. 166–186, 2017.

[34] S. Fazlollahi, N. Schuler, F. Marechal, "A solid thermal storage model for the optimization of buildings operation strategy," *Energy*, vol. 88, pp. 209–222, 2015.

[35] X. Zhang, X. Liu, X. Sun, *et al.*, "Thermodynamic and economic assessment of a novel CCHP integrated system taking biomass, natural gas and geothermal energy as co-feeds," *Energy Conversion and Management*, vol. 172, pp. 105–118, 2018.

[36] A. Ghosh, B. Norton, "Optimization of PV powered SPD switchable glazing to minimize probability of loss power supply," *Renewable Energy*, vol. 131, pp. 993–1001, 2019.

[37] S. Nourollah, G. Gharehpetian, "Coordinated load shedding strategy to restore voltage and frequency of micro-grid to secure region," *IEEE Transactions on Smart Grid*, vol. 10, no. 4, pp. 4360–4368, 2019.

[38] M. Jordan, T. Mitchell, "Machine learning: Trends, perspectives, and prospects," *Science*, vol. 349, no. 6245, pp. 255–260, 2015.

[39] T. Haarnoja, A. Zhou, K. Hartikainen, *et al.*, "Soft actor-critic algorithms and applications," 2018. [Online]. Available: https://arxiv.org/abs/1801.01290.

[40] M. Turchetta, F. Berkenkamp, A. Krause, "Safe exploration in finite Markov decision process with Gaussian processes," in *Proceedings of Advances in Neural Information Processing Systems*, 2016, pp. 4312–4320.

[41] J. Schulman, F. Wolski, P. Dhariwal, *et al.*, "Proximal policy optimization algorithm," 2017. [Online]. Available: https://arxiv.org/abs/1707.06347.

[42] V. Mnih, A. Badia, M. Mirza, *et al.*, "Asynchronous methods for deep reinforcement learning," 2016. [Online]. Available: https://arxiv.org/abs/1602.01783.

[43] T. Lillicrap, J. Hunt, A. Pritzel, *et al.*, "Continuous control with deep reinforcement learning," 2015. [Online]. Available: https://arxiv.org/abs/1509.02971.

[44] R. Sutton, A. Barto, *Reinforcement learning: An introduction*, vol. 1, Cambridge, MA, USA: MIT Press, 2005.

[45] D. Wolf, Y. Smeers, "The gas transmission problem solved by an extension of the simplex algorithm," *Management Science*, vol. 46, no. 11, pp. 1454–1465, 2000.

[46] H. Hasselt, A. Guez, D. Silver, "Deep reinforcement learning with double Q-learning," 2015. [Online]. Available: https://arxiv.org/abs/1509.06461.

[47] D. Connolly, K. Hansen, D. Drysdale, *et al.*, "Enhanced heating and cooling plans to quantify the impact of increased energy efficiency in EU member states," *Tech. Rep.*, Aalborg University, Denmark, 2015.

Chapter 7
Artificial intelligence in energy management of microgrid

Weihao Hu[1], *Guozhou Zhang*[2], *Zhenyuan Zhang*[1], *Sayed Abulanwar*[3,4] *and Frede Blaabjerg*[5]

7.1 Introduction

Fuel scarcity and escalating carbon emissions pose a dual challenge to conventional energy sources. Photovoltaic (PV) solar and wind energy systems serve as renewable substitutes for traditional energy systems due to their eco-friendliness, versatility, and green nature. Furthermore, regulations adopted by numerous governments encouraging the installation of PV and wind power systems have reduced the cost of electricity production. The development of renewable energy sources, such as solar, wind, fuel cells, and biomass, has enabled the current electric power industry to advance technologically. A smart grid is a digital technology that integrates various microgrids with the grid and monitors them with appropriate management and control to reduce or eliminate power quality concerns. The likelihood of individual microgrid stability criteria being met is increased by interconnected microgrids. Therefore, self-sustaining smart grid technology is necessary to decrease carbon emissions and provide an energy management solution. The use of passive or active filters is one method to enhance the grid's power quality. Despite their higher cost, active filters are preferred due to their size and performance advantages. Utility grid-interfaced PV and wind energy system installations as active filters have gained popularity as a significant research topic worldwide since the turn of the millennium. The active filtering capabilities of smart grid systems connected with microgrids have received significant attention in the literature. Instantaneous reactive power theory (IRPT), synchronous reference frame theory (SRFT), and the improved linear sinusoidal tracer (ILST) are some of the methods used for managing shunt active filters (SAFs). These algorithms

[1]School of Mechanical and Electrical Engineering, University of Electronic Science and Technology of China, China
[2]College of Engineering and Technology, Southwest University, China
[3]Electrical Engineering Department, Faculty of Engineering, Mansoura University, Egypt
[4]Faculty of Engineering, Horus University-Egypt, Egypt
[5]Department of Energy Technology, Aalborg University, Denmark

can reduce current harmonics on the source side and offer reactive power compensation when utilized to manage the voltage source inverter (VSI) in grid-connected systems.

A "smart microgrid" refers to an intelligent electricity distribution system that connects loads, distributed energy resources, and storage within clearly defined electrical boundaries, functioning as a single, controllable entity with respect to the main grid, according to India's Model Smart Grid Regulations. Microgrids (MGs) are small-scale power plants that can operate in both grid-connected and island modes, boasting high levels of energy security, dependability, storage capacity, and economic efficiency for demand-side and load-side control. The electricity sector can self-recover by assessing and addressing issues, as they are capable of local control via automated methods. To reduce costs and return excess energy to local microgrids, the electric industry has been revolutionized through the use of solar and wind renewable energy sources (RESs) and smart grid technologies. Smart microgrids have the potential to integrate with the grid, but they are also self-sufficient and can service a local community without relying on centralized power systems. Based on comparable electrical system research, future generations will require intelligent controller-based smart grid systems, with many microgrids integrating with the grid without affecting the quality of electricity.

A smart microgrid system comprises one or more interconnected smart microgrids, facilitated by a robust controller that can be integrated into or operated independently from the main grid. This comprehensive system encompasses diverse renewable and non-renewable energy sources, along with load centers catering to residential, commercial, and industrial sectors. Real-time management of electric power systems necessitates data analytics and artificial intelligence (AI) techniques for statistical analysis of consumer energy usage data and weather forecasts pertaining to various RESs. Furthermore, the AI-based controller incorporates modules such as tariff control and power flow management while processing customer power data.

7.2 DRL-based energy management for multi-microgrid considering voltage control and thermal flow

7.2.1 Literature review

The distribution network (DN) level of a multi-energy microgrid (MG) creates, transports, and consumes heterogeneous energy carriers, including electric and thermal energy [1]. A multi-energy microgrid typically consists of distributed energy resources (DERs), energy coupling technologies, nearby active loads, and energy storage devices at the distribution network level. Microturbines (MTs) and electric heat pumps (HPs), two recent advances in energy coupling, may be combined by various energy carriers to improve the economics and environmental sustainability of the energy system. Through increased RES use and multi-energy coordination, multi-energy MG has evolved into a dependable and affordable method of delivering multi-energy supply. However, owing to the local nature of

RES power supply, multi-energy MG and RES have restricted energy-supplying capacities [2]. Due to the capacity restrictions of each individual MG, many nearby MGs may be able to exchange energy in specific places to prevent this shortfall [3,4]. These studies support the idea that, in addition to connecting each individual MG directly to the DN, it may also be advantageous to network the various microgrids into an interconnected multi-microgrid (MMG) system in a specific location before connecting it to the DN to increase the economic benefits and power supply reliability for both the interconnected MMGs and the DN [5]. Energy management is essential for dependable and effective multi-energy MG operation and control [6], especially for multi-energy MMG systems, because consumers in a multi-energy MG often need a considerable quantity of electrical and thermal energy at the same time [7].

A specialized controller is often used in a totally centralized control system to carry out a number of tasks, such as receiving data, doing calculations and optimizations, and choosing the control actions that will be applied to all controlled units in the system. To perform all of these tasks at one location, a complex communication network must be used between the central controller and the controlled devices. For instance, in [8], the distribution system operator (DSO) and the MMGs are one entity, with the DSO directly overseeing each MG's operations. To maximize the financial advantages of networked MMGs with combined heat and power (CHP) systems through energy management, the centralization-based heuristic method is presented in [9]. The flexibility of CHP units employing electrical boilers and heat storage tanks within a centralized model is enhanced by an updated linear model for integrated energy system dispatch reported in [10]. Centralized control does not exhibit plug-and-play capabilities, which is necessary in an MG arrangement, due to the need to concurrently process massive volumes of data at a single location [11]. In general, isolated power systems that must sustain a crucial supply-and-demand balance over the long term in infrastructure that changes slowly benefit more from the adoption of centralized control [11]. A distributed system's MGs may also be owned by a variety of parties, and each MG's scheduling of its numerous distributed generators may be governed by its own set of economic and policy regulations [12]. As a result, for privacy and security considerations, information exchange between DSOs and MGs as well as MMGs may be limited.

Given the aforementioned factors, centralization-based methods are inappropriate for planning the energy of linked MMGs. Significantly, linked MMGs are increasingly using distributed and decentralized administration. By breaking the linked MMG energy scheduling issue down into smaller, simpler MG subproblems, the authors of [13] present a distributed stochastic optimum scheduling approach with little information exchange cost. A distributed transactive resource trading framework and a multilateral bargaining-based mechanism are proposed in [14] to measure the advantages of proactive resource trading among interconnected MMGs. Proactive resource trading enables different multi-energy generation and load profiles to be used to mutual advantage. To enable energy trading across island MGs, a distributed convex optimization framework is being developed in [15]. For coordinating the operation of linked MMGs within a distribution system, a decentralized, bi-level algorithm-based energy management system is proposed in [16].

Even though significant studies have been done, there are still four main problems with linked systems' energy management that need to be addressed. First, all of the methods covered above fall under the category of model-based methods. Since power flow and energy coupling connections are impacted by a wide range of parameters [17], it is very difficult to construct an accurate physical model of an interconnected MMG that is precise and efficient enough for its energy management. Second, when applied to a range of MMG contexts, model-based techniques' generalizations are constrained since they depend on particular linked MMG ecosystems [18]. Third, the amount of data that needs to be transferred, processed, and stored in the power system is quickly growing as more MGs and devices are linked to DN. As a result, typical model-based techniques dramatically increase the computing cost of energy management. Fourth, model-based techniques need constant problem-solving and are unable to adapt to constantly changing situations, which may delay real-time judgments. Therefore, it is crucial that vast volumes of data are used precisely, effectively, and efficiently.

Learning-based methods, such as deep reinforcement learning, can be used to get around the aforementioned limitations. Without being aware of the precise physical model of linked MMGs, they are capable of completely using the information in interactions with the interconnected MMG environment to learn an ideal policy. Once the learning process is over, the generated policy may be applied for end-to-end sophisticated decision-making, such as the application of quickly (e.g., within milliseconds) creating optimum actions in response to real-time system status without any prior information. To manage the energy sources of MMGs, a deep reinforcement learning strategy is put forth, where each MG is managed by an agent to provide a decentralized control mechanism. Each agent, however, lacks coordinated interaction with other agents since there is no coordinated mechanism between them. As a result, each agent is independently optimized by its own reward function. A deployment choice like this is inappropriate for the MMG system because it integrates numerous MGs to form an MMG system, which realizes energy complementarity inside the MMG system through the integration of many MGs. As a result, the MMG system's efficient functioning is supported by a high-quality and successful coordinated optimization method [12]. In light of this, Gregoratti and Matamoros [15] suggest a MADRL (multi-agent deep reinforcement learning)-based method for managing the MMG system's energy sources in concert.

7.2.2 System description

As Figure 7.1 shows, (1) each individual MG consists of a heat and electricity network where the heat network consists of a microturbine, power to heat unit (e.g., heat pump, etc.), and thermal load; and the electricity network consists of a diesel generator (DG), renewable energy (e.g., solar and wind energy, etc.), energy storage unit, and electrical load; (2) the power line (black line) and heat pipe (red line) integrate the individual MGs into an interconnected MMG system so that electrical and thermal energy can be supplied to each other between the different MGs.

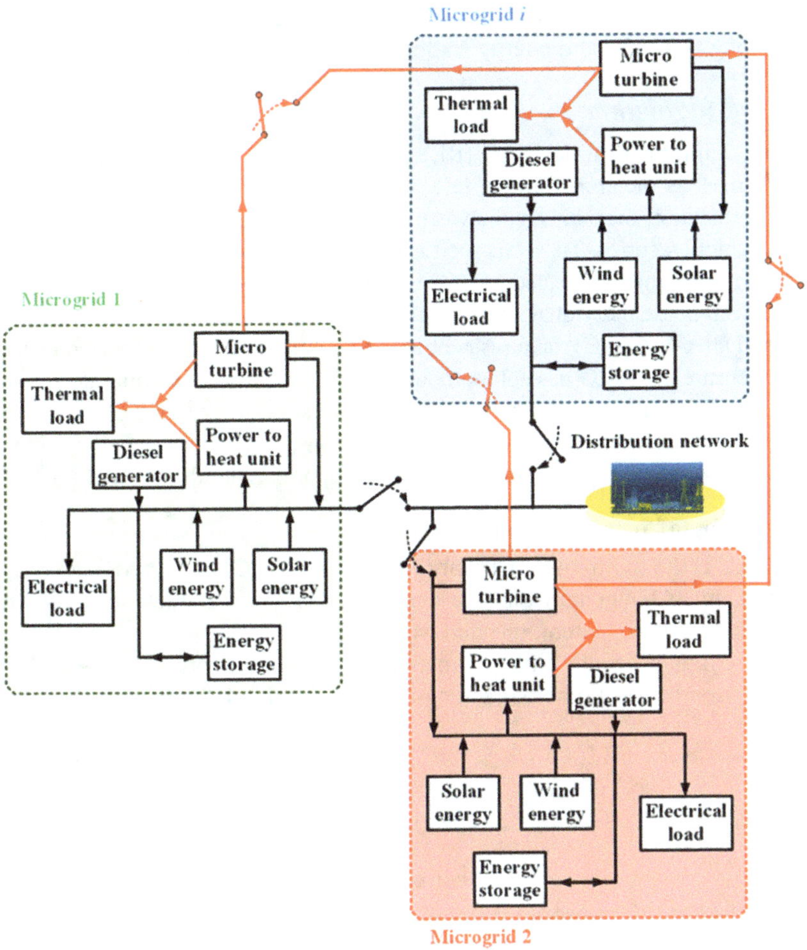

Figure 7.1 MMG architecture [19]

Individual MGs in Figure 7.1 are autonomous systems that are geographically dispersed and equipped with RES units, storage units, and DGs to sustain the local load. Energy management at the individual MG level is intended to handle the time-varying issues associated with renewable energy sources and loads by using the complimentary characteristics of energy storage system (ESS), diesel generator (DG), and RES. Individual MGs are often not intended to manage the geographically scattered RESs and loads of a region, but rather a local area. As a result, their overall energy management solutions may not be the global optimum in that region. As shown in Figure 7.1, the MGs can be linked to form an interconnected MMG system by power lines and heat pipes. In the interconnected MMG system, the purpose of the energy management program is to handle the spatial variations of

RESs and loads within a region or wider area by coordinating the exchange of power among the MGs and energy trading between the DN and individual MGs.

7.2.3 RL algorithm

Traditional reinforcement learning (RL) algorithms are difficult to deal with the Markov games, as mentioned in the last section, due to the curse of dimension and a lack of coordination mechanisms between each agent while considering the privacies and information security. To solve such a problem, a novel MADRL is proposed to solve the complex interconnected MMGs energy management. The MADRL, which is called the deep dense individual attention architecture in the multi-agent soft actor-critic algorithm (D2IA-MASAC), features centralized training and a decentralized execution manner to ensure each MG has autonomous energy management within interconnected MMGs. The proposed approach consists of two neural network-based parts. One is the actor part, which is responsible for making decisions, and the other is the critic part, which is responsible for guiding the actor part to approximate the optimal policy.

7.2.3.1 A brief introduction of MASAC

The MASAC is the multi-agent variant [20] of the SAC [21]. Each MG controller is modeled as an SAC agent within the centralized training framework to solve the energy management of interconnected MMGs.

Considering N agents with parameterized critic network $Q^j_{\phi_j}, j \in [1, N]$, the update of individual critic network for j-th agent can be defined as:

$$
\begin{aligned}
&Loss = \mathbb{E}_{X_t, A_t \sim \mathcal{D}} \left(Q^j_{\phi_j}(X_t, A_t) - y\right)^2 \\
&y = r^j_t + \gamma \mathbb{E}_{X_{t+1} \sim \mathcal{D},\ A_{t+1} \sim \mu_\theta(X_{t+1})} \left[Q^j_{\phi'_j}(X_{t+1}, A_{t+1}) - \alpha \log\left(\mu^j_{\theta'_j}(s^j_{t+1})\right)\right]
\end{aligned}
\tag{7.1}
$$

where $X_t = \left[s^1_t, \ldots, s^N_t\right]$ is the state set of time t, $A_t = \left[a^1_t, \ldots, a^N_t\right]$ represents the action set of time t, $Q^j_{\phi_j}(X_t, A_t)$ is the centralized action-value function with parameters ϕ_j, and μ_θ denotes the policy network with parameters θ .

There are N agents with parameterized actor network $\mu^j_{\theta_j}, j \in [1, N]$, the individual policies are updated by ascent in accordance with the following gradient:

$$
\nabla_{\theta_j} J(\mu^j_{\theta_j}) = \mathbb{E}_{X_t,\ A_{t j\backslash} \sim \mathcal{D}} \left[\nabla_{\theta_j} \log\left(\mu^j_{\theta_j}(s^j_t)\right) \begin{pmatrix} Q^j_{\phi_j}(X_t, A_t) - \\ \alpha \log\left(\mu^j_{\theta_j}(s^j_t)\right) \end{pmatrix} |_{a^j_t = \mu^j_{\theta_j}(s^j_t)}\right] \tag{7.2}
$$

where $A_{tj\backslash} = [a^1_t, \ldots, a^{j-1}_t, a^{j+1}_t, \ldots, a^N_t]$ is the action set of other agents of time t, $\mathcal{D}$ is the replay buffer that contains the experience of N agents, and α is the temperature ratio used to balance exploration and exploitation.

ϕ'_j and θ'_j can be updated as

$$
\begin{aligned}
\phi'_j &\leftarrow \tau \phi_j + (1 - \tau)\phi'_j \\
\theta'_j &\leftarrow \tau \theta_j + (1 - \tau)\theta'_j
\end{aligned}
\tag{7.3}
$$

where soft replacement $\tau \ll 1$.

7.2.3.2 Architectures of D2IA-MASAC networks

The architecture of the critic part used in the proposed approach is shown in Figure 7.2. Each agent has its own individual critic, where each critic is composed of a deep dense architecture in reinforcement learning (D2RL) [22] and attention mechanism [23]. Take the first agent in Figure 7.2 as an example, the input of the critic is the whole agents' states and actions. Each agent's information, i.e., states and actions are encoded by the D2RL.

Deep neural network is a powerful approach for extracting features from inputs and mapping them to desired outputs, which enhances the nonlinear capability of a critic network. It is a pity that merely increasing the depth of the dense neural network may prove ineffective because the original input information gradually disappears with further network depth [24]. D2RL offers a way to avoid this problem. The architecture of D2RL is shown in Figure 7.3. As this figure shows, the s^g is first input to h_1 dense neural network and output m_1. Next, the input of h_2 is the concatenation of s^g and m_1 instead of m_1. Loop back and forth until the final a^g is output. It makes better use of the non-linear mapping ability of deeper dense

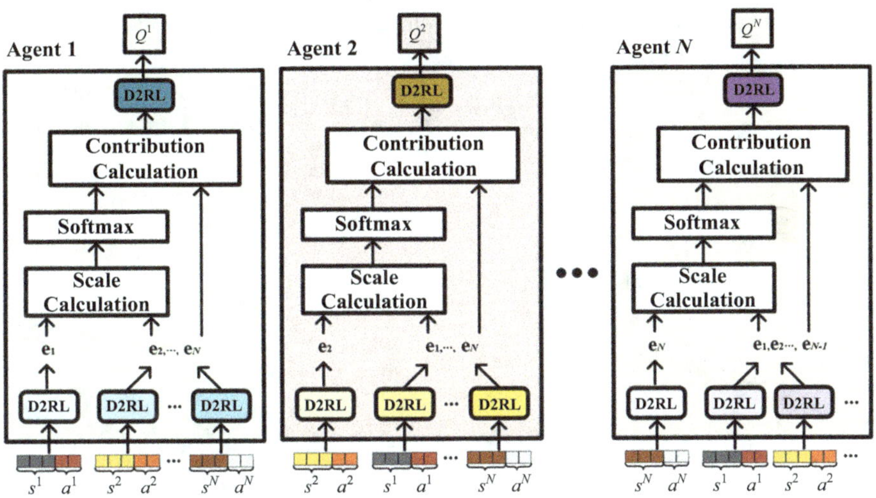

Figure 7.2 The architecture of critic network of each agent [19]

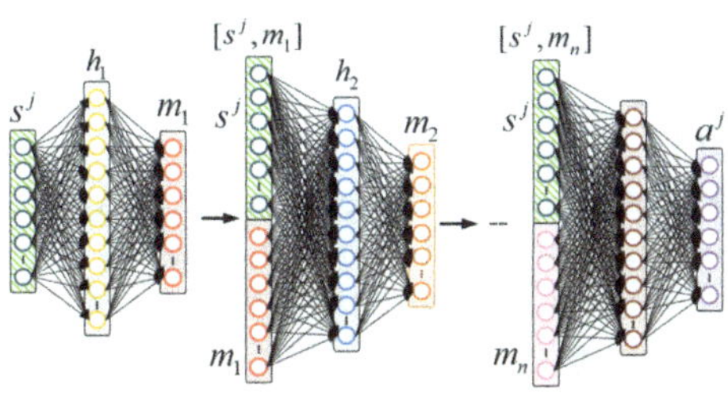

Figure 7.3 The architecture of D2RL [19]

networks by incorporating the original input s^g into each hidden layer of the network so that the s^g information can be retained to the maximum extent. Mathematically, the calculation process from the input of the critic network belonging to the first agent to e_1 is

$$\begin{aligned} I_1 &= \text{concat}(s^1, s^2, ..., s^N, a^1, a^2, ..., a^N), \\ I_{u+1} &= \text{ReLu}(\text{concat}(I_1, I_u) \cdot \mathrm{W}_u^1 + \mathrm{B}_u^1), u = 2, ..., U-1 \\ \zeta &= \text{ReLu}(I_U \cdot \mathrm{W}_U^1 + \mathrm{B}_U^1), \\ e_1 &= \zeta \cdot \mathrm{W}_O^1 + \mathrm{B}_O^1, \end{aligned} \tag{7.4}$$

where concat represents the "concatenate" operation, ReLu is the Rectified Linear Units, I_u represents the input of u-th hidden layer, W_u^1 and B_u^1 are weight and bias matrices of u-th hidden layer of the first agent, respectively, U denotes the total number of hidden layers, ζ denotes the latent feature extracted by the whole U hidden layers from the input I_1, W_O^1 and B_O^1 are weight and bias matrices of the output layer of the first agent, respectively.

When the calculations of e_i, $i \in [1, N]$ are finished, the attention mechanism will then process them to emphasize or select the important information of the other agents while suppressing the effect of irrelevant details on decision-making. The detailed calculation process is as follows:

$$[\chi_{1(2)}, \chi_{1(3)}, ..., \chi_{1(N)}] = \frac{\varpi(e_1, [e_2{}^{\mathrm{T}}, e_3{}^{\mathrm{T}}, \ldots, e_N{}^{\mathrm{T}}])}{\sqrt{d_\beta}} \tag{7.5}$$

$$\omega_i = \frac{\exp\left(\chi_{1(i)}\right)}{\sum_{k=2}^{N} \exp\left(\chi_{1(k)}\right)}, i \in [2, 3, ..., N] \tag{7.6}$$

$$\varepsilon_1 = \sum_{k=2}^{N} \omega_k e_k \tag{7.7}$$

where $\varpi(x_a, [x_b, \ldots, x_z]) = [x_a \cdot x_b, \ldots, x_a \cdot x_z]$, and d_β denotes the network dimensions.

The final expression for action-value function is

$$Q^1(\mathrm{X}, \mathrm{A}) = \mathbf{D2RL}(\mathrm{e}_1, \varepsilon_1) \tag{7.8}$$

where **D2RL** denotes the deep-blue D2RL neural network in the first agent.

7.2.4 Markov game

- *State:* The state variables reflect the current operating conditions of the individual MG system and are gathered by each MG controller within EMS and utilized to make energy management decisions. The state of the g-th MG $s_t^g \in S^g$ at time t is defined as

$$s_t^g = (t, \widetilde{EP}_t, SOC_t^g, \widetilde{TL_t^g}, \widetilde{EL_t^g}, P_{i,t-1}^{g,DG}, \widetilde{P_{i,t}^{g,WT}}, \widetilde{P_{i,t}^{g,PV}}, \widetilde{NL_t^g}, \widetilde{NV_t^g}) \tag{7.9}$$

where t denotes the time information, $\widetilde{EP_t}$ is the forecasted electricity price of current time t, $SOC_{i,t}^g$ is the SOC of energy storage system within g-th MG at node I, $\widetilde{TL_t^g}$ is the forecasted cumulative thermal load of each node, $\widetilde{EL_t^g}$ is the forecasted cumulative electrical load of each node, $P_{i,t-1}^{g,DG}$ is the working power of the DG at time $t-1$, $\widetilde{P_{i,t}^{g,WT}}$ and $\widetilde{P_{i,t}^{g,PV}}$ is the power of forecasted wind turbine (WT) and PV, respectively, $\widetilde{NL_t^g}$ is the forecasted electrical load of each node within g-th MG and NV_{t-1}^g is the voltage of each node.

- *Action:* In the Markov game, the action of g-th MG is defined as

$$a_t^g = (P_{i,t}^{g,ESS}, P_{i,t}^{g,DG}, P_{i,t}^{g,PtH}, Q_{i,t}^{g,PV}, Q_{i,t}^{g,ESS}, Q_{i,t}^{g,DG}, Q_{i,t}^{g,WT}) \tag{7.10}$$

where $P_{i,t}^{g,ESS}$, $P_{i,t}^{g,DG}$, and $P_{i,t}^{g,PtH}$ denotes the active power of ESS, DG, and PtH unit, respectively, $Q_{i,t}^{g,PV}$, $Q_{i,t}^{g,ESS}$, $Q_{i,t}^{g,DG}$, and $Q_{i,t}^{g,WT}$ is the reactive power of PV, ESS, DG, and WT, respectively.

- *Reward:* The reward function in Markov game is the objective function of the optimization problem. This mainly includes the cost of buying electricity from the distribution network, the cost of DG, the cost of buying gas, and the cost of unstable voltage.

7.2.5 Numerical results

7.2.5.1 Experimental setup

In this section, the performance of the proposed D2IA-MASAC algorithm is examined by numerical experiments on the interconnected MMG system, whose architecture is illustrated in Figure 7.4. In the figure, the interconnected MMG system includes three MGs, each of which consists of a heat and electricity network. Each MG has ES, DG, WT, PV, and microturbine (i.e., source node in Figure 7.4). The related parameters of this case study are listed in Tables 7.1 and 7.2. The electricity price offered by the DN

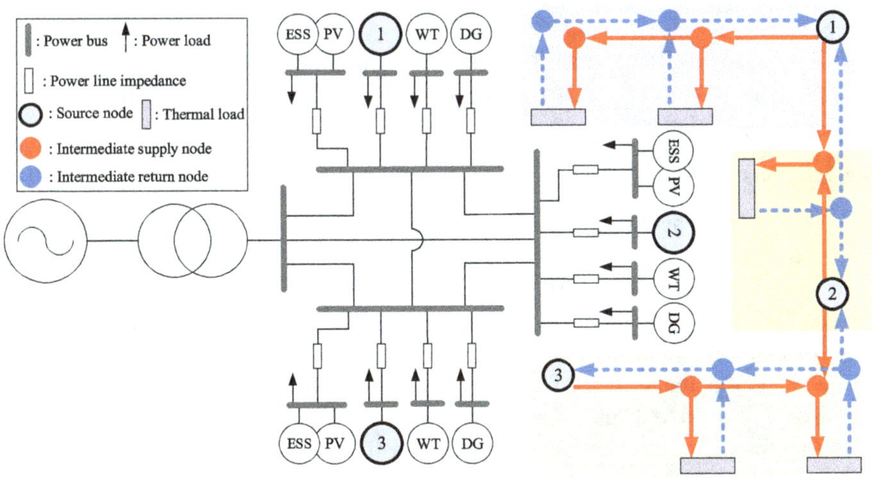

Figure 7.4 The architecture of interconnected MMG system [19]

Table 7.1 Model parameters in Section 7.2

Symbol	Value	Symbol	Value
a ($/(kW)^{2}h)	1040	b ($/MWh)	30.4
c ($)	1.3	η_{CHP}	0.36
$C^{start/stop}$ ($)	5	L_{cvng} (MWh/m^3)	9.7×10^{-3}
σ	6	–	

Table 7.2 Parameters of the interconnected MMG system

Parameter	Value			Parameter	Value		
	MG1	MG2	MG3		MG1	MG2	MG3
$P^{PV}_{i\,\max}$ (MW)	0.2	0.15	0.1	$P^{ES}_{i\,\max}$ (MW)	0.05	0.045	0.04
$P^{WT}_{i\,\max}$ (MW)	0.25	0.2	0.15	$E_{i\,\max}$ (MWh)	0.2	0.18	0.16
$P^{PtH}_{i\,\max}$ (MW)	0.005	0.005	0.005	$SOC^{ES}_{i\,\max}$	1	1	1
$P^{DG}_{i\,\max}$ (MW)	0.05	0.05	0.05	$SOC^{ES}_{i\,\min}$	0.1	0.1	0.1
$P^{DG}_{i\,\min}$ (MW)	0.02	0.02	0.02	$R^{DG}_{up} R^{DG}_{dn}$	0.025	0.025	0.025

follows the hourly pricing from PJM [25] and the natural gas prices follow the monthly Natural Gas Industrial Price from the US Energy Information Administration [26]. The training data contains the first 200 days' data, and the test data is from days 201 to 300 of 2017. The forecasted power of WT, PV, and load of the three MGs in the interconnected MMG system during a typical day is illustrated in Figure 7.5. From the figure, the regions of MG1 include largely electrical loads and have a high capacity of RES. There are fewer electrical loads in MG2 as compared to MG1, and the RES capacity of MG2 is lower as well. MG3 has the highest proportion of RES among the three MGs. In addition, the electrical loads of MG3 are relatively smaller. The thermal load of MG1 and MG2 is similar, and the thermal load of MG3 is the highest.

7.2.5.2 Comparisons with benchmark approaches

To assess the effectiveness of our proposed approach, we compare it to various benchmark approaches.

Descriptions of benchmark approaches

- TD3 [27]: The MG controllers are modeled as Twin Delayed Deep Deterministic policy gradient (TD3) agents, each of which is trained individually to maximize its own reward function without coordinating with other MGs.
- MATD3 [5]: Multi-agent Twin Delayed Deep Deterministic policy gradient (MATD3) is an optimization approach, which was proposed in [5] to optimize the energy management of the interconnected MMG system.
- MAAC [28]: MAAC uses a single common attention network to collect the whole MG information in a centralized way to construct a common gradient space for the whole MG controllers to update their parameters of the neural network.

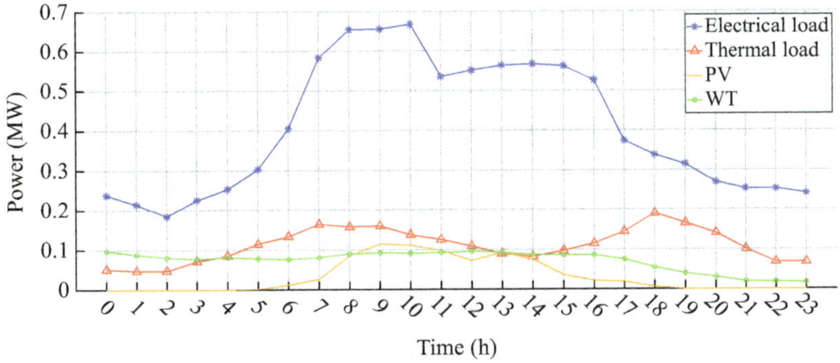

(a) The data of MG1.

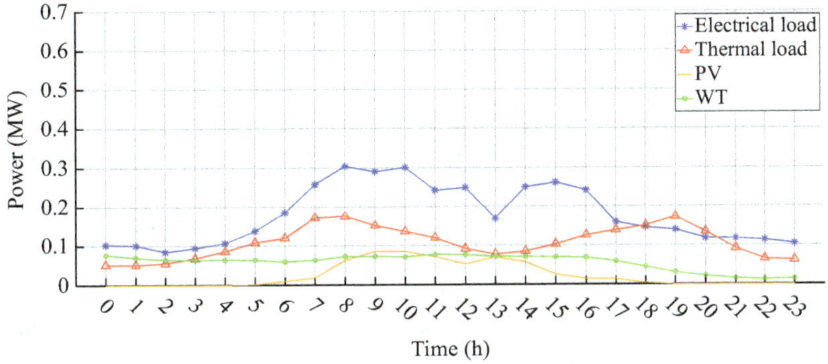

(a) The data of MG2.

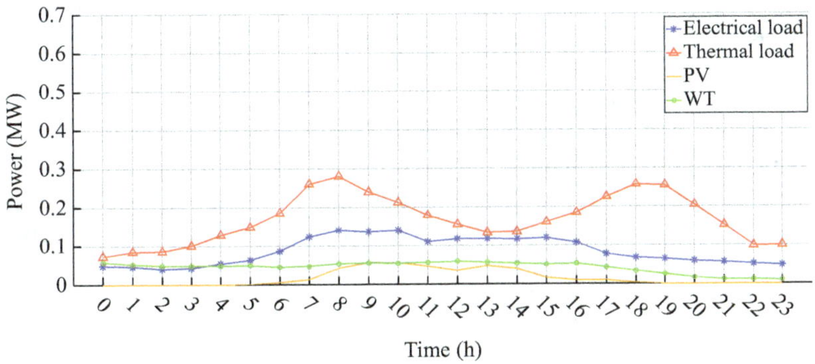

(a) The data of MG3.

Figure 7.5 The power of electrical and thermal load, PV and WT. (a) The data of MG1. (b) The data of MG2. (c) The data of MG3 [19].

Performances on training set

Figure 7.6 shows the normalized cumulative rewards (i.e., the values of the objective function) obtained by the different RL-based control strategies during the training process. In the figure, the agents do not have any knowledge of how to

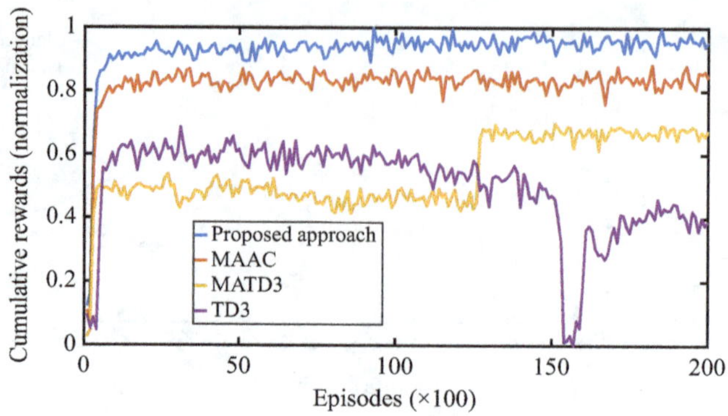

Figure 7.6 The average reward for different training approaches during each 100-episode period [19]

make decisions to obtain a high reward at the beginning. Consequently, they choose to explore the action space to collect more experience. With the progression of the training, agents gradually learn the energy management strategy of the interconnected MMG system to obtain a better reward. As the training curves come to an end, they climb slowly and gradually converge until the curve completes convergence, representing the end of training.

It is obvious that the reward curve of TD3 shows the lowest value and oscillates greatly during the training process compared to other approaches, which demonstrates that such a lack of coordination approach does not fit the energy management of interconnected MMG. Compared with TD3, the MATD3 that is proposed in [5] utilizes the multi-agent framework inspired by [20] to construct the coordination relationship between each MG for energy management in an interconnected MMG system. Obviously, MATD3 outperforms TD3 in terms of both training stability and performance due to the coordination mechanism. The MAAC proposed in [28] further enhances the coordination between each MG by using the common attention network in the critic part and, as such, achieves better performance than the MATD3. The purpose of setting the common attention network in MAAC is to make each agent's own neural network parameters include a part of the network parameters shared among the agents and give each agent an attention attribute to speed up the construction of consensus within the agents and enhance their coordination. Nevertheless, because the interconnected MMG system has complex energy network coupling characteristics in individual MGs with CHP, energy exchange among MMGs, and energy trading between MMGs and DN, the attention network design in MAAC appears not to fit the optimizer of the energy management of the interconnected MMG system. In an interconnected MMG system, each MG has its own individual system parameters (e.g., different thermal and electrical loads, the rated power of assets, heat and power flow, etc.) and as such, each MG may have negative effects on others through the common network due to the complex and individual operation mode of each system. In view of this,

Table 7.3 Performances on the test

Approach	TD3	MATD3	MAAC	Proposed approach
Average cost ($/day)	1,870.14	1,145.15	420.58	331.00
Percentage (%)	565.00	345.97	127.06	–

the proposed approach utilizes a network architecture as depicted in Section 7.3, and the training curves achieve a better performance than the MAAC, proving the effectiveness of the proposed approach.

Performances on the test set

Table 7.3 summarizes the average cost per day of the proposed and benchmark approaches over the 100 test days. The percentage column represents the average cost of benchmark approaches divided by the proposed approach. As shown in Table 7.3, the interconnected MMG system needs to spend 465.00%, 245.97%, and 27.06% more under the control of TD3, MATD3, and MAAC compared with the proposed approach. In these data, TD3 has no mechanism for one agent to coordinate with other agents and has the worst results of all approaches. This phenomenon indicates that coordination among the whole group of agents is necessary in this model. The results of MATD3, which have better results than those of TD3, further demonstrate this point. Moreover, the proposed approach is better than the MAAC, which has a common attention layer. From this point of view, using individual attention mechanisms to manage the energy of interconnected MMG systems is more suitable than the common attention mechanism. Based on these results, it is clear that the proposed approach is effective.

7.2.5.3 Pick a day on the test set for the evaluation of the proposed approach

To show the effectiveness of our proposed approach in guaranteeing secure operation, the voltage of all buses for 1 day is shown in Figure 7.7. Figures 7.7(a) and 7.9(b) show the results of the proposed approach and rule-based approach for voltage control, respectively. The red lines denote the voltage profiles of MG1, the blue lines represent the voltage profile of MG2 and the green line are the voltage profile of MG3. Therein, the line with mark '*', 'Δ', '∇', and '+' denotes the voltage profile of node 1, 2, 3, and 4, respectively. As the figure shows, the voltage can be constrained in the safe range at each bus and more stable than the rule-based approach in the system. The above results support the hypothesis that the proposed D2IA-MASAC approach can ensure the safe operation of interconnected MMG systems through the obtaining of good enough solutions.

A detailed evaluation of the proposed approach on a test day is presented in Figure 7.8 to further illustrate the effectiveness of the proposed approach. Figure 7.8(a) is the real-time electricity price for 1 day, and Figure 7.8(b) is the charging/discharging power of the MGs. MG1 has the lowest proportion of RES and the widest fluctuation

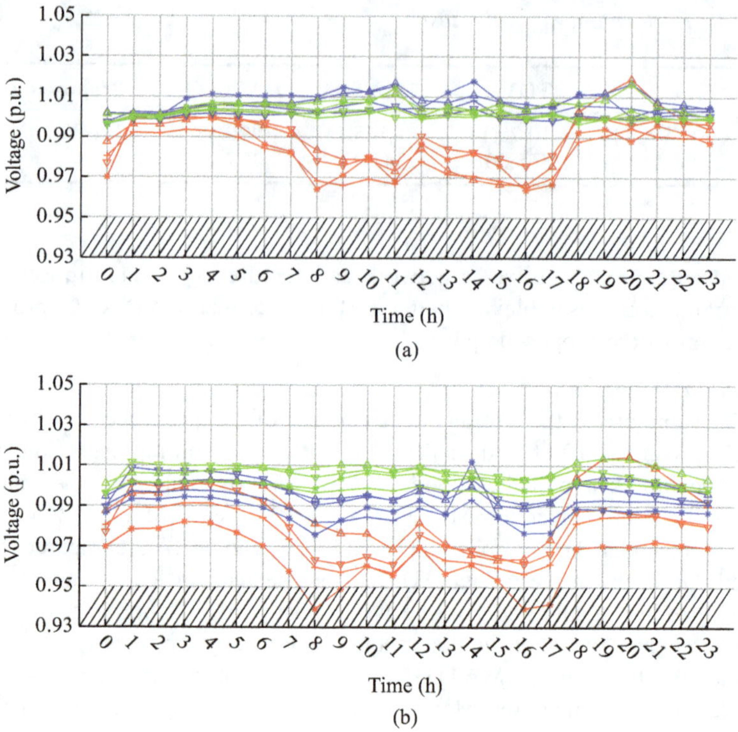

Figure 7.7 The voltage profile of each bus in the interconnected MMG system. (a) The voltage profile under the proposed approach control. (b) The voltage profile under the rule-based control, i.e., each battery, DG, and PtH, does not work. In addition, each controllable reactive power unit follows the rule that its reactive power is associated with the power factor and its active power [19].

range in its voltage profile than the other two MGs, which has the most severe undervoltage situation as shown in Figure 7.7(b). Therefore, we can observe that the MG1 has a different charging mode from that of MG2 and MG3 in Figure 7.8(b). Specifically, the MG1 has a low voltage at the early time of the period of 1 day. To raise its voltage, the ESSs of MG1 choose to discharge its power to avoid further voltage drops. On the contrary, owning the MG3 has the richest RES, which can be used to supply the electrical load to stabilize the voltage, and as such, the ESS of the MG3 can be used for energy arbitrage, i.e., charging when the electricity price is low and discharging when the electricity price is high. Similarly, the ESS of MG2 has the energy arbitrage operation, but it behaves more conservatively than MG3. Specifically, as Figure 7.8(c) shows, MG3 has a larger discharging power than that of MG2, which will result in the lowest SOC of MG3 (at 17:00) being lower than the lowest value of MG2 (at 23:00) in a period. Given that MG2 contains a lower proportion of RES than MG3, if the ESS has a low SOC, it may result in the ESS not being able to supply

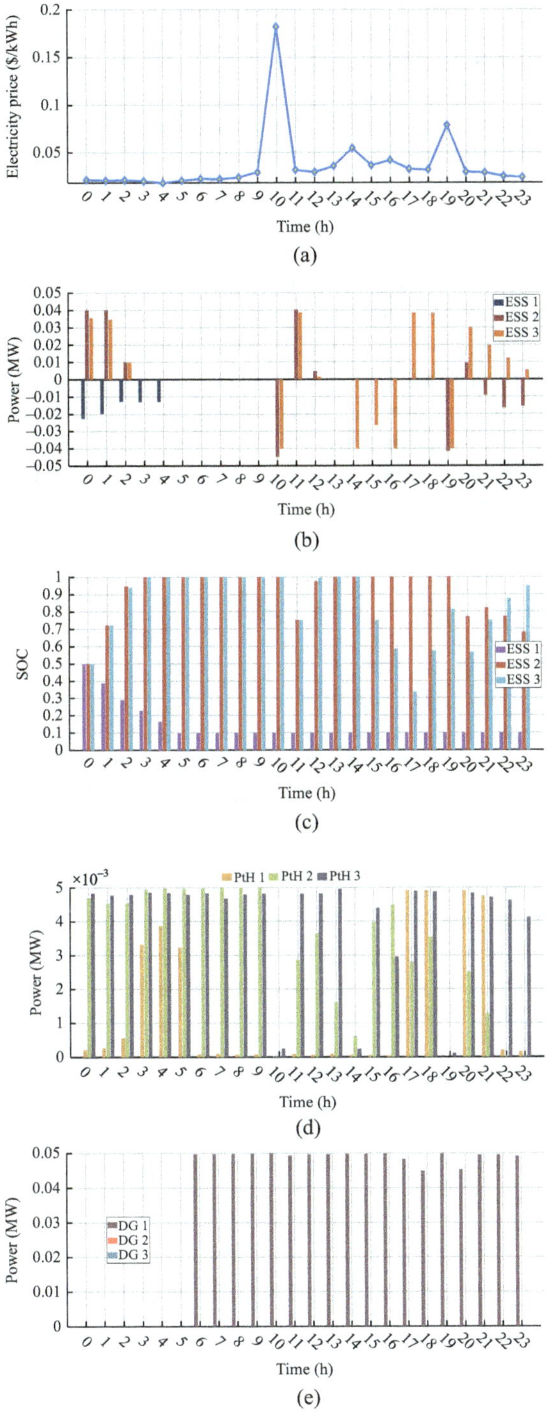

Figure 7.8 Management results of the proposed approach on a test day. (a) Real-time electricity price. (b) Charging and discharge power of the ESSs in the interconnected MMG system. (c) The energy of ESSs in the interconnected MMG system. (d) Power of the PtH units in the interconnected MMG system. (e) Power of the DG units in the interconnected MMG system [19].

enough energy to the electrical load, thus causing the undervoltage condition at MG2. Therefore, sufficient energy is required for the ESS of MG2 to stabilize the voltage profile. In other words, MG2 needs to pay more attention to voltage stability than MG3 in the charging/discharging management of ESS. As shown in Figure 7.8(b), the period from 21:00 to 23:00 is the time of day when electricity prices are low (as Figure 7.8(a) shows). At this time period, the charging power of MG3 is more reasonable, if only from the perspective of energy arbitrage. However, because MG2 has a lower proportion of RES than MG3, MG2 needs additional energy supply from ESS in this time period from 21:00 to 23:00 to achieve the purpose of stabilizing voltage. The PtH unit is a device that converts electrical energy into thermal energy. In the heat network, the thermal demand is satisfied by the microturbine and PtH units. As the PtH is also the power-consuming component, which is the same as the ESS, its power profile follows a similar rule as that of ESS. Figure 7.8(e) shows the DG working situation. The DG of the MG1 does not work from 0:00 to 5:00 and operates at almost maximum power from 6:00 to 23:00. This is due to the fact that from 0:00 to 5:00, the electrical load is not so huge, and the voltage condition of MG1 is relatively stable. To reduce the DG operation and fuel costs, it is reasonable that DG does not work when facing a more stable voltage. After 5:00, as people gradually start to get busy, the daily electricity consumption starts to increase upwards and the voltage starts to fluctuate significantly (as Figure 7.7(b) shows), which requires DG to work to a certain extent to smooth out this fluctuation. The DG of the MG1 operates at almost maximum power from 6:00 to 23:00. This is due to the fact that MG1 has the most severe undervoltage conditions at that time period, as Figure 7.7(b) shows. The voltages of MG2 and MG3 are in a more desirable stable state after overall regulation, so to avoid further cost, the DGs of MG2 as well as MG3 are chosen not to operate.

As Figure 7.9 shows, since MG1 has the lowest proportion of RES and the widest fluctuation range in its voltage profile compared to the other two MGs, MG2 and MG3 will transfer their energy to MG1 through the power line to stabilize the voltage profile of MG1.

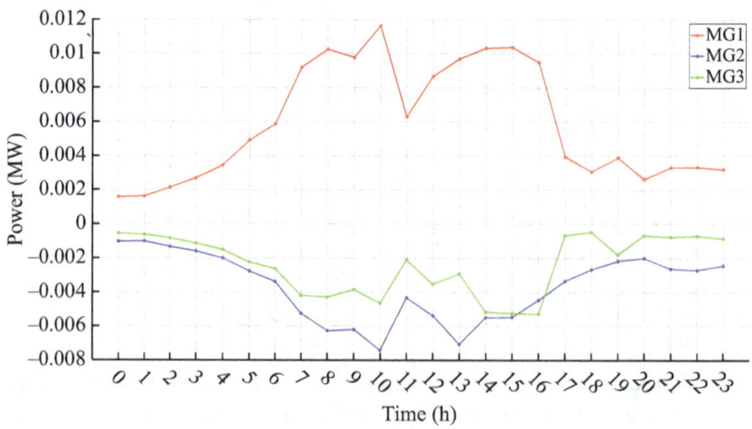

Figure 7.9 Transition power between the MGs [19]

7.3 Surrogate model-enabled DRL-based energy management for multi-microgrid

7.3.1 System description

An illustration of a grid interconnected MMG with the CHP system is shown in Figure 7.10. The number of MGs is denoted as $g \in \{1, 2, 3, \ldots, G\}$. The g-th MG is made up of a heat and electricity network. The heat network is made up of CHP, power to heat (PtH) unit (such as a heat pump), and a thermal load. The electricity network is made up of a diesel generator (DEG), renewable energy (such as solar and wind energy), an energy storage system (ESS), and an electrical load. The individual MGs are integrated into an interconnected MMG system by the power line (black line) and heat pipe (red line), allowing for the exchange of electrical and thermal energy between the different MGs. Typically, there are circuit breakers as ON or OFF switches in the MMG system. This paper assumes circuit breakers are always ON to guarantee energy exchange.

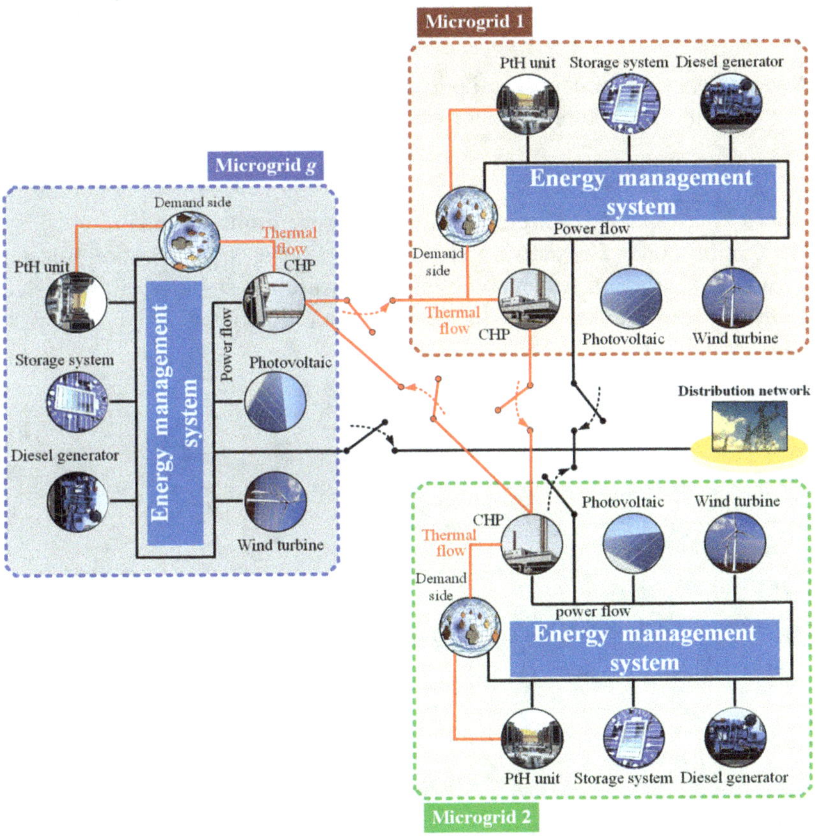

Figure 7.10 Illustration of a grid interconnected MMG with CHP system [29]

7.3.1.1 Electricity network modeling

Power flow calculation results are the basis for analyzing network planning and electricity supply scheme feasibility, safety, reliability, and economy. Power flow analysis plays an imperative role in the operation mode and design of electricity networks.

The power flow constraints are as follows:

$$P_{i,t} = V_{i,t} \sum_{j=1}^{J} V_{j,t}(G_{ij,t} \cos \theta_{ij,t} + B_{ij,t} \sin \theta_{ij,t}) \tag{7.11}$$

$$Q_{i,t} = V_{i,t} \sum_{j=1}^{J} V_{j,t}(G_{ij,t} \sin \theta_{ij,t} - B_{ij,t} \cos \theta_{ij,t}) \tag{7.12}$$

where $P_{i,t}$ is active power injected into node i at time t; $Q_{i,t}$ represents the reactive power injected into node i; J denotes the total number of nodes; $V_{i,t}$ denotes the voltage amplitude of node i; $G_{ij,t}$ and $B_{ij,t}$ are the real and imaginary part of admittance element between node i and j; $\theta_{ij,t}$ is the voltage phase difference between node i and j.

The following restrictions apply when calculating power flow:

- Node voltage amplitude: The node voltage should be in the allowed range to comply with the requirements for electricity quality and electrical supply safety.

$$V_{i\ \min} \leq V_{i,t} \leq V_{i\ \max} \tag{7.13}$$

where $V_{i\ \min}$ and $V_{i\ \max}$ are the lower and upper limits of node i.

- Generation node: The active and reactive power at the generation nodes must be within a safe and reasonable range to ensure the stability of the system. In addition, some nodes with distributed generation must satisfy the working constraints (e.g., DEG, etc.).

$$0 \leq P_{i,t}^{PV} \leq P_{i\ \max}^{PV} \tag{7.14}$$

$$0 \leq P_{i,t}^{WT} \leq P_{i\ \max}^{WT} \tag{7.15}$$

$$0 \leq P_{i,t}^{PtH} \leq P_{i\ \max}^{PtH} \tag{7.16}$$

$$\begin{aligned} P_{i,t}^{DEG} - P_{i,t-\Delta t}^{DEG} \leq R_{up}^{DEG} \Delta t \cdot s_{t-\Delta t}^{DEG} \\ + P_{i\ \min}^{DEG} \left(s_t^{DEG} - s_{t-\Delta t}^{DEG}\right) + P_{i\ \max}^{DEG} \left(1 - s_t^{DEG}\right) \end{aligned} \tag{7.17}$$

$$\begin{aligned} P_{i,t-\Delta t}^{DEG} - P_{i,t}^{DEG} \leq R_{dn}^{DEG} \Delta t \cdot s_t^{DEG} \\ + P_{i\ \min}^{DEG} \left(s_{t-\Delta t}^{DEG} - s_t^{DEG}\right) + P_{i\ \max}^{DEG} \left(1 - s_{t-\Delta t}^{DEG}\right) \end{aligned} \tag{7.18}$$

$$-P_{i\ \max}^{ESS} \leq P_{i,t}^{ESS} \leq P_{i\ \max}^{ESS} \tag{7.19}$$

$$SOC_{i\ \min}^{ESS} \leq SOC_{i,t}^{ESS} \leq SOC_{i\ \max}^{ESS} \tag{7.20}$$

$$Q_{i\ \min}^{PV} \leq Q_{i,t}^{PV} \leq Q_{i\ \max}^{PV} \tag{7.21}$$

$$Q_{i\,\min}^{WT} \leq Q_{i,t}^{WT} \leq Q_{i\,\max}^{WT} \tag{7.22}$$

$$Q_{i\,\min}^{DEG} \leq Q_{i,t}^{DEG} \leq Q_{i\,\max}^{DEG} \tag{7.23}$$

$$Q_{i\,\min}^{ESS} \leq Q_{i,t}^{ESS} \leq Q_{i\,\max}^{ESS} \tag{7.24}$$

where $P_{i,t}^{PV}$, $P_{i,t}^{WT}$, $P_{i,t}^{PtH}$, $P_{i,t}^{DEG}$, and $P_{i,t}^{ESS}$ are the active powers of PV, wind turbine (WT), PtH, DEG, and ESS of node i; R_{up}^{DEG} and R_{dn}^{DEG} denote the ramp-up and ramp-down limits of DEG; s_t^{DEG} denotes the binary variable, which represents the working state of DEG at time t (1 denotes ON, 0 represents OFF); $P_{i\,\max}^{PV}$, $P_{i\,\max}^{WT}$, $P_{i\,\max}^{PtH}$, $P_{i\,\max}^{DEG}$, and $P_{i\,\max}^{ESS}$ are the upper limits of active powers of PV, WT, PtH, DEG, and ESS of node i; $SOC_{i,t}^{ESS}$ represents the state-of-charge (SOC) of ESS of node i; $SOC_{i\,\max}^{ESS}$ is the upper limit of SOC; $Q_{i,t}^{PV}$, $Q_{i,t}^{WT}$, $Q_{i,t}^{DEG}$, and $Q_{i,t}^{ESS}$ are the reactive powers of PV, WT, DEG, and ESS at node i; $Q_{i\,\max}^{PV}$, $Q_{i\,\max}^{WT}$, $Q_{i\,\max}^{DEG}$, and $Q_{i\,\max}^{ESS}$ are the maximum reactive powers of PV, WT, DEG, and ESS at node i; $Q_{i\,\min}^{PV}$, $Q_{i\,\min}^{WT}$, $Q_{i\,\min}^{DEG}$, and $Q_{i\,\min}^{ESS}$ are the minimum values of reactive powers.

7.3.1.2 Heat network modeling

A heat network typically consists of one source node and multiple intermediate/load nodes linked by a primary network and a secondary network. The primary network is known as the generation-side network, while the secondary network is known as the consumption-side network. During the thermal energy transfer process, water is used to transport thermal energy from the supply pipes in the primary network to the thermal exchanger in the secondary network, where the thermal energy is distributed to all load nodes, causing the transmitted water to undergo a drastic temperature shift. Finally, the thermal circulation is completed when the water travels back through the return pipes [30].

7.3.1.3 Hydraulic model

Continuity of flow: As per the definition, continuous flow is when the mass flow into a node is equal to the mass flow out of the node plus the flow consumed at the node [31]. The matrix $\boldsymbol{A}$ depicts the network incidence of the nodes on the branches. The mass flow vector within a pipe and the mass flow vector through a node that is injected from a source or discharged to a load are defined as the vectors $\boldsymbol{m}$ and $\boldsymbol{m_q}$, respectively.

$$\boldsymbol{Am} = \boldsymbol{m_q} \tag{7.25}$$

Loop pressure equation: The pressure difference in meters caused by pipe friction is referred to as head loss. The loop pressure equation requires that the total head losses around a closed loop be equal to zero. The following is the loop pressure equation for the hydraulic network [32]:

$$\boldsymbol{Bh_f} = 0 \tag{7.26}$$

where $\boldsymbol{B}$ is the loop incidence matrix that relates the loops to the branches, and $\boldsymbol{h_f}$ is the vector of the head losses.

Head loss equation: Along each pipe, the relationship between the flow and the head losses is as follows [31]:

$$\boldsymbol{h}_f = \boldsymbol{K}\boldsymbol{m}|\boldsymbol{m}| \tag{7.27}$$

where $\boldsymbol{K}$ is the vector of each pipe's resistance coefficients. $\boldsymbol{K}$ typically has a strong relationship with a pipe's diameter.

7.3.1.4 Thermal model

Nodal thermal flow balance: Thermal flow follows the Kirchhoff law, which states that water flows in and out of each node in equal amounts.

$$\sum\nolimits_{p\in\mathbf{P}^{s/r}_{i,start}} m^{s/r}_{p,t} = \sum\nolimits_{p\in\mathbf{P}^{s/r}_{i,end}} m^{s/r}_{p,t} \tag{7.28}$$

where $\mathbf{P}^{s/r}_{i,start}$, $\mathbf{P}^{s/r}_{i,end}$ denotes the set of pipes starting and ending at node i in supply/return network, respectively; $m^{s/r}_{p,t}$ represents the mass flow rate of supply/return pipe p at time t.

Nodal temperature mixing: The nodal temperature is dependent on water mass and temperature, as stated by the first law of thermodynamics. A node's temperature is the same as the temperature of its matching starting node in the same pipe. Mathematically, they can be expressed as:

$$T^{s/r}_{i,t} \cdot \sum\nolimits_{p\in\mathbf{P}^{s/r}_{n,start}} m^{s/r}_{p,t} = \sum\nolimits_{p\in\mathbf{P}^{s/r}_{n,end}} T^{s/r}_{p,end,t} \cdot m^{s/r}_{p,t} \tag{7.29}$$

$$T^{s/r}_{i,t} = T^{s/r}_{p,start,t} \tag{7.30}$$

where $T^{s/r}_{i,t}$ denotes the temperature of node i of supply/return network at time t; $T^{s/r}_{p,start,t}$ and $T^{s/r}_{p,end,t}$ represent temperature at the start and end of supply/return pipe p at time t.

Internal temperature range: A pipe's water temperature boundaries for inflow and outflow are as follows:

$$T^{s/r,\,\min}_{p,start/end,t} \le T^{s/r}_{p,start/end,t} \le T^{s/r,\,\max}_{p,start/end,t} \tag{7.31}$$

Thermal transmission delay: In contrast to the instantaneous nature of electric energy, there is a noticeable delay in the transmission of heat. The water temperature at the beginning and end of a pipe may not be the same because of the sluggish flow speed of water in the pipeline compared with the electric energy transfer rate. This paper refers to [30] for more specific formulas that can be used to quantify this transmission delay.

Thermal energy balance: There should be a balance between thermal energy production and consumption. The total amount of thermal energy needed in the thermal system is equal to the energy difference between the start of the supply pipe and the end of the return pipe, taking into account that hot and cold water are used to transfer thermal energy between the heat source node and the load node via a thermal pipe.

$$H_{p,t} = C_{WH} \cdot \left(m^{s}_{p,t} \cdot T^{s/r}_{p,start,t} - m^{r}_{p,t} \cdot T^{s/r}_{p,end,t} \right), p \in \mathbf{N}_{SN} \tag{7.32}$$

where $H_{p,t}$ denotes the thermal demand of pipe p at time t; C_{WH} represents water heat capacity; and $\mathbf{N}_{SN}$ is the set of source nodes.

7.3.1.5 Optimization problem

Considering the above models, real-time energy management of MMG systems can be viewed as an optimization problem involving minimizing the following objective function.

$$\min \left\{ \sum_{t}^{T} \left(C^{EP}_{t} + C^{Voltage}_{t} + \sum_{g}^{G} \left(C^{g,DEG}_{t} + C^{g,Gas}_{t} \right) \right) \right\} \tag{7.33}$$

where T denotes the length of a period; G denotes the number of MGs; C^{EP}_{t} is the cost of buying electricity from the distribution network (DN); $C^{Voltage}_{t}$ is the cost of unstable voltage $C^{g,DEG}_{t}$ represents the cost of DEG; $C^{g,Gas}_{t}$ denotes the cost of buying gas.

$$C^{EP}_{t} = P^{DN}_{t} \cdot EP_{t} \cdot \Delta t \tag{7.34}$$

$$C^{g,Voltage}_{i,t} = \begin{cases} 1 - 100 \cdot \left|1 - v^{g}_{i,t}\right|, v^{g}_{i,t} \in [0.95, 1.05] \\ -100 \cdot \left|1 - v^{g}_{i,t}\right|, v^{g}_{i,t} \in [0.8, 0.95] \cup v^{g}_{i,t} \in [1.05, 1.25] \\ -1000, \qquad \text{Otherwise.} \end{cases} \tag{7.35}$$

$$C^{Voltage}_{t} = \sum_{i=1}^{I} C^{g,Voltage}_{i,t} \tag{7.36}$$

$$C^{g,DEG}_{t} = C^{g,Fuel}_{t} + C^{g,Operation}_{t} \tag{7.37}$$

$$C^{g,Fuel}_{t} = \sum_{i=1}^{I} s^{DEG}_{i,t} \cdot \left(a \cdot \left(P^{g,DEG}_{i,t} \right)^{2} + b \cdot P^{g,DEG}_{i,t} + c \right) \cdot \Delta t \tag{7.38}$$

$$C^{g,Operation}_{t} = \sum_{i=1}^{I} C^{g,start/stop}_{i} \left| s^{g,DEG}_{i,t} - s^{g,DEG}_{i,t-\Delta t} \right| \tag{7.39}$$

$$C^{g,Gas}_{t} = \sum_{i=1}^{I} \frac{P^{g,CHP}_{i,t} \cdot \Delta t}{\eta_{CHP} \cdot L_{cvng}} \cdot GP_{t} \tag{7.40}$$

where P^{DN}_{t} symbolizes the power energy exchange between MMG system and DN at time t; $v^{g}_{i,t}$ represents the node voltage; the MMG system's overall node count is indicated by I; the fuel cost for the g-th MG's DEG is $C^{g,Fuel}_{t}$, while its operating cost is $C^{g,Operation}_{t}$; a, b, and c are coefficients of the DEG fuel cost function; $C^{g,start/stop}_{i,t}$ is the startup or shutdown cost of DEG at node i in g-th MG; $s^{g,DE}_{i,t}$ denotes the working state of DEG; η_{CHP} is the energy generation efficiency of microturbine; L_{cvng} represents the low calorific value of natural gas; and GP_{t} denotes the gas price.

7.3.2 Proposed physics-model-free framework for energy management of MMG system

This paper proposes a physics-model-free control framework for the energy management of MMGs with heat-electricity energy.

Conventional optimization strategies rely on well-defined electrical and thermal network parameters and topological structure. In actuality, they are often unknown to the operators. In such cases, it is hard to model the premise power and thermal flow models to calculate realistic flow results (i.e., P_t^{DN}, $v_{i,t}^g$, and $P_{i,t}^{g,CHP}$). To solve this problem, surrogate models are proposed to obtain realistic calculation results for power and thermal flow. Later, the trained surrogate models are integrated into the environment of the proposed MADRL approach to construct the proposed physics-model-free control framework.

7.3.2.1 Proposed surrogate model

Given the load requirements of each node, the power introduced by renewable energy and CHP and the controllers' decision, the electrical surrogate model attempts to estimate the power exchange between the MMG system and DN P_t^{DN}, and the voltages of each node of the MMG system $v_{i,t}^g$. Accounting for the losses in the thermal flow and given the heat output of the heat pump, the goal of the thermal surrogate model is to figure out how much heat the CHP needs to produce $P_{i,t}^{g,CHP}$ to meet the heating load. The thermal and electrical surrogate models' mapping correlations are learned from historical data. Unlike deep learning technologies, the proposed surrogate model can construct ideal mapping relationships between input data and output power and thermal flow results in relatively few data points, which has a few-shot learning characteristic, considering the historical data of the MMG system used to build the surrogate model is limited. This eliminates the need for precise model parameters of electricity networks and heat networks, which are notoriously hard to come by and are subject to a wide range of practical errors.

Assume that there is a relationship in the training data between the input $X \in R^{N\times Q}$ and the output $Y \in R^{N\times 1}$ is subject to the Gaussian Process (GP):

$$Y = f(X) + \varepsilon \tag{7.41}$$

where $\varepsilon \sim \mathcal{N}(0, \sigma_n^2 I)$ stands for Gaussian white noise, σ_n^2 represents variance; I denotes an identity matrix with appropriate dimensions, and $f(\cdot)$ represents Gaussian process mapping between input and output:

$$f(x) \sim GP(m(x), K(x, x')) \tag{7.42}$$

where $x, x' \in X$, the mean and covariance matrices of $f(x)$ are represented, respectively, by $m(x)$ and $K(x, x')$. Most commonly, $m(x)$ is assumed to be equal to zero if there are few priors, and elements in the covariance matrix $k(\cdot)$, which is also called the kernel function, can be expressed as follows:

$$k(x, x') = \sigma_k^2 e^{-0.5(x-x')^2} \tag{7.43}$$

where σ_k denotes covariance.

After knowing m and k, the prior distribution of y can be expressed as:

$$y \sim \mathcal{N}(0, K(X,X) + \sigma_n^2 I) \tag{7.44}$$

When new input x^* in test data is given, the joint prior distribution of y and f_* is:

$$\begin{bmatrix} y \\ f_* \end{bmatrix} \sim \mathcal{N}\left(0, \begin{bmatrix} K(X,X) + \sigma_n^2 I & K(X,x^*) \\ K(x^*,X) & K(x^*,x^*) \end{bmatrix}\right) \tag{7.45}$$

Thus, the posterior distribution of the predicted value f_* can be obtained by [32]:

$$f_*|X,y,x^* \sim \mathcal{N}\left(\mu^*, \sum\nolimits^*\right) \tag{7.46}$$

where $\mu*$ and $\sum *$ represent the mean and variance of test data, respectively, and they can be calculated as [32]:

$$\mu^* = K(x^*,X)\left(K(X,X) + \sigma_n^2 I\right)^{-1} y \tag{7.47}$$

$$\begin{aligned} \sum\nolimits^* = K(x^*,x^*)- \\ K(x^*,X)\left(K(X,X) + \sigma_n^2 I\right)^{-1} K(X,x^*) \end{aligned} \tag{7.48}$$

When the parameters $\theta = \{\sigma_n, \sigma_k\}$ are known, the mean and variance of test data can be calculated. The parameters θ can be updated by maximizing the negative log-likelihood function $-\log p\ (Y|X\ )$ [32].

The calculation complexity of the above-mentioned step is $O(N^3)$ where the N is the input $X \in R^{N\times Q}$ or output $Y \in R^{N\times 1}$ data's data size. When dealing with enormous amounts of data, GP may face a significant computational strain. To reduce computational complexity, a variational inference approach known as sparse variational Gaussian processes (SVGP) is used, in which M ($M \ll N$) auxiliary input $Z \in R^{M\times Q}$ and corresponding outputs $u \in R^{M\times 1}$ are introduced. After the above improvement, the computational complexity is reduced to $O(NM^2)$. The joint density of y, f, and u is:

$$p(y,f,u) = p(f,u;X,Z)p(y|f) \tag{7.49}$$

where $p(f,u;X,Z) = p(u|Z)p(f|u;X,Z)$ and $p(u|z) = N(u|0,K(Z,Z))$. If $p(f|u)$ is known, the joint density of the above can be obtained.

The relationship between u and f is given as follows since the u and f are generated by the same GP:

$$p(f|u;X,Z) = \mathcal{N}(f|\alpha(f),\beta(f)) \tag{7.50}$$

where the $\alpha(f)$ and $\beta(f)$ represent the mean and variance of f, respectively. They can be expressed as:

$$\alpha(f) = K(X,Z)k(Z,Z)^{-1}u \tag{7.51}$$

$$\beta(f) = K(X,X) - K(X,Z)K(Z,Z)^{-1}K(Z,X) \tag{7.52}$$

To obtain the true posterior $p(f, u)$ and lessen the computing overhead, variational inference is implemented. Its goal is to minimize the Kullback-Leibler divergence between the true posterior $p(f, u)$ and the variational posterior q to get an approximation posterior $q(f, u) = p(f|u)q(u)$. The minimization Kullback-Leibler divergence is equivalent to maximizing the evidence lower bound (ELBO) [33]:

$$\mathbb{L} = \log p(y) \geq \mathrm{E}_{q(f,u)} \log \frac{p(y, f, u)}{q(u, f)} \tag{7.53}$$

Further, the ELBO can be simplified as [33]:

$$\mathbb{L} \geq \mathrm{E}_{q(f)} \log p(y|f) - KL(q(u)||p(u)) \tag{7.54}$$

Therein, the marginal likelihood $q(f)$ is:

$$q(f) = \int p(f|u)q(u)\mathrm{d}u = \mathcal{N}\left(f|\mu^l, \sum\nolimits^l\right) \tag{7.55}$$

where $q(u) = \mathcal{N}\left(\mu^l, \sum^l\right)$. The μ^l and $\sum^l$ is expressed as:

$$\mu^l = K(X, Z)K(Z, Z)^{-1}u \tag{7.56}$$

$$\begin{aligned} \sum\nolimits^l &= K(X, X) + \\ &K(X, Z)K(Z, Z)^{-1}(\sum - K(Z, Z))K(Z, Z)^{-1}K(Z, X) \end{aligned} \tag{7.57}$$

These allow for the knowledge of every component of the simplified ELBO presented in (7.54).

Summarized, the parameters of SVGP that need to be optimized are $\Theta = \left\{\sigma_n, \sigma_k, \mu^l, \sum^l, Z\right\}$. They are optimized by maximizing (7.54).

After the optimized parameters are obtained by maximizing the ELBO, input a new test point x^* in the test stage, and the posterior distribution of the predicted value can be calculated:

$$q(f^*) = \int p(f^*|u)q(u)\mathrm{d}u \tag{7.58}$$

The heat demand supplied by CHP, the voltages of each node in the MMG system, and the exchange power between the MMG system and DN can be obtained based on the distribution calculated by (7.58).

7.3.2.2 Decision-making approach based on MADRL

The thermal and electrical surrogate models eliminate the need for precise model parameters for the electricity network and heat network. This means that the calculation of objective functions can get rid of the model-based power and thermal flow calculation processes. Later, the thermal and electrical surrogate models integrate into the environment of the proposed MADRL approach, consisting of the whole proposed physics-model-free control framework.

The optimization problem is reformulated as a Markov game, which is solved by the proposed MADRL approach. There are several parts to model in the Markov game.

7.3.2.3 Problem reformulation

▪ *State:* State represents the current operating characteristics of the individual MG system and is collected by each MG controller and used to make energy management decisions. The state of the g-th MG at time t is defined as:

$$s_t^g = (t, \widetilde{\mathcal{P}}_t, SOC_t^g, \widetilde{TL_t^g}, \widetilde{EL_t^g}, \\ P_{i,t-1}^{g,DEG}, \widetilde{P_{i,t}^{g,WT}}, \widetilde{P_{i,t}^{g,PV}}, NV_t^g) \tag{7.59}$$

where t denotes the time information; $\widetilde{\mathcal{P}}_t$ is the forecasted electricity price of current time t; $SOC_{i,t}^g$ is the SOC of energy storage system within g-th MG at node i; $\widetilde{TL_t^g}$ is the forecasted cumulative thermal load of each node $\widetilde{P_{i,t}^{g,TL}}$; $\widetilde{EL_t^g}$ is the forecasted cumulative electrical load of each node $\widetilde{P_{i,t}^{g,EL}}$; $P_{i,t-1}^{g,DEG}$ is the working power of the DEG at time t-1; $\widetilde{P_{i,t}^{g,WT}}$ and $\widetilde{P_{i,t}^{g,PV}}$ is the power of forecasted WT and PV, respectively, and NV_t^g is the voltage of each node.

▪ *Action:* The action of g-th MG is defined as

$$a_t^g = (P_{i,t}^{g,ESS}, P_{i,t}^{g,DEG}, P_{i,t}^{g,PtH}, Q_{i,t}^{g,PV}, Q_{i,t}^{g,WT}, Q_{i,t}^{g,ESS}, Q_{i,t}^{g,DEG}) \tag{7.60}$$

where $P_{i,t}^{g,ESS}$, $P_{i,t}^{g,DEG}$, and $P_{i,t}^{g,PtH}$ denote the active power of ESS, DEG, and PtH unit, respectively; $Q_{i,t}^{g,PV}$, $Q_{i,t}^{g,ESS}$, $Q_{i,t}^{g,DEG}$, and $Q_{i,t}^{g,WT}$ is the reactive power of PV, ESS, DEG, and WT, respectively.

▪ *Reward:* The reward function can be defined as:

$$r_t^g = -\left(\frac{\left(C_t^{EP} + C_t^{Voltage}\right)}{G} + C_t^{g,DEG} + C_t^{g,Gas}\right) \tag{7.61}$$

where $\sum_t^T \sum_g^G r_t^g$ denotes the objective function of this paper.

▪ *Transition function:* The transition function $s_{t+1}^g = \mathcal{T}(s_t^g, a_t^g, \mathfrak{S})$ maps the current state s_t^g to the next state s_{t+1}^g based on the action a_t^g and the environment's unpredictability $\mathfrak{S}$. The deterministic elements of the s_t^g are the ones most affected by the action, as shown by the equation $SOC_{t+1}^g = SOC_t^g + a_t^g \Delta t / E_{\max}^g$, where $E_{\max}^g$ is the ESS capacity of the g-th MG. The $\widetilde{\mathcal{P}}_t$, $\widetilde{TL_t^g}$, $\widetilde{EL_t^g}$, $\widetilde{P_{i,t}^{g,WT}}$, and $\widetilde{P_{i,t}^{g,PV}}$ in s_t^g depends on $\mathfrak{S}$ for their state transition, and the NV_t^g is influenced by both a_t^g and $\mathfrak{S}$.

7.3.2.4 Proposed MADRL approach

The proposed MADRL approach consists of two neural network-based components: an actor part that responds to make decisions and a critic part that responds to help the actor part approximate the optimal control strategy.

Actor network: The state design of this paper contains various TSDs (e.g., $\widetilde{\mathcal{P}}_t$, TL_t^g, EL_t^g, etc.). These TSDs are unaffected by action during the state transition process, but their trend with certain laws can influence decision-making. Therefore, the strategy's performance will be significantly impacted by TSD due to the lack of knowledge about their trend. If the historical trend characteristic of TSD can be captured, it can mitigate the problem of TSD data having a negative impact on decision-making [34]. From this point of view, the historical trend representation of current time TSD is critical for the agent to make a decision. In addition, the actor part of the proposed approach only utilizes local information to make decisions. Agents can benefit from conditioning based on the entire history of their actions and states in such cases [35]. The actor network of each agent uses the state of the current time and the action from the previous time as input into a recurrent neural network (RNN)-like network (e.g., gated recurrent unit (GRU)) to get a representation of trajectory history information τ^g for g-th agent. Mathematically, the τ_t^g can be calculated as follows:

$$\tau_t^g = GRU\left(s_t^g, a_{t-1}^g\right) \tag{7.62}$$

where GRU denotes the GRU network, the technical details about it can be referred to [36].

Any increase in the number of agents in multi-agent systems will result in an exponential expansion of the action space. An environment with a large-scale action space requires DRL to deploy exploration mechanisms throughout its optimization process in order to avoid premature convergence. Action space noise (ASN) technology is often used in conventional DRL to avoid the local optimum. For instance, π denotes the actor network, the action can be mathematically described as: $a_t^g = \pi(\tau_t^g) + \vartheta$ when taking into account the continuous action space impacted by the Gaussian noise ϑ . Instead of introducing noise into the action space, as is done in ASN, parameter space noise (PSN) does it into the actor network's parameter space, which can be expressed as: $a_t^g = \widehat{\pi}(\tau_t^g)$ where the π is affected by the PSN to get $\widehat{\pi}$. The actor network with ASN will be independent of the input τ_t^g since, despite having a fixed input τ_t^g, the output a_t^g is different because of the ASN. Even though training can create a mapping relationship between input τ_t^g and output a_t^g, such a procedure is likely to weaken the connection between them in situations where ASN has a significant impact on actions, leading to subpar performance in complicated contexts [37]. The PSN is an effort to meet this demand. At the beginning of each episode, the PSN keeps itself fixed until the arrival of the next episode. Following a pattern like this ensures consistency in actions and introduces a direct dependency between the τ_t^g and the a_t^g [38]. A perturbed actor network (PAN) and an adaptive actor network (AAN) are used to implementing PSN in actor networks. Importantly, PAN and AAN do not update parameters through gradients. The parameters of PAN and AAN networks are only updated once per episode in accordance with the actor network parameters and the sampled noise and then remain fixed until the next episode rolls around. A linear layer of a neural network is used to characterize the PSN in detail, as shown below:

$$y = \mathrm{W}x + \mathrm{B} \tag{7.63}$$

where $y \in \mathbb{R}^q$, $x \in \mathbb{R}^p$, $\mathrm{W} \in \mathbb{R}^{q \times p}$ denotes the parameters, and $\mathrm{B} \in \mathbb{R}^q$ represents the bias. The corresponding layer with PSN can be defined as:

$$y = (\mathrm{W} + \vartheta^{\omega})x + \left(\mathrm{B} + \vartheta^{b}\right) \tag{7.64}$$

where the ϑ denotes the Gaussian noise variables, $\vartheta^{\omega} \in \mathbb{R}^{q \times p}$ and $\vartheta^{b} \in \mathbb{R}^q$.

In light of the fact that PSN's impact on outcomes is largely determined by the network structure and that parameters' sensitivity to PSN changes throughout training, it is not recommended that PSN be sampled from a fixed noise distribution in actor networks [38]. In an alternative metric, the parameter space noise can be adaptively increased or decreased based on whether it falls below or above a threshold depending on the difference in output between the PAN and the actor network. This type of metric can be defined as follows:

$$\sigma_{t+1} = \begin{cases} \lambda\sigma_t, \mathcal{D}(\pi, \widehat{\pi}) \leq \xi \\ \sigma_t/\lambda, \mathcal{D}(\pi, \widehat{\pi}) > \xi \end{cases} \tag{7.65}$$

$$\mathcal{D}(\pi, \widehat{\pi}) = \sqrt{N^{-1} \cdot \sum_{i=1}^{N} \mathbb{E}_s\left[\left(\pi(\tau)_i - \widehat{\pi}(\tau)_i\right)^2\right]} \tag{7.66}$$

where σ is the standard deviation of Gaussian noise; the λ is the scaling factor; the ξ is the threshold value and N represents the dimensions of action.

The actor network is utilized to establish the mapping relationship from the input state to the output decision-making, and this process is crucial to the model's effectiveness. To map features from the input to the desired output, neural networks with multiple layers are important because deep neural networks can improve the nonlinear mapping capability of actor networks to extract features from the input and map them to the desired output. As the network becomes deeper, the original information from the input data may gradually disappear, so simply increasing the depth of the network cannot be effective [22]. To maximize information retention and take greater advantage of the deeper neural network's non-linear mapping capabilities, D2RL handles the challenges deftly by linking the original input to each hidden layer. Mathematically, the mapping relationship from the g-th agent's input to the output using the D2RL neural network architecture can be defined as:

$$\begin{aligned} I_1 &= \tau^g, \\ I_{u+1} &= \mathrm{ReLu}\left([I_u, \tau^g] \cdot \mathrm{W}_u^g + \mathrm{B}_u^g\right), u = 1, \ldots, U-1 \\ \zeta &= \mathrm{ReLu}\left(I_U \cdot \mathrm{W}_U^g + \mathrm{B}_U^g\right), \\ a^g &= \tanh\left(\zeta \cdot \mathrm{W}_O^g + \mathrm{B}_O^g\right), \end{aligned} \tag{7.67}$$

where W_u^g and B_u^g are the weight and bias matrices of the u-th hidden layer of the g-th agent, respectively; $[I_u, \tau^g]$ represents the concatenation of I_u and τ^g, which is the critical operation of D2RL; U denotes the total number of hidden layers; ζ

denotes the latent feature extracted by all the U hidden layers; and W_O^g and B_O^g are the weight and bias matrices of the output layer, respectively.

Summarized, the advantages of combining PSN and D2RL technologies are: The parameter-level operations of PSN and the complex and efficient parametric connection relationship built by D2RL introduce more exploratory activity than does the combination of shallow dense architecture and PSN. In addition, D2RL has better non-linear representational capability than shallow dense neural network architecture to fast-track the link between input and optimal output decision-making, thereby helping the MADRL approach surmount the instability and poor convergence speed of the training process brought on by the introduction of PSN.

Critic network: By utilizing the attention mechanism, the critic network effectively manages the MMG system's energy by coordinating strategies among actor networks.

The g-th agent action-value function Q_g can be obtained according to the following steps:

$$\mathrm{e}_1 = \mathcal{F}_1(\tau^1, a^1), ..., \mathrm{e}_g = \mathcal{F}_g(\tau^g, a^g), ..., \mathrm{e}_G = \mathcal{F}_G(\tau^G, a^G) \tag{7.68}$$

$$\left[\alpha_{g(1)}, ..., \alpha_{g(g-1)}, \alpha_{g(g+1)}, ..., \alpha_{g(G)}\right] = \frac{f\left(\mathrm{e}_g \mathrm{w}_Q^h + \mathrm{b}_Q^h, [\mathrm{e}_1 \mathrm{w}_K^h + \mathrm{b}_K^h, ..., \mathrm{e}_G \mathrm{w}_K^h + \mathrm{b}_K^h]\right)}{\sqrt{d_k}} \tag{7.69}$$

$$\omega_i = \frac{\exp\left(\alpha_{g(i)}\right)}{\sum_{i \neq g} \exp\left(\alpha_{g(i)}\right)} \tag{7.70}$$

$$\mathrm{O}_h = \sum_{i \neq g} \omega_i \left(\mathrm{e}_i \mathrm{w}_V^h + \mathrm{b}_V^h\right), h \in [1, H] \tag{7.71}$$

$$Q_g = \mathcal{G}_g\left(\mathrm{e}_g, \mathrm{O}_1, ..., \mathrm{O}_h, ..., \mathrm{O}_H\right) \tag{7.72}$$

where e_1, e_g, and e_G represent the encoders of the (τ^1, a^1), (τ^g, a^g), and (τ^G, a^G) pairs, respectively; $\mathcal{F}_g$ denotes the feed-forward neural network of the g-th agent with an input layer, two hidden layers of rectified linear units function neurons, and a linear output layer; $f(\mathcal{U}, [\mathcal{V}, ..., \mathcal{W}]) = [\mathcal{U} \cdot \mathcal{V}, ..., \mathcal{U} \cdot \mathcal{W}]$, d_k represents the network dimensions and attention layer with H heads are used in the proposed approach; in each head, there are six optimizable parameters $\mathrm{w}_Q^h, \mathrm{b}_Q^h, \mathrm{w}_K^h, \mathrm{b}_K^h, \mathrm{w}_V^h, \mathrm{b}_V^h$, $h \in [1, H]$; $\mathcal{G}_g$ denotes the feed-forward neural network similar to $\mathcal{F}_g$.

7.3.2.5 Training of the proposed physical-model-free control framework

To begin, the surrogate model is trained in a supervised manner. The proposed MADRL-based control approach is then combined with surrogate models. The ongoing interaction between the two can generate the desired control strategy. The flowchart of the training process is shown in Figure 7.11. The following also includes a detailed description of training processes.

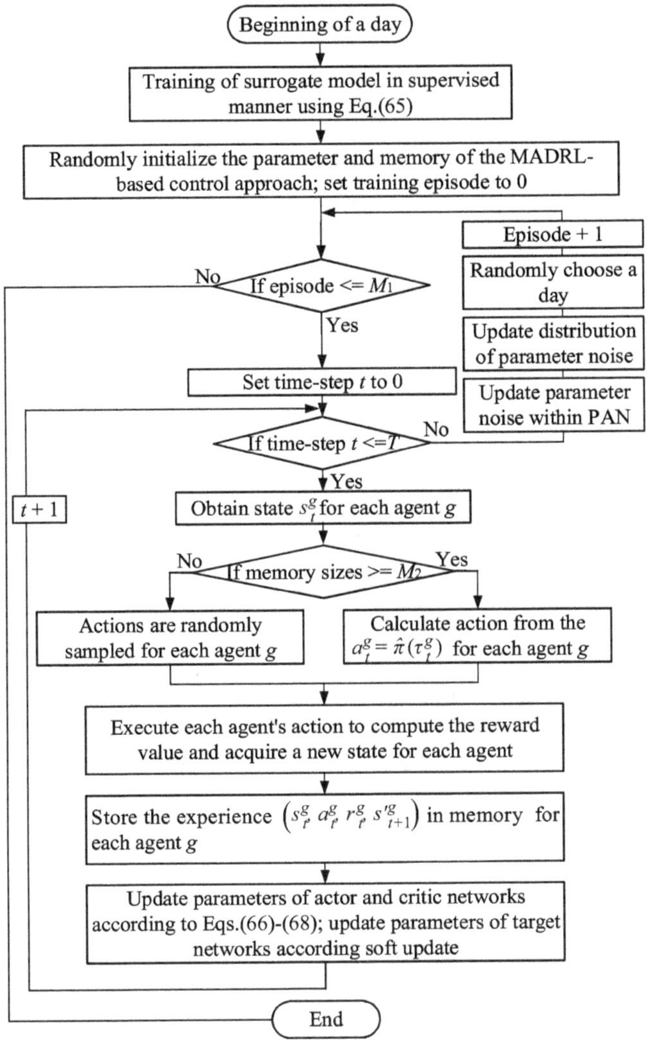

Figure 7.11 Flowchart for the training of proposed physical-model-free control framework [29]

At the beginning of training, surrogate models are trained in a supervised manner by minimizing the following metric:

$$MAE = \frac{1}{N}\sum_{i=1}^{N} \left| f_{true} - f_{predict} \right| \tag{7.73}$$

where f_{true} and $f_{predict}$ denote the true and predicted value. When surrogate models finish their training, the parameters and memory of MADRL-based approach are initialized. After initializing, the training of MADRL begins and ends until the training episode is equal to M_1. Each episode represents a random sample of a day of operation from a training set, in which T time steps make up an operation day. In each time step, the action of each agent is chosen at random to explore the action space. When the memory size exceeds M_2, each agent performs its action in accordance with $a_t^g = \widehat{\pi}(\tau_t^g)$. When all agents execute their actions a_t^g, each one is immediately rewarded r_t^g, and the state of system s_t^g moves on to the next state $s'_{t+1}g$. Following that, each agent stores the experienced pairings $(s_t^g, a_t^g, r_t^g, s'_{t+1}g)$ in their memory. After storing the pairs, the parameters of the actor network can be updated for it to ascend the gradients shown below:

$$\nabla J = \mathbb{E}\left[\nabla \log(\pi^g(a_t^g|\tau_t^g))\begin{pmatrix} Q^g(\Gamma_t, A_t) \\ -\alpha \log \pi^g(a_t^g|\tau_t^g) \end{pmatrix}\right] \tag{7.74}$$

where $\Gamma_t = [\tau_t^1, ..., \tau_t^G]$ and $A_t = [a_t^1, ..., a_t^G]$; the use of temperature parameter α is intended to balance exploration and exploitation in the process of training. The parameters of critic network can be updated to minimize the following equation:

$$\mathcal{L} = \sum\nolimits_{g=1}^{G} \mathbb{E}\left[(Q^g(\Gamma_t, A_t) - \mathcal{Y}^g)^2\right] \tag{7.75}$$

$$\mathcal{Y}^g = r_t^g + \gamma\mathbb{E}\left[Q_{\text{tar}}^g(\Gamma_t, A_t) - \alpha \log\left(\pi_{\text{tar}}^g\left(a_{t+1}^g|\tau_{t+1}^g\right)\right)\right] \tag{7.76}$$

where γ is the reward discount factor; Q_{tar}^g and π_{tar}^g are the target critic and actor networks that are used to stabilize the training, whose parameters are updated according to the soft update mechanism [39]; $-\log\left(\pi_{\text{tar}}^g\left(a_{t+1}^g|\tau_{t+1}^g\right)\right)$ indicates the entropy regularization. Once time step t equals T, it means that an episode is over, and the parameter noise within PAN and the noise distribution are updated for the preparation of a new episode.

7.3.3 *Case studies*

In this section, the detailed parameter settings are first provided, followed by the performance evaluation of the proposed physics-model-free control framework, which consists of a surrogate model and a MADRL-based control model.

7.3.3.1 Case study setup

Three MGs make up the MMG system in the case study. Each MG is represented by a modified IEEE 13-bus network [40] that includes a heat network and an electricity network. Figure 7.12 shows the architecture of MMG system in a case study. In the figure, each MG has ESS, DE, WT, PV, CHP, and PtH units. The used datasets for electrical load, thermal load, electricity price, wind energy, and solar energy, respectively, are [18,41–44]. The 200-day training set (from January 1 to July 19) and the 100-day test set (July 20 to October 27) are created from these datasets. The related parameters of the optimization model are listed in Tables 7.4 and 7.5. The parameters of the proposed control framework are listed in Table 7.6.

7.3.3.2 Evaluation of the proposed surrogate model

Three surrogate models need to be applied in the optimization of MMG system energy management: voltage, exchange power, and CHP-generated energy

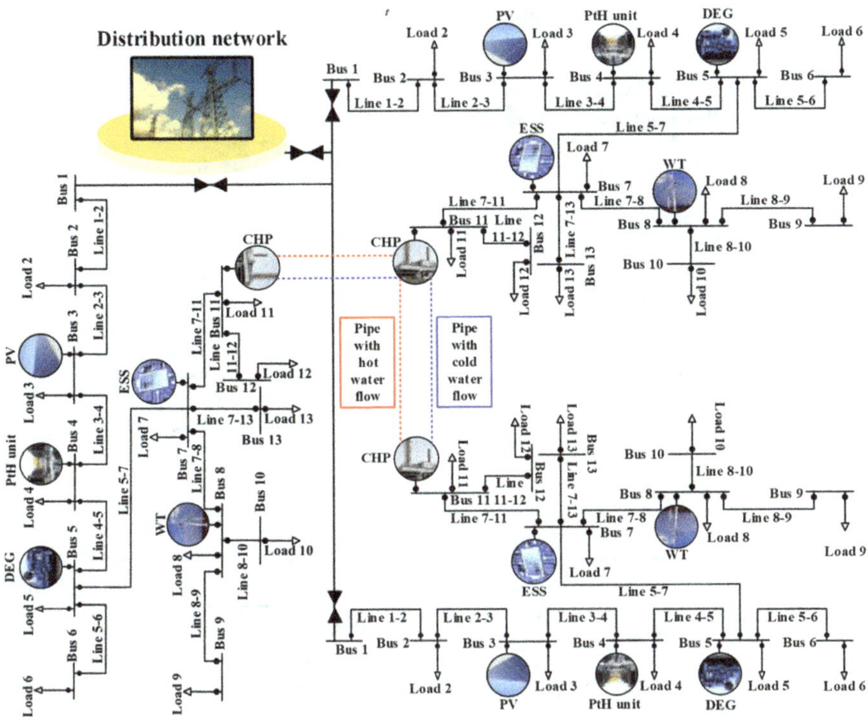

Figure 7.12 The MMG system in a case study [29]

Table 7.4 Model parameters

Symbol	Value	Symbol	Value
a (\$/(kW)^{2}h)	1,040	$C^{start/stop}$ (\$)	5
b (\$/MWh)	30.4	η_{CHP}	0.36
c (\$)	1.3	L_{cvng} (MWh/m^3)	9.7×10^{-3}

Table 7.5 Parameters of the MMG system

	$P_{i\,\max}^{PV}$	$P_{i\,\max}^{WT}$	$P_{i\,\max}^{PtH}$	$P_{i\,\max}^{DEG}$	$P_{i\,\min}^{DEG}$
MG1	0.2 MW	0.25 MW	0.005 MW	0.05 MW	0.02 MW
MG2	0.15 MW	0.2 MW	0.005 MW	0.05 MW	0.02 MW
MG3	0.1 MW	0.15 MW	0.005 MW	0.05 MW	0.02 MW
	$P_{i\,\max}^{ES}$	$E_{i\,\max}$	$SOC_{i\,\max}^{ES}$	$SOC_{i\,\min}^{ES}$	$R_{up/dn}^{DEG}$
MG1	0.05 MW	0.2 MWh	1	0.1	0.025
MG2	0.045MW	0.18 MWh	1	0.1	0.025
MG3	0.04 MW	0.16 MWh	1	0.1	0.025

Table 7.6 Parameters of the proposed control framework

Parameters	Value
Threshold value	0.1
Scaling factor	1.001
Temperature parameter	0.1
Reward discount factor	0.95
Memory capacity	1e6
Learning rate of actor	3e-4
Learning rate of critic	3e-4
Soft replacement	1e-3
Batch sizes for updating	256
Optimizer	Adam

Table 7.7 Mapping relationships of different surrogate models

	Input	Output
Voltage	$TL_t^g, EL_t^g, P_{i,t-1}^{g,DEG}, P_{i,t}^{g,WT}, P_{i,t}^{g,PV}, NV_t^g, P_{i,t}^{g,ESS}, P_{i,t}^{g,DEG}, P_{i,t}^{g,PtH}, Q_{i,t}^{g,PV}, Q_{i,t}^{g,WT}, Q_{i,t}^{g,ESS}, Q_{i,t}^{g,DEG}$	NV_{t+1}^g
Exchange power	$TL_t^g, EL_t^g, P_{i,t-1}^{g,DEG}, P_{i,t}^{g,WT}, P_{i,t}^{g,PV}, NV_t^g, P_{i,t}^{g,TL}, P_{i,t}^{g,EL}, P_{i,t}^{g,ESS}, P_{i,t}^{g,DEG}, P_{i,t}^{g,PtH}, Q_{i,t}^{g,PV}, Q_{i,t}^{g,WT}, Q_{i,t}^{g,ESS}, Q_{i,t}^{g,DEG}$	P_{t+1}^{DN}
CHP	$P_{i,t}^{g,TL}$	$P_{i,t+1}^{g,CHP}$

surrogate models, respectively. The mapping relationship of training of these three surrogate models is summarized in Table 7.7. To assess the effectiveness of the proposed surrogate model, a comparison is conducted between linear regression (LR), deep neural networks (DNN), and the proposed SVGP as surrogate models to evaluate their fit accuracy when there are limited data points available. Therein, DNN contains two hidden layers with neuron counts of 100 and 100, respectively; 32 and 0.001 are chosen as the batch size and learning rate; 2,000 episodes are used to train the DNN. Tables 7.8–7.10 summarize the comparison results for the voltage, exchange power, and CHP-generated energy surrogate models, respectively, using mean absolute error (MAE) as the metric. As Tables 7.8–7.10 show, when the limited instances, i.e., 20, 40, and 80 instances of training data are used to train the surrogate model, the proposed SVGP-based approach can achieve a lower MAE than that obtained by the LR and DNN-based surrogate models. Therein, the LR performance shown in Table 7.8 has the poorest performance (SVGP outperformed it by 98.4% at 20 instances), the DNN performance shown in Table 7.9 has the poorest performance (SVGP outperformed it by 93.6% at 80 instances), and all three approaches can achieve similar performance in the CHP-generated energy surrogate model shown in Table 7.10. In the complex mapping relationship

Table 7.8 MAE of voltage surrogate models with different numbers of training instances

Instances	LR	BPNN	SVGP
20	0.2923	0.0703	0.0048
40	0.1598	0.0181	0.0028
80	0.1390	0.0133	0.0026

Table 7.9 MAE of exchange power surrogate models with different numbers of training instances

Instances	LR	BPNN	SVGP
20	0.0608	0.3503	0.0459
40	0.0536	0.2797	0.0281
80	0.0213	0.1539	0.0098

Table 7.10 MAE of CHP generated energy surrogate models with different numbers of training instances

Instances	LR	BPNN	SVGP
20	$1.570e^{-4}$	$1.755e^{-4}$	$1.310e^{-4}$
40	$1.400e^{-4}$	$1.328e^{-4}$	$1.152e^{-4}$
80	$1.233e^{-4}$	$1.205e^{-4}$	$1.145e^{-4}$

surrogate model, i.e., the voltage and power exchange surrogate model, the LR and DNN cannot do a good job of taking into account different complex mapping relationships at the same time. This means that the LR and DNN-based surrogate models find it difficult to achieve the desired results in the different flow calculations, but the proposed SVGP-based surrogate model can achieve that. The above results demonstrate the effectiveness of the proposed SVGP-based surrogate model.

7.3.3.3 Evaluation of the proposed physics-model-free-based control framework

We evaluate the effectiveness of the proposed control framework against several baselines.

Descriptions of baselines

- TD3 [27]: The MG controllers are modeled as TD3 agents, with each one being trained to maximize its own reward function while disregarding procedures for coordinating with other MGs.

- MATD3 [5]: MATD3 is an optimization technique described in [5] to optimize energy management in an MMG system with integrated energy.
- MASAC [45]: The MASAC approach is described in [45], which aims to improve the economic dispatch and stability of the MMG system by optimizing energy management.
- MAAC [28]: multi-actor-attention-critic (MAAC) is a MADRL approach that integrates the attention layer to enhance agents' coordination.
- Proposed MADRL approach: As opposed to the proposed physics-model-free control framework that utilizes surrogate models to obtain power and thermal flow results, this approach utilizes mathematical power and thermal flow models to obtain calculation results. Setting this baseline will allow us to evaluate the effectiveness of the surrogate model proposed.
- Optimization of deterministic information scenario with NLopt [46]: NLopt's optimization setting is based on perfect future information. The perfect future information means that the future electricity price, electrical load, etc. can be known beforehand, i.e., NLopt yields optimized results based on deterministic and accurate information.

Performances on the test set

The convergence processes of the proposed and other MADRL approaches are shown in Figure 7.13, in which we evaluate the performance on the test set one time every 10 episodes of parameter updates, i.e., there are 30,000 episodes of training and thus 3,000 evaluations on the test set. In comparison to other approaches, the reward curve for TD3 shows the lowest value and oscillates greatly during training, demonstrating that such a lack of coordination is not appropriate for energy management in the MMG system. The MATD3 is an improvement over the TD3 in that it uses the multi-agent framework to build the coordination relationship between each MG for energy management in an MMG system. There is little doubt that MATD3's coordinated mechanism gives it a performance and stability edge over

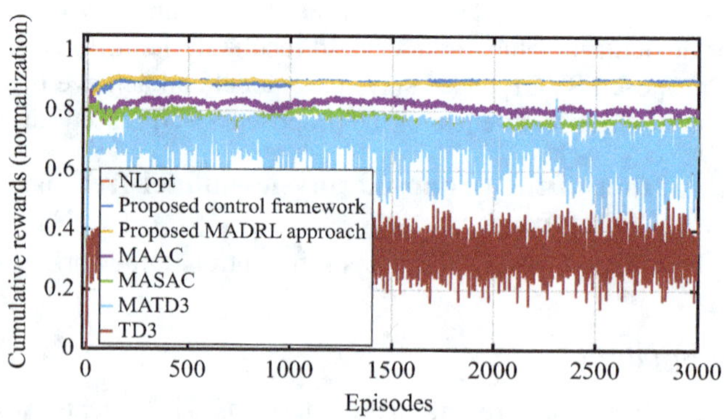

Figure 7.13 The convergence process of proposed control framework and baselines on the test set [29]

TD3. The MADRL with the entropy-based approach, MASAC, can obtain better performance than the MATD3 approach. To improve upon the MASAC's already impressive performance, the authors of [28] presented the MAAC, which uses the attention layer in the critic network to significantly improve coordination between MGs. Setting the attention network in MAAC is meant to speed up the building of consensus among the agents and improve their coordination by including some of the network parameters shared across the agents in each agent's own neural network parameters. Furthermore, the proposed MADRL approach first uses GRU to generate historical information representation in order to overcome the negative impact of the TSD on the decision-making process. Additionally, due to the exponential increase in action space in multi-agent environments, premature convergence to locally optimal solutions is much more likely than in single-agent environments. The proposed MADRL approach introduces PSN technology into the neural network to overcome the negative influence brought by the local optimal solutions. Then, the proposed MADRL approach leverages D2RL to get rid of the noise-related instability and inefficiency of PSN exploration while also providing richer parameter connection relationships to construct a more powerful exploration mechanism overall. The proposed MADRL approach achieves better performance and stability than MAAC through an efficient exploration mechanism incorporating historical information representation. The proposed MADRL approach relies on the accurate physical information of the MMG system and uses a mathematical power and thermal flow model to calculate power and thermal flow results, whereas the proposed control framework uses surrogate models to acquire results for power and thermal flow. This eliminates the need for precise model parameters for electricity networks and heat networks, which are notoriously hard to come by and are subject to a wide range of practical errors. Although these two ways of calculating power and thermal flow are different, the proposed physics-model-free control framework has a similar performance to the proposed MADRL approach. As a result, the surrogate model can provide a precise reward signal for the training of DRL agents, thus demonstrating the effectiveness of the proposed control framework. For the energy management of MMG to be stated as a deterministic optimization problem and addressed by NLopt, we assume that the unknown variables are known in advance. The research objectives of this paper include two parts: one is economic cost, and the other is voltage stability. NLopt is a traditional model-based centralized optimization algorithm based on perfect future information that outperforms the proposed control framework. NLopt spends 8.95% less on the economic cost compared to the proposed control framework, 38.27% less on the voltage cost compared to the proposed control framework, and 13.80% less on the total cost compared to the proposed control framework. However, because there is unpredictability, it cannot be used in a realistic context. The proposed control framework not only gets rid of the premise that one has accurate knowledge of the parameters and topology of the electricity and heat networks but also has the closest performance to that of NLopt compared with other baselines. The results further demonstrate the effectiveness of the proposed control framework.

Optimizing stability

One phenomenon is that, despite having the same number of updates each time, a DRL-based approach's final performance will not be the same in these times. In other words, the excellent performance of a DRL-based approach may be accidental. Due to the nature of DRL optimization, there is sometimes a situation where it is difficult to escape after falling into a local optimal trap. In the process of the agent falling into the trap for a long time, gradient information that may have a persistent negative impact on the update of parameters will accumulate. The negative impacts brought about by gradient information in parameter updates are difficult to erase, even when new gradient information that may have a positive impact on parameter updates is introduced. Similar deceptive traps are very common in complex problem optimization, which is why the reward curve changes from stable convergence to difficult convergence when DRL with poor optimization ability is applied in environments with increasing optimization difficulty. The research in this paper involves internal multi-energy allocations within individual MGs, external multilateral multi-energy interactions among interconnected MMGs, and energy trading between MMGs and the distribution network. In this context, any decision will have multiple impacts on the MMG system, which increases the difficulty of the agent's seeking optimization in the decision space. Thus, the above-mentioned phenomenon is more serious in the multi-agent environments considered in this paper. To verify the performance stability of the proposed control framework, box plots of the proposed control framework and baselines are provided in Figure 7.14. As the figure shows, the TD3 has the worst stability due to its lack of coordination. With the introduction of a coordinated mechanism, the MATD3 is capable of achieving better and more stable performance than the TD3. The experiment results show that the multi-agent control problem in the coordinated environment needs to introduce the mechanism of coordinated agent interaction into the control algorithm. The advantage of this is that it can avoid the negative impact on the parameter update of the control algorithm caused by falling into the local optimal trap frequently and for a long time. Further, the MASAC approach can have better performance than the MATD3 approach due to the introduction of entropy. Through the comparison of the performance of MASAC and MAAC, it is known that the use of the attention mechanism can enhance the coordinated relationships between agents, thus reducing the accumulation of negative gradient information caused by falling into the local optimal trap, and further improving the performance and stability of the control algorithm. Furthermore, the proposed MADRL approach and physics-model-free control framework not only achieve the lowest cost but also have the most stable performances among various approaches, which demonstrates that (1) the desired results achieved by the proposed MADRL approach are not accidental. The efficiently designed exploration mechanism incorporating historical information representation can ensure the MADRL approach gets rid of the deceptive trap as much as possible. Thus, the stability of the desired performance achieved by the proposed MADRL approach can be guaranteed; (2) in the training of DRL agents, the surrogate model can provide a precise reward signal for the agents, thus resulting in the proposed control framework having similar

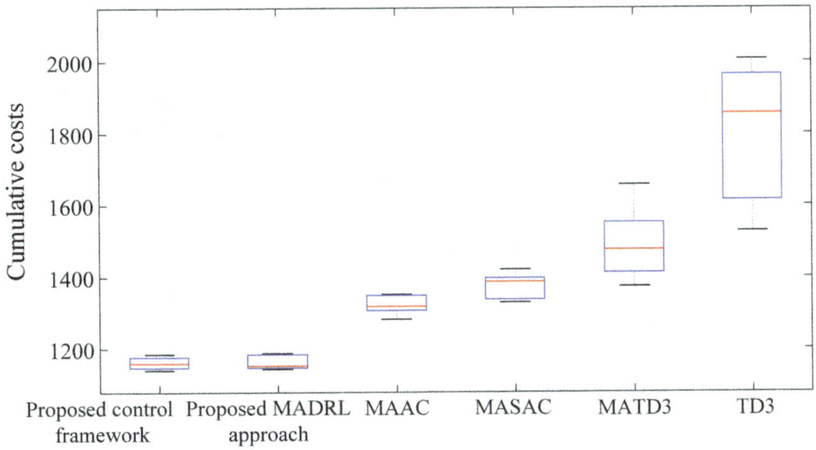

Figure 7.14 Box plots of the performance of various approaches under the 100-day test set [29]

performance and stability to the proposed MADRL approach that relies on physics models. It is evident from this that the proposed control framework is effective.

Solution quality

Figure 7.15(a) shows the real-time electricity price for 1 day. Figure 7.15(b) shows the voltage profile of the MMG system under control, in which the red tone, green tone, and blue tone lines denote the voltages of MG1, 2, and 3, respectively. The gradient color in the same tone represents the voltage curve of different nodes in the same MG. Figure 7.15(c) shows the charging (positive value)/discharging (negative value) powers of ESSs of different MGs. The ESS can leverage the temporal differences in electricity prices to obtain energy arbitrage and utilize charge/discharge behavior to maintain voltage stability. (1) When the electricity prices are low, ESS2 and 3 execute charging behavior for the energy arbitrage objective at the intervals of 4:00, 5:00. (2) When the voltage is high, ESS1 utilizes charge behavior to store the rich electric energy in the grid to reduce the high voltage at the intervals of 7:00–10:00, in which charging power gradually decreases with the rise in electricity prices, considering the energy arbitrage target. (3) When the electricity prices are low and the voltages are high, the ESS1-3 have charging behavior at intervals of 21:00–23:00 in order to suppress voltage climbing while also purchasing electrical energy storage at a low cost, in which the charging behavior of ESS considers the economic cost at the same time, and as such, as the electricity price decreases (21:00–23:00), the charging power will gradually increase (21:00–23:00). (4) When the electricity prices are high and the voltages are low, ESS1 discharges energy at the intervals of 11:00–19:00, and ESS2 and 3 discharge energy at the intervals of 9:00–19:00.

Figure 7.15(d) depicts the operational status of the PtH units. This unit transforms electrical energy into heat energy. As the PtH is also a component that

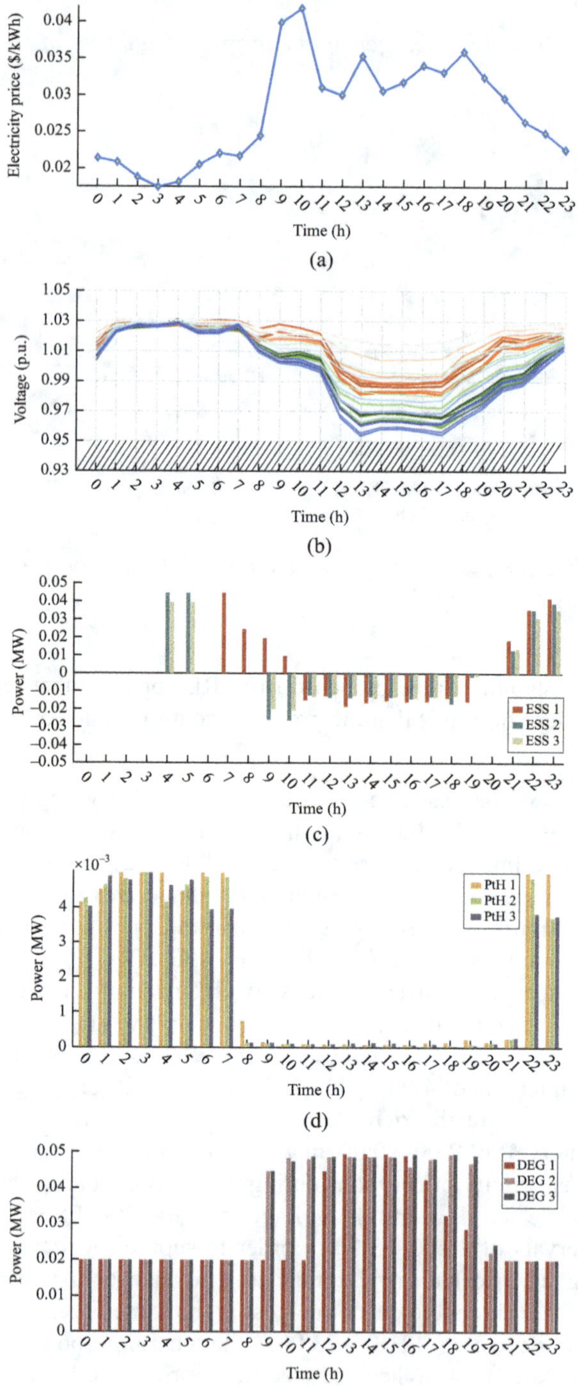

Figure 7.15 Energy management results of the proposed control framework on a test day. (a) Real-time electricity price. (b) The voltage profile under the proposed approach control. (c) Charging and discharge powers of the ESSs. (d) Power of the PtH units. (e) Power of the DEG units [29].

consumes power, much like the ESS, its power profile follows similar rules to that of the ESS. CHP and PtH units supply the thermal demand inside the heat network. When electricity prices are low (i.e., between 0:00 and 7:00, 22:00 and 23:00), thermal energy is mostly supported by the PtC using electrical energy. When electricity prices are high (i.e., between 8:00 and 21:00), thermal energy is predominantly supported by the CHP.

Figure 7.15(e) shows the DEG's working situation. The DEG1, 2, and 3 operate at minimum power at intervals of 0:00–8:00. This is due to the fact that electricity prices are low at these intervals, so purchasing electricity directly from the DN to satisfy demand is more cost-effective. In addition, from 0:00 to 8:00, the voltage of MG1, 2, and 3 is relatively stable. To reduce DEG operation and fuel costs, it is reasonable that DEG operates at the minimum power when facing a more stable voltage. After 8:00, as people gradually start to get busy, the daily electricity consumption starts to increase. This leads to the electricity price starting to rise and the voltage starting to fluctuate significantly, which requires DEG to work to a certain extent to reduce the electricity purchased from the DN for the cost-effective objective and supply the electrical energy for the voltage-stable objective. For example, the DEG2 and 3 operate at almost maximum power from 9:00 to 19:00. This is due to the fact that (1) there are high electricity prices and (2) MG2 and MG3 have the most severe undervoltage conditions during that time period. DEG1's working power is similar to DEG2 and DEG3, but the phenomenon of DEG1's power changing with voltage is more pronounced due to MG1's greater voltage stability than MG2 and MG3 (e.g., during 12:00–20:00, when the voltage decreases, the power of DEG1 gradually increases to achieve the purpose of raising the voltage, and when the voltage rises, the power of DEG1 gradually decreases to slow down the trend of voltage continuing to rise).

Ablation study

To further demonstrate the effectiveness of the proposed control framework, an ablation study is given. Table 7.11 summarizes the different combinations. The proposed physics-model-free control framework, i.e., M1, consisted of BasicMADRL, GRU, PSN, D2RL, and SurrogateModel. The other three combinations, M2, M3, and M4, have one less component each than M1. The convergence processes of the four combinations are shown in Figure 7.16, in which we evaluate the performance on the test set one time every 10 episodes of parameter updates, i.e., there are 30,000 episodes of training and thus 3,000 evaluations on the test set. Obviously, M1, i.e., the proposed control framework, has the best performance compared with others. This is due to the introduction of GRU and the exploratory mechanisms consisting of D2RL and PSN. M2 removes the GRU compared with M1 and obtains worse performance compared with M1. This is due to the TSD having a negative impact on the M2 and influencing its decision-making. The simulation results demonstrate that if the state contains TSD, its trend with certain laws can have a negative influence on the decision-making process, and GRU can mitigate the negative impact. M3 removes the D2RL, resulting in worse performance than M1. Additionally, the reward curve of M3 exhibits the largest fluctuation in the process of convergence compared with other

Table 7.11 Different components for ablation study

	Components
M1	BasicMADRL+GRU+PSN+D2RL+SurrogateModel
M2	BasicMADRL+GRU+PSN+D2RL+SurrogateModel
M3	BasicMADRL+GRU+PSN+D2RL+SurrogateModel
M4	BasicMADRL+GRU+PSN+D2RL+SurrogateModel

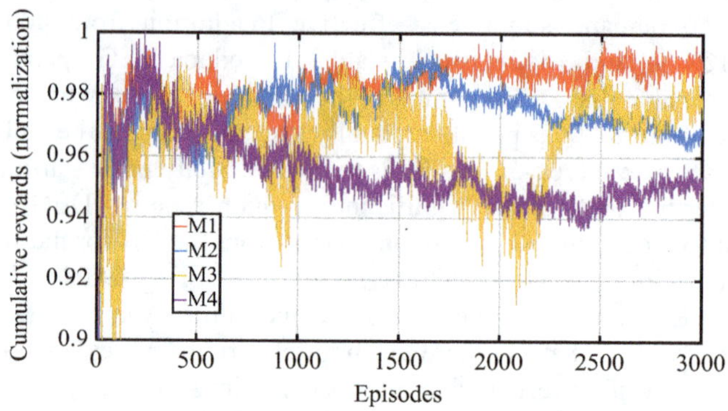

Figure 7.16 The convergence process of M1, M2, M3, and M4 on the test set [29]

approaches. The simulation results demonstrate that, compared with conventional parametric link relationships, D2RL establishes a more complicated and effective parametric link relationship that is more suitable to combine with PSN. This increases the potential for exploratory action while making sure that training is stable and efficient. There is a decrease in performance in M4 due to the removal of the PSN. Based on the simulation results, it is evident that only applying GRU and D2RL cannot achieve the desired results in the multi-agent scenario due to the large action space that needs to be optimized, and as such, the effective exploratory mechanism is critical for the optimization of the multi-agent scenario.

Using the results of the ablation study, the proposed control framework is analyzed to show why it performs well. The experimental results show that by using a representation of historical information to counteract the bad effects of TSD and an effective exploration mechanism to find the desired solution, the proposed control framework solves the given control problem and yields good performance gains.

References

[1] C. Guo, X. Wang, Y. Zheng, *et al.*, "Optimal energy management of multi-microgrids connected to distribution system based on deep reinforcement learning," *International Journal of Electrical Power & Energy Systems*, vol. 131, 2021.

[2] B. Zhao, X. Wang, D. Lin, *et al.*, "Energy management of multiple microgrids based on a system of systems architecture," *IEEE Transactions on Power Systems*, vol. 33, no. 6, pp. 6410–6421, 2018.

[3] N. Liu, J. Wang and L. Wang, "Hybrid energy sharing for multiple microgrids in an integrated heat–electricity energy system," *IEEE Transactions on Sustainable Energy*, vol. 10, no. 3, pp. 1139–1151, 2019.

[4] M. Fathi and H. Bevrani, "Statistical cooperative power dispatching in interconnected microgrids," *IEEE Transactions on Sustainable Energy*, vol. 4, no. 3, pp. 586–593, 2013.

[5] T. Chen, S. Bu, X. Liu, *et al.*, "Peer-to-peer energy trading and energy conversion in interconnected multi-energy microgrids using multi-agent deep reinforcement learning," *IEEE Transactions on Smart Grid*, vol. 13, no. 1, pp. 715–727, 2022.

[6] Y. Guo and C. Zhao, "Islanding-aware robust energy management for microgrids," *IEEE Transactions on Smart Grid*, vol. 9, no. 2, pp. 1301–1309, 2018.

[7] J. Wang, S. You, Y. Zong, *et al*, "Optimal dispatch of combined heat and power plant in the integrated energy system: A state of the art review and case study of Copenhagen," *Energy Procedia*, vol. 158, pp. 2794–2799, 2019.

[8] N. Nikmehr and SN. Ravadanegh, "Reliability evaluation of multi-microgrids considering optimal operation of small scale energy zones under load-generation uncertainties," *International Journal of Electrical Power & Energy Systems*, vol. 78, pp. 80–87, 2016.

[9] N. Nikmehr and S. Najafi-Ravadanegh, "Optimal operation of distributed generations in micro-grids under uncertainties in load and renewable power generation using heuristic algorithm," *IET Renewable Power Generation*, vol. 9, no. 8, pp. 982–990, 2015.

[10] X. Chen, *et al.*, "Increasing the flexibility of combined heat and power for wind power integration in China: Modeling and implications," *IEEE Transactions on Power System*, vol. 30, no. 4, pp. 1848–1857, 2015.

[11] Y. S. Foo. Eddy, H. B. Gooi and S. X. Chen, "Multi-agent system for distributed management of microgrids," *IEEE Transactions on Power Systems*, vol. 30, no. 1, pp. 24–34, 2015,

[12] M. N. Alam, S. Chakrabarti and A. Ghosh, "Networked microgrids: state-of-the-Art and future perspectives," *IEEE Transactions on Industrial Informatics*, vol. 15, no. 3, pp. 1238–1250, 2019.

[13] D. Xu, B. Zhou, K. Chan, *et al.*, "Distributed multienergy coordination of multi-microgrids with biogas-solar-wind renewables," *IEEE Transactions on Industrial Informatics*, vol. 15, no. 6, pp. 3254–3266, 2019.

[14] D. Xu, B. Zhou, N. Liu, *et al.*, "Peer-to-peer multienergy and communication resource trading for interconnected microgrids," *IEEE Transactions on Industrial Informatics*, vol. 17, no. 4, pp. 2522–2533, 2021.

[15] D. Gregoratti and J. Matamoros, "Distributed energy trading: The multiple-microgrid case," *IEEE Transactions on Industrial Electronics*, vol. 62, no. 4, pp. 2551–2559, 2015.

[16] Z. Wang, B. Chen, J. Wang, *et al.*, "Decentralized energy management system for networked microgrids in grid-connected and islanded modes," *IEEE Transactions on Smart Grid*, vol. 7, no. 2, pp. 1097–1105, 2016.

[17] D. Cao, W. Hu, J. Zhao, *et al.*, "Reinforcement learning and Its applications in modern power and energy systems: a review," *Journal of Modern Power Systems and Clean Energy*, vol. 8, no. 6, pp. 1029–1042, 2020.

[18] Q. Zhang, K. Dehghanpour, Z. Wang, *et al.*, "Multi-agent safe policy learning for power management of networked microgrids," *IEEE Transactions on Smart Grid*, vol. 12, no. 2, pp. 1048–1062, 2021.

[19] S. Li, D. Cao, W. Hu, Q. Huang, Z. Chen, and F. Blaabjerg, "Multi-energy management of interconnected multi-microgrid system using multi-agent deep reinforcement learning," *Journal of Modern Power Systems and Clean Energy*, vol. 11, no. 5, pp. 1606–1617, 2023.

[20] R. Lowe, Y. Wu, A. Tamar, *et al.*, "Multi-agent actor-critic for mixed cooperative-competitive environments," in *Proceedings of Advances in Neural Information Processing Systems (NeurIPS)*, California, USA, pp. 6379–6390, 2017.

[21] T. Haarnoja, A. Zhou, P. Abbeel, *et al.*, "Soft actor-critic: off-policy maximum entropy deep reinforcement learning with a stochastic actor," in *Proceedings of International Conference on Machine Learning (ICML)*, Stockholm, Sweden, 2018.

[22] S. Sinha, H. Bharadhwaj, A. Srinivas, *et al.*, "D2RL: Deep dense architectures in reinforcement learning," *arXiv*:2010.09163, 2021.

[23] S. Li, W. Hu, D. Cao, *et al.*, "EV charging strategy considering transformer lifetime via evolutionary curriculum learning-based multi-agent deep reinforcement learning," *IEEE Transactions on Smart Grid*, doi:10.1109/TSG.2022.3167021.

[24] S. Li, W. Hu, D. Cao, *et al.*, "A multi-agent deep reinforcement learning-Based approach for the optimization of transformer life using coordinated electric vehicles," *IEEE Transactions on Industrial Informatics*, doi:10.1109/TII.2021.3139650.

[25] "Historical hourly electricity price from PJM," 2017. [Online]. Available: https://www.pjm.com/.

[26] "Historical gas price from U.S. Energy Information Administration," 2018. [Online]. Available: https://www.eia.gov/.

[27] S. Yao, J. Gu, H. Zhang, *et al.,* "Resilient load restoration in microgrids considering mobile energy storage fleets: A deep reinforcement learning approach," *2020 IEEE Power & Energy Society General Meeting (PESGM)*, pp. 1–5, 2020.

[28] L. Yu, Y. Sun, Z. Xu, *et al.*, "Multi-agent deep reinforcement learning for HVAC control in commercial buildings," *IEEE Transactions Smart Grid*, vol. 12, no. 1, pp. 407–419, 2021.

[29] S. Li, W. Hu, D. Cao, *et al.*, "Physics-model-free heat-electricity energy management of multiple microgrids based on surrogate model-enabled multi-agent deep reinforcement learning," *Applied Energy*, vol. 346, 2023.

[30] Z. Li, L. Wu and Y. Xu, "Risk-averse coordinated operation of a multi-Energy microgrid considering voltage/var control and thermal flow: An adaptive stochastic approach," *IEEE Transactions on Smart Grid*, vol. 12, no. 5, pp. 3914–3927, 2021.

[31] X. Liu, J. Wu, N. Jenkins, *et al.*, "Combined analysis of electricity and heat networks," *Applied Energy*, vol. 162, pp. 1238–1250, 2016.

[32] C. E. Rasmussen, C. K. I. Williams. *Gaussian Processes for Machine Learning*. Cambridge, MA: MIT Press, 2005.

[33] H. Salimbeni and M. Deisenroth, "Doubly stochastic variational inference for deep Gaussian processes," in *Proceedings of Neural Information Processing Systems (NeurIPS)*, pp. 4589–4600, 2017.

[34] Z. Wan, H. Li, H. He, *et al.*, "Model-free real-time EV charging scheduling based on deep reinforcement learning," *IEEE Transactions on Smart Grid*, vol. 10, no. 5, pp. 5246–5257, 2019.

[35] T. Rashid, M. Samvelyan, C. Schroeder, *et al.*, "QMIX: Monotonic value function factorisation for deep multi-agent reinforcement learning," in *Proceedings of the 35th International Conference on Machine Learning*, Stockholm, Sweden, 2018.

[36] J. Chung, C. Gulcehre, K. Cho, *et al.*, "Empirical evaluation of gated recurrent neural networks on sequence modeling," *arXiv preprint arXiv*:1412.3555, 2014.

[37] I. Osband, B. Van Roy, D. J. Russo, *et al.*, "Deep exploration via randomized value functions," *Journal of Machine Learning Research*, vol. 20, no. 124, pp. 1–62, 2019.

[38] M. Plappert, R. Houthooft, P. Dhariwal, *et al.*, "Parameter space noise for exploration," in *Proceedings of International Conference on Learning Representations (ICLR)*, Vancouver, Canada, 2018.

[39] S. Li, W. Hu, D. Cao, *et al.*, "Electric vehicle charging management based on deep reinforcement learning," *Journal of Modern Power Systems and Clean Energy*, vol. 10, no. 3, pp. 719–730, 2022.

[40] K. D. Shinde, and P. B. Mane, "Analysis of radial distribution test feeders in presence of solar photovoltaic systems using PowerFactory," in *Proceedings of IEEE International Conference in Power Engineering Application (ICPEA)*, 2022.

[41] B. Zhang, W. Hu, D. Cao, *et al.*, "Deep reinforcement learning–based approach for optimizing energy conversion in integrated electrical and heating system with renewable energy," *Energy Conversion and Management*, vol. 202, 2019.

[42] S. Li, W. Hu, D. Cao, *et al.*, "Energy management of multiple microgrids considering missing measurements: a novel MADRL approach," *IEEE Transactions on Smart Grid*, vol. 14, no. 5, pp. 4133–4136, 2023.

[43] X. Wu, W. Hu, Q. Huang, *et al.*, "Optimizing the layout of onshore wind farms to minimize noise," *Applied Energy*, vol. 267, 2020.

[44] D. Cao, W. Hu, J. Zhao, *et al.*, "A multi-agent deep reinforcement learning based voltage regulation using coordinated PV inverters," *IEEE Transactions on Power Systems*, vol. 35, no. 5, pp. 4120–4123, 2020.

[45] Y. Xia *et al.*, "A safe policy learning-based method for decentralized and economic frequency control in isolated networked-microgrid systems," *IEEE Transactions on Sustainable Energy*, vol. 13, no. 4, pp. 1982–1993, 2022.

[46] "Library for nonlinear optimization", [Online]. Available: https://nlopt.readthedocs.io/en/latest/.

Chapter 8

Artificial intelligence in renewable energy systems small signal stability control

Weihao Hu[1], *Guozhou Zhang*[2], *Zhenyuan Zhang*[1], *Sayed Abulanwar*[3,4] *and Frede Blaabjerg*[5]

8.1 Introduction

With the increasing development of multiple renewable energy sources (wind, photovoltaic, and hydropower), the nonlinear, strongly time-varying, and high-dimensional characteristics exhibited by the power system are reinforced. This brings several stability phenomena into practice: first, the critical modes within the system are not limited to low-frequency oscillation modes, but different generation structures and different operating conditions will make the system exhibit ultra-low frequency and sub-synchronous oscillation modes. Second, the original stability performance is weakened. Third, the operating conditions of the system have become complex due to the volatility of renewable energy output. This makes it difficult for traditional control methods and strategies to perform well in such complex scenarios. Here, control methods based on deep reinforcement learning can give computers very good performance on various complex tasks. Therefore, this chapter applies it to the adaptive optimization of multiple control devices in power systems to solve different types of stability problems. From the viewpoint of the control object, stability problems are broadly classified into voltage stability, power angle stability, and frequency stability, and from the viewpoint of the dynamic behavior, they can be classified into electromagnetic oscillations and electromechanical oscillations. The research in this chapter focuses on improving the dynamic performance of electromechanical oscillations.

In Section 8.2, a novel sparsity-promoting adaptive control method is proposed for online self-tuning of the power system stabilizer (PSS) parameter settings. The proposed method combines a deep deterministic policy gradient (DDPG) algorithm and sensitivity analysis theory to train an agent to learn the sparse coordinated control

[1]School of Mechanical and Electrical Engineering, University of Electronic Science and Technology of China, China
[2]College of Engineering and Technology, Southwest University, China
[3]Electrical Engineering Department, Faculty of Engineering, Mansoura University, Egypt
[4]Faculty of Engineering, Horus University-Egypt, Egypt
[5]Department of Energy Technology, Aalborg University, Denmark

policy of multi-PSS. After training, a well-trained agent can be employed for online sparse coordinated adaptive control, and the control signal is only applied when it is required and only to the key PSS parameters that have the maximum influence on the system's stability. In Section 8.3, an artificial neural network-based estimator for system identification that can identify the equivalent transfer function of the whole system in real-time is used. Then a DDPG algorithm is adopted to train the agent on learning an adaptive, robust control strategy for the static synchronous compensator additional damper controller (STATCOM-ADC). In Section 8.4, a robust and computationally efficient control method for damping ultra-low frequency oscillations in a hydropower-dominated system is proposed. The proposed method reformulates the control problem into a bi-level, robust optimization model. This allows us to consider a wide range of extreme system operating conditions.

8.2 Deep reinforcement learning enabled sparsity promoting adaptive control method to improve the stability of power systems with wind energy penetration

The integration of wind power also poses a great challenge to the stability of the power system due to the randomness of the wind speed. This may result in unsatisfactory performance of the conventional PSS (CPSS) within the system under actual operating conditions. To solve this problem, this chapter applies a deep reinforcement learning algorithm to finally realize CPSS adaptive control by building an agent that contains the mapping relationship between the wind speed under actual working conditions and the optimal control strategy.

8.2.1 Stability analysis of conventional power system stabilizer

In this section, a CPSS model widely used in power systems is adopted. The CPSS is composed of three parts: a washout filter, a gain model, and two leading-lag dynamic compensators. The transfer function is given as follows [1]:

$$\Delta u_{\text{pss}} = K_{STAB}\frac{sT_w}{1+sT_w}\times\frac{1+sT_1}{1+sT_2}\times\frac{1+sT_3}{1+sT_4}\times\Delta x \tag{8.1}$$

where Δu_{pss} is the output signal of PSS; K_{STAB} is the controller gain; T_w is the washout filter time constant; T_1, T_2, T_3, and T_4 are time constants of the dynamic compensator; Δx is the input signal of PSS. In general, the time constants T_w, T_2, and T_4 are predetermined. T_w is set to 10. T_2 and T_4 are both set to 0.05. Other variables must meet the following restrictions:

$$\begin{aligned}&K_{STAB}^{\min}\le K_{STAB}\le K_{STAB}^{\max}\\&T_1^{\min}\le T_1\le T_1^{\max}\\&T_3^{\min}\le T_3\le T_3^{\max}\end{aligned} \tag{8.2}$$

In general, the stability of using the system is strongly related to the eigenvalues of the system, and these eigenvalues can be described by the Prony method.

Without losing generality, assume the eigenvalues are identified and constructed as a diagonal matrix Λ_i:

$$\begin{aligned} \Lambda_i &= \mathrm{diag}\left(\lambda_{i1}, \lambda_{i2}, \cdots \lambda_{ij} \cdots, \lambda_{in}\right) \\ \lambda_{ij} &= \alpha_{ij} + j\sigma_{ij}; \xi_{ij} = -\alpha_{ij} / \sqrt{\alpha_{ij}^2 + \sigma_{ij}^2} \end{aligned} \tag{8.3}$$

where λ_{ij} is the jth eigenvalues at the ith operating point; n is the number of eigenvalues; a_{ij} and σ_{ij} are the real part and imaginary part of the λ_{ij}; β_{ij} is the damping ratio of the λ_{ij}.

The objective of the PSS parameter adjustment is to enhance the stability of the system. Therefore, it is necessary to formulate the evaluation criterion to quantify the performance of the PSS parameters in terms of the stability of the system. An eigenvalue-based stability function is used to serve as a performance indicator, and it can be defined as follows [2]:

$$J_{STB-i} = \sum_{j=1}^{n} \sum_{\xi_{ij} < \xi_{\mathrm{set}}} \left| \xi_{ij} - \xi_{\mathrm{set}} \right| + \sum_{j=1}^{n} \sum_{\sigma_{ij} > \sigma_{\mathrm{set}}} \left| \sigma_{ij} - \sigma_{\mathrm{set}} \right| \tag{8.4}$$

Based on Equation (8.4), a small $J_{STB\text{-}i}$ corresponds to a good damping of the whole system. By minimizing $J_{STB\text{-}i}$, the eigenvalues can be shifted to a stable region ($\xi > \xi_{\mathrm{set}}, \sigma < \sigma_{\mathrm{set}}$, σ_{set} and ξ_{set} are the desired real part and damping ratio of the eigenvalues. In this chapter, ξ_{set} is set to 0.05, σ_{set} is set to -0.15 [3]). When $J_{STB\text{-}i} = 0$, all the eigenvalues are in a stable region and with satisfactory damping.

As mentioned above, the effects of different PSS parameters on system stability are significantly different. Some PSS parameters have a much smaller impact on the system's stability than other PSS parameters. Therefore, it is not necessary to adjust all the PSS parameters, but only some key ones. A function sensitivity is introduced to evaluate the influence of each PSS parameter on the stability of the system.

Function sensitivity is one of the most useful sensitivity analysis tools, and it quantifies the variation in a system function corresponding to a change in a parameter. It is used to quantify the effects of the PSS parameters on system stability, which can be defined as follows:

$$\frac{\partial J_{STB-i}(\theta_{\mathrm{PSS}-ij})}{\partial \theta_{\mathrm{PSS}-ij}} = \lim_{\Delta\theta_{\mathrm{PSS}-ij} \to 0} \frac{J_{STB-i}(\theta_{\mathrm{PSS}-ij} + \Delta\theta_{\mathrm{PSS}-ij}) - J_{STB-i}(\theta_{\mathrm{PSS}-ij})}{\Delta\theta_{\mathrm{PSS}-ij}} \tag{8.5}$$

where $\theta_{\mathrm{PSS}\text{-}ij}$ is the jth PSS parameter in the ith wind speed condition; $\Delta\theta_{\mathrm{PSS}\text{-}ij}$ is the increment of $\theta_{\mathrm{PSS}\text{-}ij}$. A high function sensitivity of a parameter implies that it has a significant influence on the stability of the system.

For a power system, it is difficult to calculate the function sensitivity directly using (8.5). Therefore, the Lagrangian interpolation method is introduced to solve this problem:

The interpolation points are obtained through multiple perturbations:

$$\begin{cases} \theta_{\mathrm{PSS}-ij}^{k} = \theta_{\mathrm{PSS}-ij} + \Delta\theta_{\mathrm{PSS}-ij} \cdot k \\ J_k = J_{STB-i}(\theta_{\mathrm{PSS}-ij}^{k}) \end{cases} k = (0, \pm 1, \pm 2, \cdots, \pm n) \tag{8.6}$$

The interpolation basis function can be constructed according to the above interpolation points:

$$L_k(\theta_{\text{PSS}-ij}) = \prod_{\substack{l=-n \\ l \neq k}}^{n} \frac{\theta_{\text{PSS}-ij} - \theta^l_{\text{PSS}-ij}}{\theta^k_{\text{PSS}-ij} - \theta^l_{\text{PSS}-ij}} \tag{8.7}$$

The corresponding interpolation polynomial is:

$$\begin{aligned} L(\theta_{\text{PSS}-ij}) &= \sum_{k=-n}^{n} L_k(\theta_{\text{PSS}-ij}) \cdot J_k\left(\theta^k_{\text{PSS}-ij}\right) \\ &= \sum_{k=-n}^{n} \left(J_{STB-i}\left(\theta^k_{\text{PSS}-ij}\right) \cdot \prod_{\substack{l=-n \\ l \neq k}}^{n} \frac{\theta_{\text{PSS}-ij} - \theta^l_{\text{PSS}-ij}}{\theta^k_{\text{PSS}-ij} - \theta^l_{\text{PSS}-ij}} \right) \end{aligned} \tag{8.8}$$

The function sensitivity of θ_{PSS} is:

$$\begin{aligned} \frac{\partial J_{STB-i}(\theta_{\text{PSS}-ij})}{\partial \theta_{\text{PSS}-ij}} &= L(\theta_{\text{PSS}-ij})' \\ &= \sum_{k=-n}^{n} \left(J_{STB-i}\left(\theta^k_{\text{PSS}-ij}\right) \cdot \prod_{\substack{l=-n \\ l \neq k}}^{n} \frac{\theta_{\text{PSS}-ij} - \theta^l_{\text{PSS}-ij}}{\theta^k_{\text{PSS}-ij} - \theta^l_{\text{PSS}-ij}} \right)' \end{aligned} \tag{8.9}$$

Based on the above method, the sensitivity of each PSS parameter can be calculated. Subsequently, the PSS parameters can be classified into high- and low-sensitivity parameters using a threshold, as follows:

$$Se(\theta_{\text{PSS}-ij}) = \begin{cases} 0, if & \dfrac{\partial J_{STB-i}(\theta_{\text{PSS}-ij})}{\partial \theta_{\text{PSS}-ij}} \leq \sigma_{\text{set}} \\ 1, if & \dfrac{\partial J_{STB-i}(\theta_{\text{PSS}-ij})}{\partial \theta_{\text{PSS}-ij}} > \sigma_{\text{set}} \end{cases} \tag{8.10}$$

where

$$\sigma_{\text{set}} = \frac{1}{m} \sum_{j=1}^{m} \frac{\partial J_{STB-i}(\theta_{\text{PSS}-ij})}{\partial \theta_{\text{PSS}-ij}} \tag{8.11}$$

where $Se(\theta_{\text{PSS-}ij})$ is the sensitivity marker of parameter $\theta_{\text{PSS-}ij}$, σ_{set} is the threshold, which is set as the mean value of the sum of the sensitivity of all the PSS parameters, and m is the total number of adjustable PSS parameters.

If $Se(\theta_{\text{PSS-}ij}) = 1$, it implies that at the i-th operating point, $\theta_{\text{PSS-}ij}$ is a high-sensitivity parameter. If $Se(\theta_{\text{PSS-}ij}) = 0$, it signifies that $\theta_{\text{PSS-}ij}$ is a low-sensitivity parameter, and typically it is sufficient to adjust only the high-sensitivity PSS parameters.

8.2.2 DDPG-based agent for multi-PSS sparse adaptive control

The parameter self-tuning of PSS is a decision-making problem in an uncertain environment, and the DDPG has become a good alternative with better robustness to solve this problem. For a self-tuning of the multi-PSS parameter problem, the related elements of the Markov decision process (MDP) are listed as follows:

1. **Agent**: An agent is the producer of the action, and it gradually learns how to identify the high-sensitivity parameters and provide their optimal settings by repeatedly interacting with the environment to gain experience.
2. **State**: The state is the information obtained by the agent perception environment, and it is used to determine the status of the environment. In the *k*th time step, the state can be defined as $(P_{\text{wind-}k}, \xi_{k1}, \cdots \xi_{kj}, \cdots \xi_{k10})$. where $P_{\text{wind-}k}$ is the output power of the wind farm in the *k*th time step; ξ_{kj} is the damping of the *j*th eigenvalue in the *k*th time step.
3. **Action**: The action is the output of the agent. In this section, for each PSS parameter, the action outputs two types of information: parameter sensitivity and optimal settings for each PSS parameter. To this end, two neurons (Neuron-I and Neuron-II) are designed to control each PSS parameter. Neuron-I outputs the value of the PSS parameter, and Neuron-II outputs the sensitivity marker of the parameter. The output of Neuron-II is only 0 or 1. When Neuron-II outputs 1 and 0, it signifies that the agent identifies this parameter as a high- and low-sensitivity parameter, respectively. Therefore, the action can be defined as $\left[\theta_{\text{PSS}-1k}, \cdots \theta_{\text{PSS}-mk}; Se^{'}(\theta_{\text{PSS}-1k}), \cdots Se^{'}(\theta_{\text{PSS}-mk})\right]$. where $\theta_{\text{PSS}-mk}$ is the output of the Neuron-I for the *m*th PSS parameter in the *k*th time step; $Se^{'}(\theta_{\text{PSS}-mk})$ is the output of the Neuron-II for the *m*th PSS parameter in the *k*th time step.It should be noted that, in the process of interacting with the environment, based on the output of Neuron-II, the agent will filter out the low-sensitivity PSS parameters and only send the high-sensitivity PSS parameters to the environment.
4. **Reward**: A reward reflects the evaluation after an agent takes an action at each time step. The reward consists of two parts: a sensitivity function, which is used to evaluate whether high-sensitivity PSSs are accurately identified, and a stability function, which is used to evaluate the parameter settings for the stability of the system. It can be defined as follows:

$$r(s_k, a_k) = -(J_{STB-k} + \beta J_{SEN-k}) \tag{8.12}$$

where $J_{STB\text{-}k}$ is the stability function of the power system at the *k*th time step; β is the weight coefficient.

$$\begin{aligned} J_{SEN-k} &= \sum_{i=1}^{m} f\left(Se(\theta_{\text{PSS}-ik}), Se^{'}(\theta_{\text{PSS}-ik})\right) \\ f\left(Se(\theta_{\text{PSS}-ik}), Se^{'}(\theta_{\text{PSS}-ik})\right) &= \begin{cases} 0, \textit{if } Se(\theta_{\text{PSS}-ik}) = Se^{'}(\theta_{\text{PSS}-ik}) \\ 1, \textit{if } Se(\theta_{\text{PSS}-ik}) \neq Se^{'}(\theta_{\text{PSS}-ik}) \end{cases} \end{aligned} \tag{8.13}$$

where $J_{SEN\text{-}k}$ is the sensitivity function of PSS parameters at *k*th time step; $Se(\theta_{\text{PSS}-ik})$ is the real marker of sensitivity of parameter $\theta_{\text{PSS-}i}$ in the *k*th time step. $Se^{'}(\theta_{PSS-ik})$ is the output of Neuron-II, and it is the sensitivity identification result of the agent to the $\theta_{PSS\text{-}i}$ in the kth time step.

Based on (8.4) and (8.13), Equation (8.14) can be further formulated as:

$$r(s_k, a_k) = \sum_{j=1}^{n} \sum_{\xi_{kj} < \xi_{set}} \left| \xi_{kj} - \xi_{set} \right| + \sum_{j=1}^{n} \sum_{\sigma_{kj} > \sigma_{set}} \left| \sigma_{kj} - \sigma_{set} \right| + \beta \sum_{i=1}^{m} f\left(Se(\theta_{PSS-ik}), Se^{'}(\theta_{PSS-ik}) \right) \quad (8.14)$$

In reality, the sparse adaptive control of multi-PSS is a decision-making problem in an uncertain environment. It can be described as an MDP: At each step k, the agent perceives the current state of the environment s_k and takes an action a_k based on a policy $\pi(a_k|s_k)$ (π is the policy function that maps states to actions). After that, this action and transition probability affect the environment and make it transfer to the next state s_{k+1}. Finally, the agent can receive a reward $r(s_k, a_k)$ immediately. The interactions between the agent and environment are repeated until the agent learns the optimal sparse adaptive control policy π that maximizes the expected discounted reward.

In general, a value function $Q^{\pi}(s_k, a_k)$ can be used to map state-action pairs (s_k, a_k) to the expected cumulative discounted reward following the policy π, and it can be expressed as follows:

$$Q^{\pi}(s_k, a_k) = \mathbb{E}_{\pi}[R_k | s_k, a_k] \quad (8.15)$$

where

$$R_k = r_k(s_k, a_k) + \gamma r_{k+1}(s_{k+1}, a_{k+1}) + \cdots = \sum_{i=k}^{\infty} \gamma^{i-k} r_i(s_i, a_i) \quad (8.16)$$

and γ is the discounting rate ($\gamma \in [0, 1]$), which reflects the impact of future rewards on the current action.

Since Equation (8.15) satisfies the recursive relationship, it can be transformed into the Bellman expectation equation:

$$Q^{\pi}(s_k, a_k) = \mathbb{E}_{\pi}\left[R_k + \gamma \mathbb{E}_{a_{k+1} \sim \pi}[Q^{\pi}(s_{k+1}, a_{k+1})]\right] \quad (8.17)$$

Therefore, the key to solving the multi-PSS sparse adaptive control problem is to obtain a policy π that maximizes the value function $Q^{\pi}(s_k, a_k)$ for the agent.

An actor-critic architecture DRL algorithm, DDPG can be adopted to solve the above MDP to learn the optimal sparse adaptive control policy π. To deal with the nonlinearity of the MDP problem, two deep neural networks are applied to approximate the actor and critic, named actor network and critic network, respectively. Among them, critic network Q (which is parameterized by θ^{Q}) is used to approximate the value function, and it takes state-action pairs (s_k, a_k) as the input, and outputs a scalar estimate of the value function. Actor network μ (which is parameterized by θ^{μ}) is used to approximate the policy π, and it takes a state s_k as the input, and outputs an action, a_k. During the training process, the parameters of these two networks are updated by training against each other. Specifically, the actor network would first interact with the environment to obtain the state. Then, based on the obtained state, the actor network would produce the next-state action, $\mu(s_{k+1}|\theta^{\mu})$. Subsequently, the critic network evaluates this action and provides an estimate Q (Q is the estimated value of the value function by the critic network). To improve the accuracy of the estimate, the critic network is trained by minimizing the loss function [4]:

$$L(\theta^Q) = \mathbb{E}_{\mu'}\left[\left(y_k - Q^{\pi}(s_k, a_k|\theta^Q)\right)^2\right] \tag{8.18}$$

where

$$y_k = r_k + \gamma Q^{\pi}(s_{k+1}, \mu(s_{k+1}|\theta^{\mu})) \tag{8.19}$$

In reality, the training of the critic networks is a process of minimizing the difference between $Q^{\pi}(s_k, a_k|\theta^Q)$ and y_k. It is a supervised learning process.

To make the actor network provide the action that maximizes the estimated Q value, the actor network can be updated by the policy gradient theorem [5]:

$$\nabla_{\theta^{\mu}} J(\theta^{\mu}) = \nabla_a Q(s, a|\theta^Q)|_{s=s_k, a=\mu(s_k)} \nabla_{\theta^{\mu}} \mu(s|\theta^{\mu})|_{s=s_k} \tag{8.20}$$

where

$$\mu'(s_k) = \mu(s_k|\theta_k^{\mu}) + \mathcal{N} \tag{8.21}$$

where $\mathcal{N}$ represents Ornstein-Uhlenbeck (OU) process, and is added to the action strategy μ to form the exploratory strategy μ', which contributes to making the agent explore the continuous action domain more effectively.

However, if the networks are updated directly by the above equations, which early makes the training process unstable. To this end, both the target networks (which include the target actor and critic networks and can be represented by $\mu'(s_k|\theta^{\mu'})$ and $Q'(s_k, a_k|\theta^{Q'})$, respectively) and experience replay technique are introduced. Specifically, in each iteration, all the observed information is saved as an experience $e = (s_t, a_t, r_t, s_{t+1})$, and stored in a buffer $D = \{e_1, e_2, \cdots, e_M\}$. After the buffer is full, the new generating experience replaces the old one. During the training process, an N-size mini-batch of experience is sampled from the buffer for the gradient calculation and the network parameter update. The loss function Equation (8.18) can be rewritten:

$$L(\theta^Q) = \frac{1}{N}\sum_{k=1}^{N}\left[y_k - Q^{\pi}(s_k, a_k|\theta^Q)\right]^2 \tag{8.22}$$

where

$$y_k = r_k + \gamma Q^{\pi}(s_{k+1}, \mu(s_{k+1}|\theta^{\mu})) \tag{8.23}$$

where y_k denotes the target value and it is calculated by the added target critic network.

Similar to the other DRL methods, the policy gradient is applied to train the neural network, which can be described as follows:

$$\nabla_{\theta^Q} L(\theta^Q) = \frac{1}{N}\sum_{i=1}^{N} \left(y_k - Q^\pi\left(s_k, a_k|\theta^Q\right)\right)\nabla_{\theta^Q} Q^\pi\left(s_k, a_k|\theta^Q\right) \tag{8.24}$$

Based on the above equation, the critic network can be updated:

$$\theta^Q_{k+1} = \theta^Q_k + \alpha^Q \cdot \nabla_{\theta^\mu} L\left(\theta^Q_k\right) \tag{8.25}$$

where a^Q represents the critic network learning rate.

Substituting (8.22) to (8.20), it can be obtained that:

$$\begin{gathered} \nabla_{\theta^\mu} J(\theta^\mu) = \frac{1}{N}\sum_{i=1}^{N} \nabla_a Q(s, a|\theta^Q)|_{s=s_k, a=\mu(s_k)} \nabla_{\theta^\mu} \mu(s|\theta^\mu)|_{s=s_k} \\ \theta^\mu_{k+1} = \theta^\mu_k + \alpha^\mu \cdot \nabla_{\theta^\mu} J\left(\theta^\mu_k\right) \end{gathered} \tag{8.26}$$

where α^μ represents the actor network learning rate.

The target networks can be updated by a soft update method and are shown as follows:

$$\underset{\tau=0.0001}{soft\ update} \begin{cases} \theta^{Q'} \leftarrow \tau\theta^Q + (1-\tau)\theta^{Q'} \\ \theta^{\mu'} \leftarrow \tau\theta^\mu + (1-\tau)\theta^{\mu'} \end{cases} \tag{8.27}$$

The procedure of the DDPG algorithm is listed in Table 8.1, and the computational process of the DDPG-based agent is shown in Figure 8.1.

Table 8.1 Multi-PSS self-tuning policy learning based on DDPG

Input: Output of generators and eigenvalues of the test system
Output: High-sensitivity PSS parameters
1: **Initialize**
Initialize the parameters of critic network and actor network with θ^Q and θ^μ, respectively
The parameters of the target critic network and target actor network are updated by $\theta^{Q'} \leftarrow \theta^Q$ and $\theta^{\mu'} \leftarrow \theta^\mu$, respectively.
Initialize the M-sized replay buffer
2: **for** episode = 1 **to** max-episode **do**
3: Initialize the OU process $\mathcal{N}$ for action domain exploration
4: Initialize the test system state
5: Observe the initial state of the test system s_1
6: **for** time step = 1 **to** max-step size **do**
7: Select high-sensitivity PSS parameters based on (8.10)
8: Send high-sensitivity PSS parameters to the test system and the simulations are carried out.
9: Calculate the reward a_t based on (8.21) and observe the next state s_{t+1}
10: Store the experience (s_t, a_t, a_t, s_{t+1}) into the replay buffer
11: If the replay buffer is full of experiences:
12: Pickup mini-batch of experiences by sampling from the replay buffer
13: Update critic network and actor network according to (8.25) and (8.26)
14: Update the target networks according to (8.27)
15: **end**
16: **end**

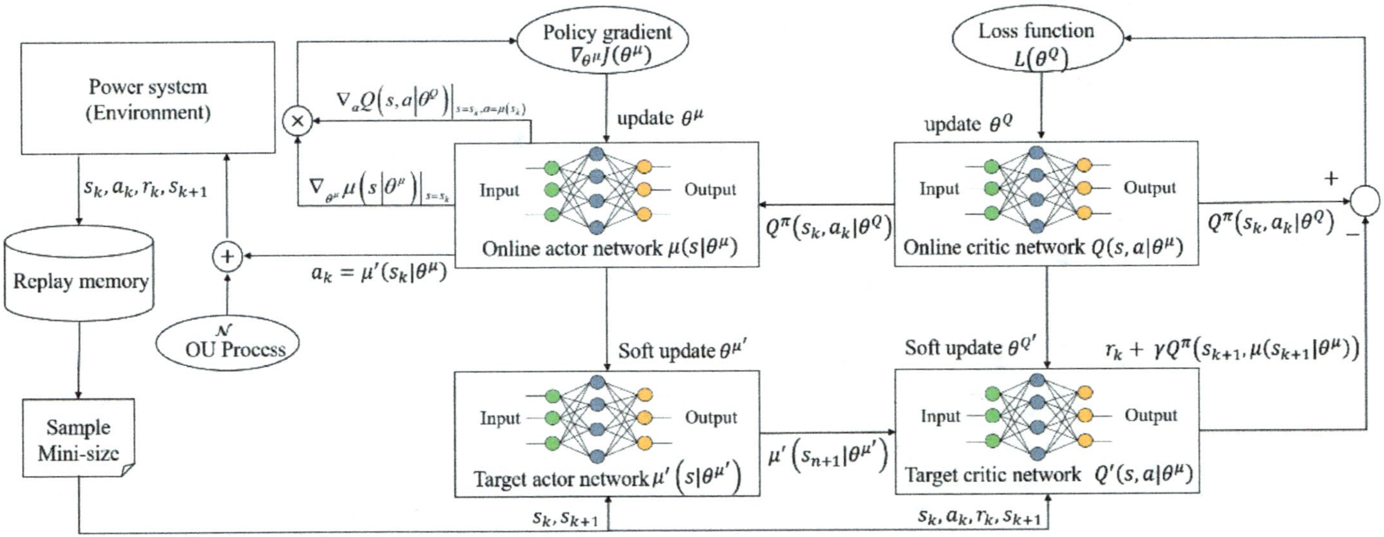

Figure 8.1 DDPG-based agent training diagram [6]

8.2.3 Numerical simulation results

Figure 8.2 shows the IEEE 10-machine 39-bus system [7], which is used to investigate the proposed method. In the 10-machine 39-bus system, the total load and generator are 5,620 MW and 6,140 MW, respectively. The reference value of the system is set to 1,000 MW. Each synchronous generator in the system adopts the fifth-order model and is equipped with the excitation system and turbine governing system. CPSSs are configured on the generators strongly related to the dominant oscillation modes, including G1, G2, G6, and G8. A 5-row 50-column rectangular layout wind farm is located on bus 9. The total capacity of the wind farm is 1,250 MW. The penetration level of wind is about 22%. Each WT in the wind farm adopts the direct-driven permanent magnet synchronous generator model, and the distance between each WT is seven rotor diameters. More details about the WT can be found in Table 8.2.

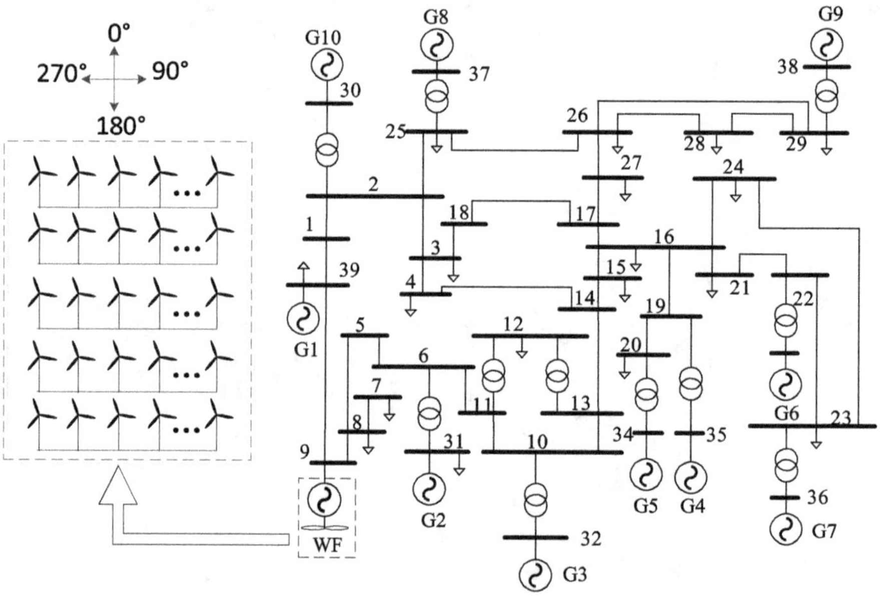

Figure 8.2 Transmission network for a 10-machine 39-bus system with a wind farm [6]

Table 8.2 Wind turbine specification

Parameter	Value	Parameter	Value
Cut-in wind speed	4 m/s	Cut-out wind speed	25 m/s
Rated wind speed	11.4 m/s	Rotor diameter	126 m
Rated rotor speed	12.1 rpm	Rated power	5 MW

Table 8.3 shows the distribution of test system eigenvalues at different wind conditions. It can be seen that the change in wind will change the distribution of eigenvalues. Moreover, it would place Mode 1 and Mode 4 out of the stable region, and the system has a risk of oscillations. To this end, a DDPG algorithm is applied to train an agent to learn the optimal control policy for the sparsity adaptive control of multi-PSS.

Before the training, the number of layers and the number of neurons need to be predetermined. In this study, all four networks adopt the same structure. Each network contains two hidden layers, one has 256 neurons, and the other has 64 neurons. The hyperparameters in the DDPG algorithm are taken from [8,9] and listed in Table 8.4. Annual wind speed data is shown in Figure 8.3. Moreover, annual data of synchronous generator and load are produced by randomly selecting from the range [0.85 p. u.–1.15 p. u.]. Both of these data are used to form a training set and a test set: 90% of the data are assigned to the training set for the agent training; the remaining data are assigned to the test set to verify the validity of the trained agent. During the training, the weight coefficient β is set to 0.04. For the PSS parameters, the range for K_{STAB} is [5, 100] and for T_1 and T_3 are [1, 10].

Table 8.3 Distribution of the electromechanical oscillation eigenvalues at different wind conditions (no PSS is used)

Modes	**Wind speed: 0 m/s**			**Wind speed: 5 m/s; wind direction: 30°**		
	Real	**Imag**	**Damping**	**Real**	**Imag**	**Damping**
Mode 1	−0.161	3.13	5.15%	**−0.077**	**3.10**	**2.48%**
Mode 2	−0.302	5.23	5.76%	−0.279	5.22	5.36%
Mode 3	−0.262	3.25	8.03%	−0.252	3.21	7.81%
Mode 4	−0.364	3.43	10.5%	−0.351	3.39	10.3%
Mode 5	−3.699	3.60	71.69%	−0.578	3.53	16.2%
Mode 6	−0.662	2.75	23.38%	−0.633	2.75	22.41%
Mode 7	−2.087	8.71	23.22%	−2.080	8.69	23.27%
Mode 8	−0.770	2.40	30.49%	−0.727	2.40	29.02%
Mode 9	−0.659	1.60	38.08%	−0.660	1.58	38.41%
Modes	**Wind Speed: 10 m/s; wind direction: 45°**			**Wind Speed: 15 m/s; wind direction: 90°**		
	Real	**Imag**	**Damping**	**Real**	**Imag**	**Damping**
Mode 1	**−0.076**	**3.09**	**2.45%**	**−0.062**	**3.10**	**2.00%**
Mode 2	−0.281	5.22	5.37%	−0.280	5.22	5.36%
Mode 3	−0.236	3.20	7.37%	−0.228	3.22	7.08%
Mode 4	−0.291	3.38	8.59%	**−0.041**	**3.37**	**1.22%**
Mode 5	−0.377	3.52	10.64%	−0.357	3.45	10.31%
Mode 6	−0.589	2.66	21.61%	−0.551	2.45	22.02%
Mode 7	−2.069	8.71	23.12%	−2.056	8.72	22.94%
Mode 8	−0.732	2.38	29.43%	−0.699	2.35	28.53%
Mode 9	−0.648	1.55	38.50%	−0.641	1.54	38.50%

The bold values denotes the information of the critical modes (whose damping ratios are less than 5%).

Table 8.4 Hyperparameters of DDPG

Parameter	Value	Parameter	Value
Step size of each episode	10	Learning rate for actor network (α^{μ})	0.001
Soft update coefficient (τ)	0.0001	Learning rate for critic network (α^{Q})	0.002
Experience memory capacity	8,000	Mini-batch size	40

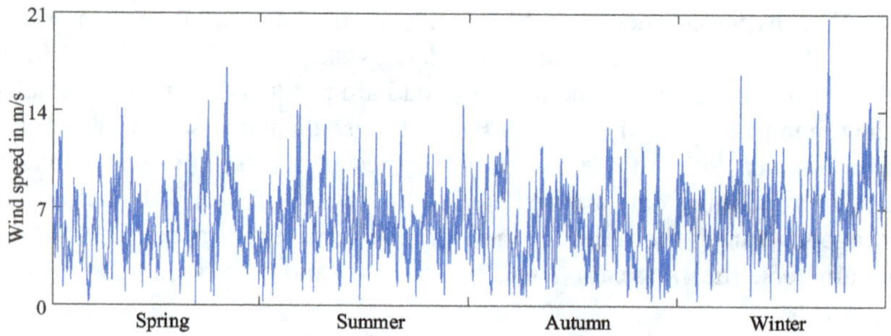

Figure 8.3 One-year wind speed data for case study [6]

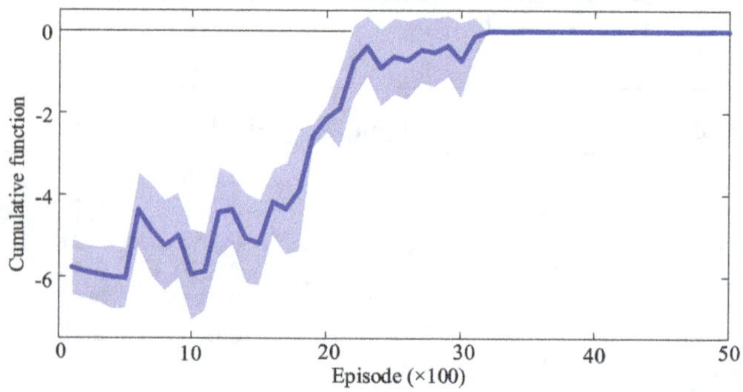

Figure 8.4 Cumulative reward changes with episodes during the training process [6]

A workstation with a NVIDIA GeForce 1080Ti GPU and an Intel Xeon E5-2630 v4 CPU is used for the training.

For the training process of the agent, the average values and the standard deviation of the accumulated reward are illustrated in Figure 8.4, and they are indicated by solid lines and shaded areas, respectively (the average accumulated reward of each episode is the average value over 100 evaluation episodes). It can be seen from Figure 8.4 that during the initial stage of the training process, the reward is low. This is because the agent cannot make the optimal decision that satisfies the correlated constraint in the beginning. With the increase in training episodes, the parameters of the DDPG networks are optimized according to the policy gradient, and the cumulative reward is increased gradually. Following 3,000 episodes, the cumulative reward converges to 0. This indicates that the agent has successfully learned the control policy to obtain higher rewards than earlier.

The cumulative reward, as presented in Figure 8.4, consists of two parts: stability function and sensitivity function, and both are shown in Figure 8.5. It can be seen that with the increase in episodes, the agent attempts to explore the environment and gain experience. Following 3,000 episodes, the values of both the stability function and the sensitivity function converge to 0. This suggests that the agent is well-trained. Under each operating condition, the agent can accurately identify the high-sensitivity parameters of the system and provide the optimal settings of these parameters to ensure that the eigenvalues are in the stable region.

Following the offline training with a test database, the agent can be applied to the test system for the PSS parameter setting self-tuning. However, it should be noted that not each wind speed change requires adjusting the PSS parameters (although the trained agent can provide the optimal parameter settings for the multi-PSS under each wind speed condition). The PSS parameters provided by the agent under one wind speed condition may still present a good performance under the next wind speed condition. Hence, the following online application strategy for the agent is proposed:

Step 1: Based on the Prony method [10], the eigenvalues of the system can be identified in real-time. Subsequently, the identified eigenvalues can be used to calculate the stability function to determine whether the PSS parameters need to be adjusted or not. If not, the above steps are repeated, and if necessary, then the next step is followed.

Step 2: Upload the measured real-time wind speed, and the agent provides the corresponding action according to it.

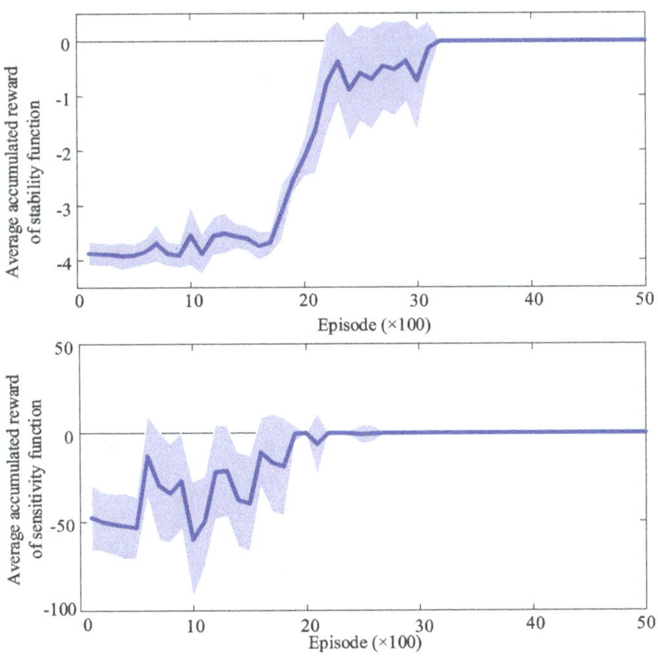

Figure 8.5 Functions change with an episode during the training process [6]

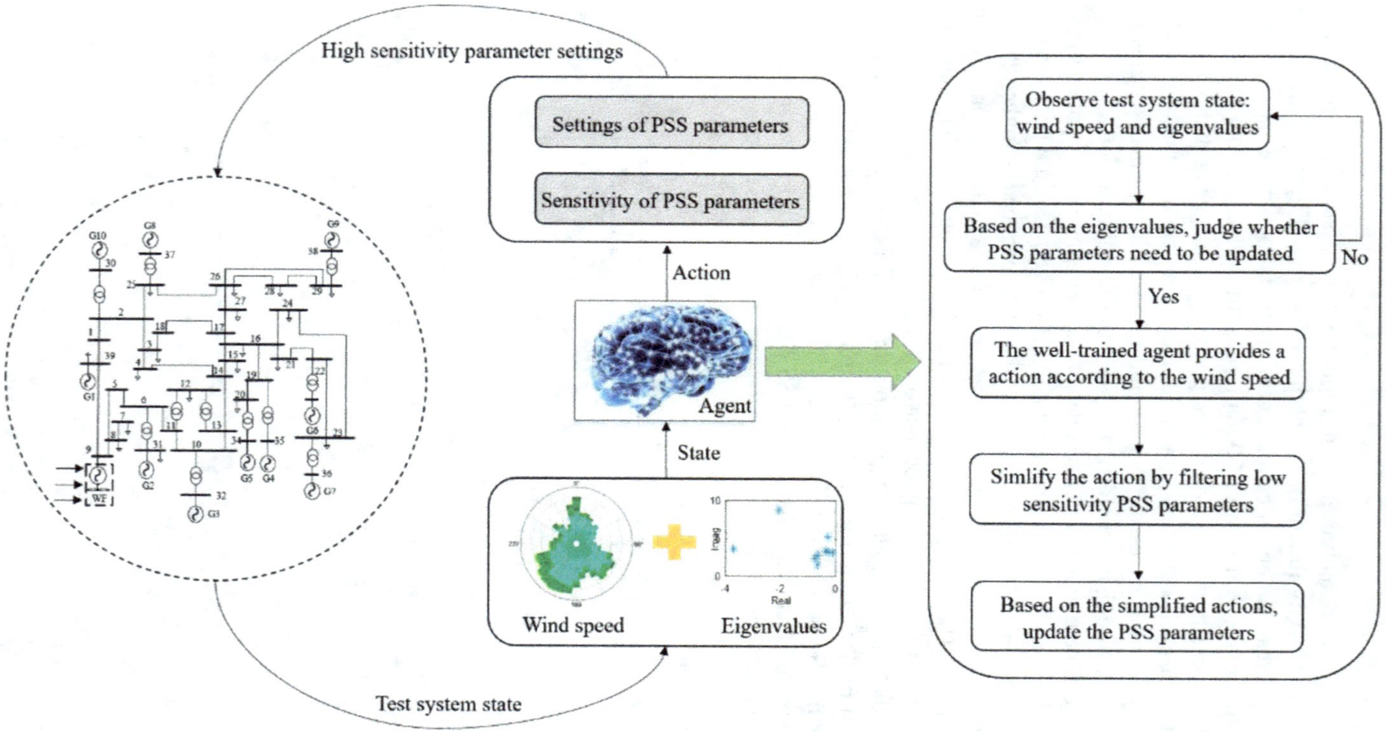

Figure 8.6 Online implementation architecture of DDPG-based agent [6]

Step 3: Simplify the action by filtering the PSS parameters with a sensitivity marker $Se'(\theta_{\mathrm{PSS}-i}) = 0$ (These PSS parameters are of low sensitivity at this wind speed condition and have little influence on the stability of the power systems).

Step 4: The simplified action is sent to the power system, and the corresponding PSS parameters are adjusted according to the action, while the other PSS parameters remain unchanged.

The above online application strategy can not only ensure a real-time stable operation of the system but also reduce the number of PSS parameter adjustments. Accordingly, the control cost of the agent can be reduced. The detailed implementation of the agent is shown in Figure 8.6.

To illustrate the effectiveness of the proposed agent, three different wind speed conditions (Table 8.5) selected from the test database were used as comparison scenarios. In each case, based on the wind speed condition, the well-trained agent provided the corresponding optimal parameter settings for the PSSs. Moreover, two other benchmark methods were also tested for comparison.

Robust method: The Taguchi method [11] is applied to tune the parameter settings of the PSS. Specifically, the parameter setting tuning problem of multi-PSS can be modeled as a nonlinear optimization problem. Then, by optimizing the signal-noise ratio function, the parameters that are less affected by the change in operation conditions are obtained.

Model predictive control (MPC): An artificial neural network (ANN)-based prediction model is proposed to predict the wind speed one time-step in advance, and it is trained by the data in the test database. Based on the prediction wind speed data, the parameter settings of PSS are optimized by minimizing the critical damping index. More details can be found in [12].

So far, three methods have been proposed and tested in each case shown in Table 8.5 (The wind direction of the three cases is fixed at 180°). To show the performance of parameter settings provided by the different methods, the time domination is carried out under Fault 1 (a three-phase short circuit fault with a 200 ms fault duration in bus 39). The simulation results are shown in Figure 8.7.

It can be observed from Figure 8.7 that during three different operating conditions, the system with the proposed DDPG-based tuning strategy can make the power angle curves of generators return to a stable state in the shortest time in comparison with the other two methods. It means that the PSS parameter setting provided by the proposed method has a better ability to suppress oscillation during different operating conditions, and it has better robustness than the robust method and MPC method in cases of uncertainty. Taking Mode 1 and Mode 4 as examples, during the three cases, the identification results are shown in Table 8.6. By comparing the damping changes of these modes under the three strategies, it can be concluded that the proposed method

Table 8.5 Three different wind speed conditions

Cases	Wind	Cases	Wind	Cases	Wind
Case 1	2 m/s	Case 2	4 m/s	Case 3	18 m/s

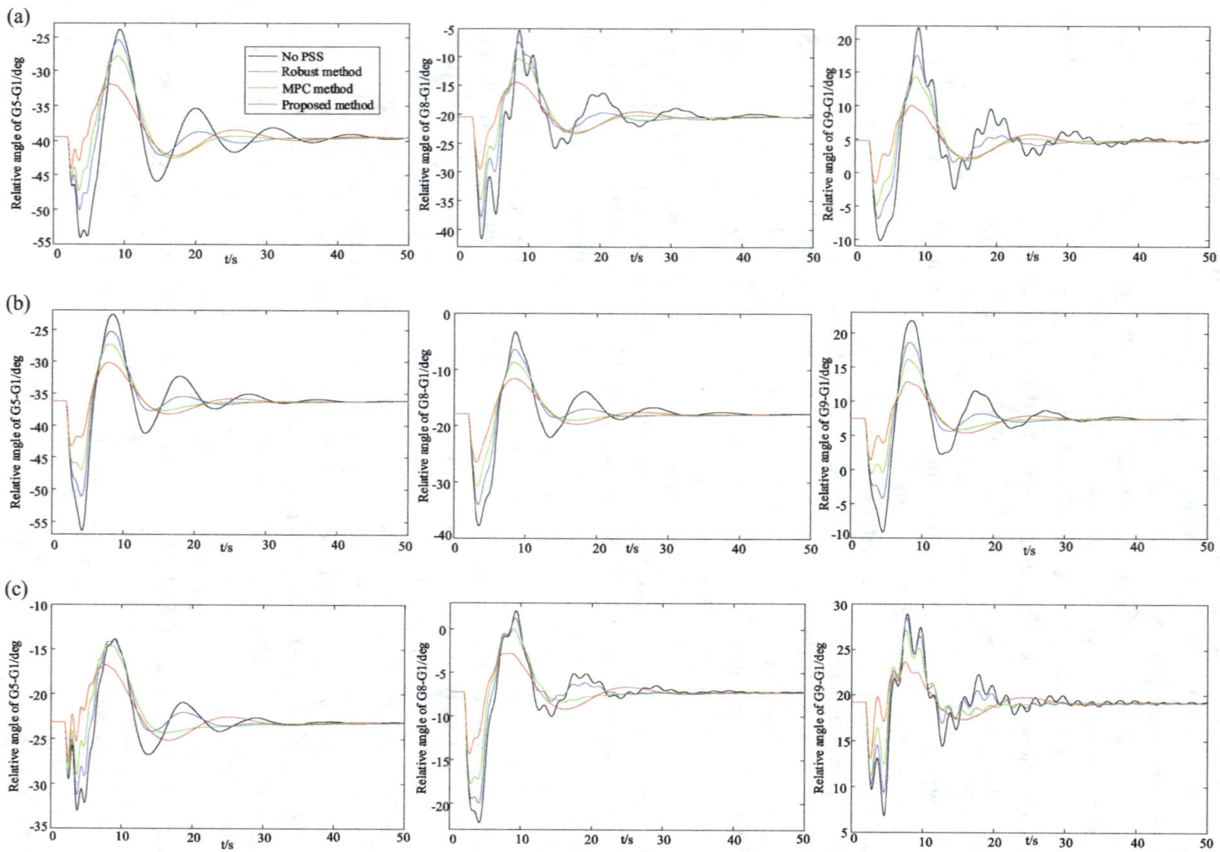

Figure 8.7 *Power oscillation curves under short-circuit faults. (a) Power oscillations of units in Case 1 during Fault 1. (b) Power oscillations of units in Case 2 during Fault 1. (c) Power oscillations of units in Case 3 during Fault 1 [6].*

Table 8.6 Comparison of modes by using different damping strategies

Modes	Strategy	Cases								
		Case 1 (2 m/s)			Case 2 (4 m/s)			Case 3 (18 m/s)		
		Real	Imag	Damp	Real	Imag	Damp	Real	Imag	Damp
Mode 1	Robust method	−0.158	3.16	5.00%	−0.181	3.14	5.75%	−0.262	3.16	8.28%
	MPC method	−0.216	3.18	6.76%	−0.247	3.16	7.78%	−0.187	3.19	5.88%
	Proposed method	−0.366	3.21	11.32%	−0.416	3.20	12.90%	−0.377	3.22	11.61%
Mode 4	Robust method	−0.413	3.32	12.34%	−0.410	3.33	12.25%	−0.227	3.36	6.75%
	MPC method	−0.462	3.24	14.14%	−0.376	3.32	11.25%	−0.377	3.44	10.90%
	Proposed method	−0.66	3.33	19.57%	−0.438	3.25	13.32%	−0.415	3.56	11.58%

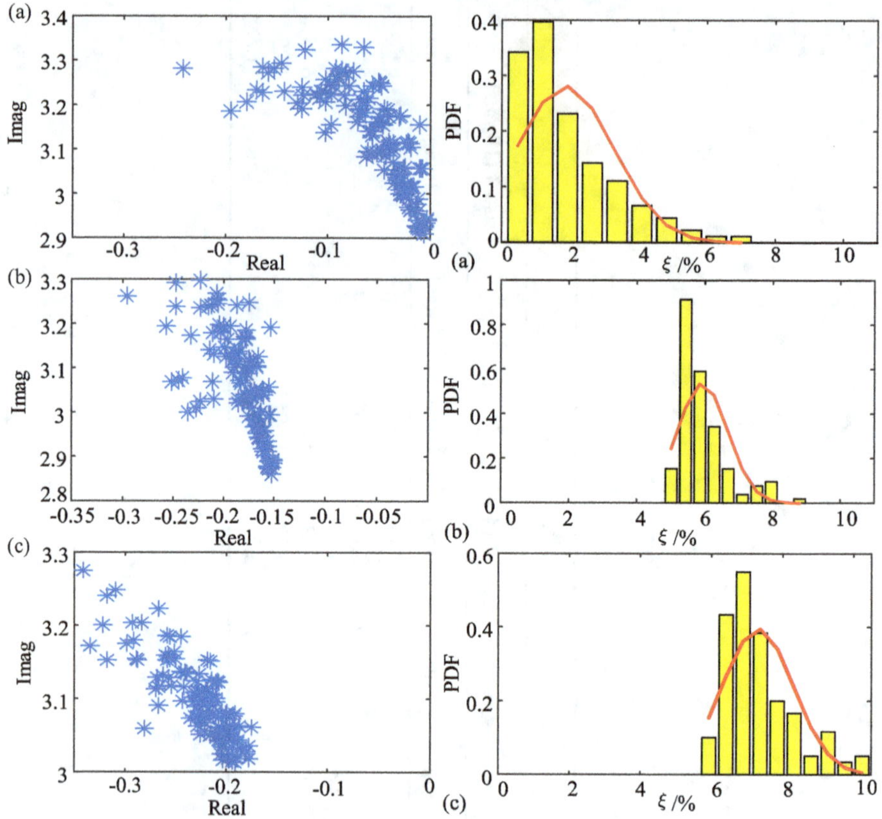

Figure 8.8 Mode 1 distribution under different cases: (a) Robust method; (b) MPC method; (c) Proposed method [6]

enhances the damping of Mode 1 and Mode 4 better than other strategies in the three cases.

To further test the robustness of the proposed method for different operating conditions, 150 operating conditions were randomly selected from the test database as comparison cases. For each case, Mode 1 and its probability density function (PDF) are calculated and plotted in the complex plane, which is shown in Figure 8.8.

Figure 8.8 shows the distribution of Mode 1 with different strategies. It can be seen from Figures 8.8(a) and 8.8(b) that the damping of partial Mode 1 is less than 5% and they are not located in the stable region, which means that the system has a risk of oscillation. Figure 8.8(c) shows the results of Mode 1 with the proposed method, respectively. In all cases, Mode 1 is located in the left plane of the real axis. It can be concluded that, compared with the robust method and MPC method, the proposed method makes Mode 1 move more to the left, and the system has a larger stable margin. Finally, the control parameter self-tuning strategy of the proposed method depends on the wind speed conditions of the wind farm, including real-time data and historical datasets, which are easily accessible in practice. Therefore, the proposed method is operational.

8.3 A data-driven approach for designing STATCOM additional damping controller for wind farms

In the power system, for the operation and regulation of wind farms, the flexible AC transmission system (FACTS) controller can be connected at the point of common coupling (PCC) to maintain a stable voltage. The additional damper controller (ADC) of FACTS also contributes to the stability of the power systems. In this section, a data-driven approach is investigated to design STATCOM-ADC. Specifically, based on DDPG and the estimation model, a novel robust adaptive control agent is proposed to adjust STATCOM-ADC parameters online.

8.3.1 Modeling of a STATCOM-ADC

As a representative of the second generation of FACTS, STATCOM is widely used in power systems with different functions. A STATCOM, which consists of a transformer, three three-phase gate turn-off thyristor (GTO)-based voltage source converter (VSC), and a DC capacitor as energy storage, is proposed and used to dampen oscillations in the power system [13]. The specific structure is shown in Figure 8.9.

For the STATCOM shown in Figure 8.9, m is the modulation ratio of the voltage source VSC control signal; ψ is the phase angle of the VSC control signal; V_{DC} is the voltage across the DC capacitor; and C_{DC} is the capacitance of the DC capacitor.

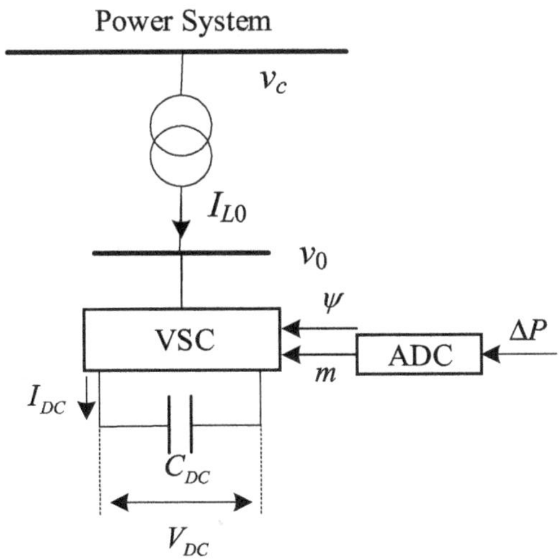

Figure 8.9 A STATCOM install in the power system [9]

The AC voltage generated by the VSC is as follows:

$$v_0(t) = mV_{DC}\sin(\omega t - \psi) \tag{8.28}$$

It can be seen from (8.28) that $v_0(t)$ is related to m and ψ. By adjusting the m and ψ, $v_0(t)$ can be controlled. The phase angle ψ is determined by the DC voltage regulator and the specific structure is shown in Figure 8.10.

In a STATCOM, the DC voltage regulator plays the role of active power exchange between the STATCOM and the power system. The dynamic equation of the DC voltage regulator dynamic is given as follows:

$$\psi = \left(K_{pdc} + \frac{K_{idc}}{s}\right)\left(\frac{ST_{dc}}{1 + ST_{dc}}\right)(V_{\mathrm{DCref}} - V_{\mathrm{DC}}) \tag{8.29}$$

where K_{pdc} and K_{idc} are the proportional and integral gains of the STATCOM DC voltage regulator, respectively; T_{dc} is the time constant of the STATCOM DC voltage regulator.

The modulation ratio m is determined by AC voltage regulator and the specific structure is shown in Figure 8.11. In general, an ADC is attached to the STATCOM AC voltage regulator to improve the stability of the whole system.

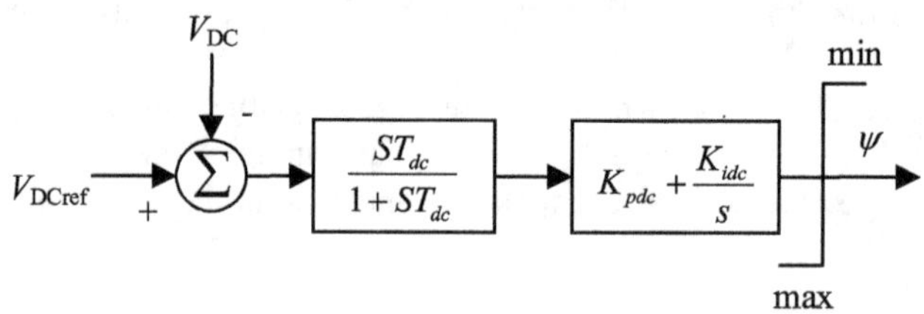

Figure 8.10 DC voltage regulator of the STATCOM [9]

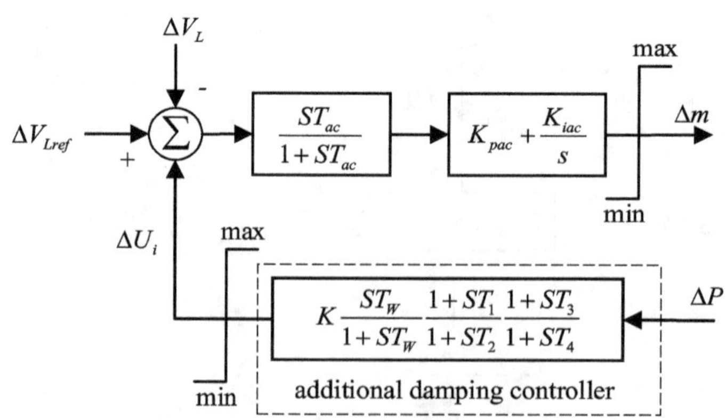

Figure 8.11 STATCOM AC voltage regulator and its ADC [9]

In a STATCOM, the AC voltage regulator plays the role of reactive power exchange between the STATCOM and the power system. The AC voltage regulator dynamic equation is shown as follows:

$$m = \left(K_{pac} + \frac{K_{iac}}{s}\right)\left(\frac{ST_{ac}}{1 + ST_{ac}}\right)(\Delta V_{Lref} - \Delta V_L + \Delta U_i) \tag{8.30}$$

where K_{pac}, K_{iac} are the proportional and integral gains of the STATCOM AC voltage regulator, respectively; T_{ac} is the time constant of the STATCOM AC voltage regulator.

As shown in Figure 8.11, the structure of the STATCOM-ADC is the same as the conventional PSS, which consists of a washout filter, a gain model, and two lead-lag dynamic compensators. The transfer function is shown as follows:

$$\Delta U_i = K \frac{ST_w}{1 + ST_W} \frac{1 + ST_1}{1 + ST_2} \frac{1 + ST_3}{1 + ST_4} \Delta P \tag{8.31}$$

where K is the controller gain; T_w is the washout filter time constant and it is set to 10; T_1, T_2, T_3, and T_4 are time constants of the dynamic compensator; ΔP is the active power deviation at PCC and it is taken as the input signal in this section.

The introduction of wind power increases the uncertainty of the power system. The design of STATCOM-ADC should ensure the stability of the whole system under different operating conditions. Therefore, it is necessary to model the uncertainty of the power system. An inverse additive perturbation model [14] is applied to represent the system uncertainties. The feedback control system with the inverse additive perturbation is shown in Figure 8.12.

In Figure 8.12, G is the actual system; Δ_A represents the system uncertainties. For the system shown in Figure 8.12, based on the small gain signal theory, if Equation (8.32) holds, the system is stable.

$$\left\| \frac{\Delta_A G(s)}{1 + G(s)G_{\text{ADC}}(s)G_{\text{STATCOM}}(s)} \right\|_\infty < 1 \tag{8.32}$$

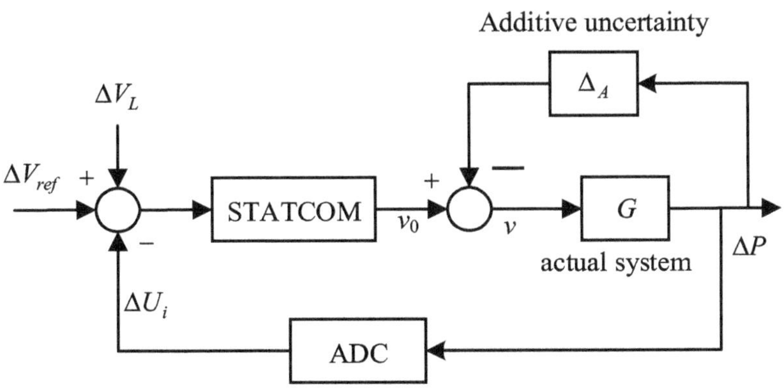

Figure 8.12 Power system with inverse additive perturbation [9]

where $G(s)$ is the transfer function of the actual system; $G_{ADC}(s)$ is the transfer function of STATCOM-ADC; $G_{STATCOM}(s)$ is the transfer function of the STATCOM. The infinity signs in (8.32) mean the gain of the equation.

For (8.32) can be transformed into:

$$\|\Delta_A\| < \frac{1}{\|G(s)/(1 + G(s)G_{ADC}(s)G_{STATCOM}(s))\|_\infty} \tag{8.33}$$

It can be seen from (8.33) that, by minimizing $\|G(s)/(1 + G(s) G_{ADC}(s)G_{STATCOM}(s))\|_\infty$, the system has the best robustness. During this process, the stability of the system should be guaranteed. It means that the STATCOM-ADC parameters can be selected to place all the eigenvalues (electromechanical modes) on the left side of the *s*-plane. To ensure the system has an adequate stability margin, all the eigenvalues should be limited by both $\xi_i \geq \xi_{spec}$ and $\sigma_i \leq \sigma_{spec}$. Moreover, each parameter of the controller must meet the parameter bounds. In this way, the STATCOM-ADC not only ensures the stability of the system but also has good robustness considering different operation conditions.

Hence, the above problem can be formulated as the following objective function:

$$\begin{aligned} &\text{Minimize}\|G(s)/(1 + G(s)G_{ADC}(s)G_{STATCOM}(s))\|_\infty \\ &\text{Subject } to\xi_i \geq \xi_{spec}, \sigma_i \leq \sigma_{spec} \\ &K_{min} \leq K \leq K_{max}, T_{j,\,min} \leq T_j \leq T_{j,\,max} \end{aligned} \tag{8.34}$$

where ξ_i and σ_i are the actual damping ratio and real part of the *i*th electromechanical mode; ξ_{spec} and σ_{sepc} are the desired damping ratio and real part of the *i*th electromechanical mode, respectively; K_{min} is the minimum limit of the STATCOM-ADC gains; K_{max} is the maximum limit of STATCOM-ADC gains; $T_{j,min}$ is the minimum limit of the STATCOM-ADC time constants T_j; $T_{j,max}$ is the maximum limit of STATCOM-ADC time constants T_j. For the $\xi_i \geq \xi_{spec}, \sigma_i \leq \sigma_{spec}$, it can be regarded as a stable region of the system.

8.3.2 Adaptive parameter adjustment method for robust STATCOM-ADC

For the problem of STATCOM-ADC parameter settings (Equation (8.34)), G_{ADC} (s) and $G_{STATCOM}(s)$ are easily obtained, where the eigenvalues of the system can be identified by the Prony algorithm. But the transfer function of the actual system $G(s)$ is normally not exactly known or in some cases even unknown. Moreover, due to the change in WT power production, the transfer function also changes in real-time. To solve this problem, an adaptive identification method is used to estimate the transfer function of the actual system. More specifically, a higher-order nonlinear model is used as an estimator and its parameters are self-tuned by the model networks.

The ANNs that make up the model networks are trained by minimizing the output error between the actual and estimation models. The computation process of the ANN-based model identification is shown in Figure 8.13.

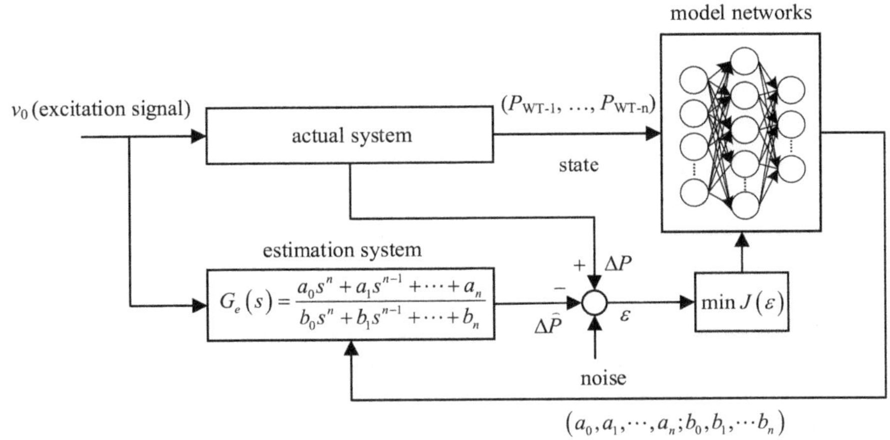

Figure 8.13 ANN-based model identification computation process [9]

In Figure 8.13, $P_{\text{WT-}n}$ is the power production of the nth WT. For each WT in the wind farm (WF), the power production can be calculated by the method proposed in [15]. $G_e(s)$ is the transfer function of the estimation system. Moreover, to excite the characteristics of the system, an excitation signal should be attached to the v_0. ΔP is the active power deviation at PCC under the excitation signal, $\Delta\hat{P}(t)$ is the output of the estimation model under the excitation signal.

The output error ε is calculated by the equation as follows:

$$\varepsilon(t) = \Delta P(t) - \Delta\hat{P}(t) \tag{8.35}$$

Based on the least square method, the performance index of the system identification is:

$$J = \frac{1}{T}\int_0^T \left[\Delta P(t) - \Delta\hat{P}(t)\right]^2 = \frac{1}{T}\int_0^T [\varepsilon(t)]^2 \tag{8.36}$$

Considering the discontinuity of measurement points, Equation (8.36) can be discretized:

$$J = \frac{1}{N}\sum_{i=0}^{N} (\varepsilon_i)^2 \tag{8.37}$$

where i represents the ith sampling time, N is the total number of samples, ε_i is the sampling value of the curve $\varepsilon(t)$ at the ith sampling time.

To reduce the influence of noise on the identification, a robust performance index can be used, which is a combination of exponential function and Huber estimation:

$$J = 1 - \frac{1}{N}\sum_{i=0}^{N} e^{-\left[v_i^2|_{|v_i|\le\rho} + \left(2\rho|v_i| - \rho^2\right)|_{|v_i|>\rho}\right]} \tag{8.38}$$

where

$$v_i = \frac{\varepsilon_i}{\sigma_0}$$
$$\delta_0 = \sqrt{\frac{\sum_{i=1}^{N}(\varepsilon_i - \bar{\varepsilon})}{N-1}}\bar{\varepsilon} = \frac{\sum_{i=1}^{N}\varepsilon_i}{N} \tag{8.39}$$

Here ρ is the core parameter which is chosen by trial and error. σ_0 is the standard deviation.

In fact, the training of the model networks is a process of minimizing (8.38) and it is a supervised learning process. The back error propagation learning strategy [16] is adopted to train the model network, and the details are as follows:

During the training, the weights of the model networks are updated by the following equation:

$$W(k) = W(k-1) + \Delta W(k) \tag{8.40}$$

where $W(k)$ is the weight of model networks in the kth iteration; $\Delta W(k)$ is the change of weight in the kth iteration and it can be calculated as:

$$\Delta W(k) = -\eta J(k)\frac{\partial J(k)}{\partial W(k)} + \alpha \Delta W(k-1) \tag{8.41}$$

where η is the learning rate of the model networks; α is the momentum factor of the model networks; $J(k)$ is the performance index of the system identification in the kth iteration.

DDPG is the state-of-the-art DRL algorithm, which combines the perception of deep learning with the decision-making ability of reinforcement learning [17]. DDPG is introduced to train an agent that can real-time adjust the STATCOM-ADC parameters.

For the problem of STATCOM-ADC parameter settings, it can be regarded as a decision-making problem under a stochastic environment and modeled as a Markov Decision Process (MDP). In general, MDP can be presented as a tuple, namely <S, A, P, R>.

- S is the state set. In this section, the state s_t at time step t is defined as ($P_{\text{WT-1}}$, ..., $P_{\text{WT-}n}$, $G_{\text{STATCOM}}(s)$, $G_{\text{ADC}}(s)$, ζ_i, σ_i).
- A is the action set. In this section, the action a_t at time step t is defined as (K, T_1, T_2, T_3, T_4), which are the parameters of STATCOM-ADC.
- P is the transition probability function and it is related to action and system uncertainty. In general, the system moving from one state s_t to another state s_{t+1} can be given by $s_{t+1} \sim P(s_{t+1}|s_t, a_t)$.
- R is the reward function r (a_t, s_t) and it is the reward that the agent obtains when action a_t is taken in state s_t. The reward is used to evaluate the STATCOM-ADC parameter setting provided by the agent. In this section, (8.34) can be used to achieve this goal.

At one step k, the agent perceives the state of the environment $s_k \in S$ and takes an action $a_k \in A$ based on the policy π (π is the function that maps states to actions).

Then it can receive a reward $r_k \sim R(s_t, a_t)$ immediately. After that, the environment transfers to the next state $s_{k+1} \in S$.

In DDPG, the cumulative discounted reward can be defined as follows:

$$R_t = r_{(t+1)} + \gamma r_{(t+2)} + \gamma^2 r_{(t+3)} + \cdots = \sum_{k=0}^{\infty} \gamma^k r_{t+1+k} \tag{8.42}$$

where γ is the discounting rate ($\gamma \in [0, 1]$) reflects the impact of future rewards on the current action.

The final purpose of DDPG can be described as a process of learning the policy π by maximizing the expected cumulative discounted reward from the start state. Like most other DRL algorithms, the action-value function is used to describe the expected cumulative reward after taking the action a_t in state s_t under the policy π.

$$Q^{\pi}(s_t, a_t) = \mathbb{E}_{\pi}[R(s_t, a_t) + \gamma \mathbb{E}_{a_{t+1} \sim \pi}[Q^{\pi}(s_{t+1}, a_{t+1})]] \tag{8.43}$$

For the DDPG algorithm, two neural networks are used to construct actor networks and critic networks, respectively. The actor networks (which are represented by Q^{u}, parameterized by μ) are used to approximate the policy function π and the critic networks (which are represented by Q^{w}, parameterized by w) are used to approximate the action-value function $Q^{\pi}(s_t, a_t)$.

For the critic networks, it can be optimized by minimizing the loss:

$$\begin{aligned} L(\theta^{w}) &= \mathbb{E}_{w'}\left[(Q(s_t, a_t|\theta^{w}) - y_t)^2\right] \\ w_{t+1} &= w_t + n_c \nabla_{\mu} L(\theta^{w}) \end{aligned} \tag{8.44}$$

where

$$y_t = r(s_t, a_t) + \gamma Q(s_{t+1}, u(s_{t+1})|\theta^{w}) \tag{8.45}$$

where n_c is the learning rate of critic networks. As it is shown in (8.44), the training of critic networks is a process of minimizing the difference y_t and $Q(s_t, a_t|\theta^{w})$.

The actor networks can be optimized by policy gradient [18]:

$$\begin{aligned} \nabla_{\theta^{\mu}} J &= E_{s_t \sim \rho^{u}}\left[\nabla_{\theta^{\mu}} Q(s, a|\theta^{w})|_{s=s_i, a=\mu(s_i)}\right] \\ &= E_{s_t \sim \rho^{u}}\left[\nabla_a Q(s, a|\theta^{w})|_{s=s_i, a=\mu(s_i)} \cdot \nabla_{\theta^{\mu}} \mu(s|\theta^{w})|_{s=s_i}\right] \mu_{t+1} = \mu_t + n_a \nabla_{\theta^{w}} J \end{aligned} \tag{8.46}$$

where n_a is the learning rate of critic networks. As it is shown in (8.46), the training of the actor networks is the process of maximizing $Q(s_t, a_t|\theta^{w})$.

To make the neural networks more stable in the training process, the target networks (target actor networks and target critic networks), which share the same structure and neuron numbers with actor-critic networks are adopted in the DDPG. In each iteration, the soft update is used to synchronize its parameters to the actor networks and critic networks and it can be described as follows:

$$\text{soft update}\begin{cases} \theta^{w'} \leftarrow \tau\theta^{w} + (1-\tau)\theta^{w'} \\ \theta^{\mu'} \leftarrow \tau\theta^{\mu} + (1-\tau)\theta^{\mu'} \end{cases} \tag{8.47}$$

where $\theta^{\mu'}$ represents the parameters of target actor networks; $\theta^{w'}$ represents the parameters of target critic networks; τ is the update rate and it is very small.

In the process of training, the information $e = \{s_t, a_t, r_t, s_{t+1}\}$ generated by each iteration is stored in a *N*-sized memory $D = \{e_1, e_2, \cdots e_N\}$. The principle of this memory is similar to a queue: when the memory is full, the new information produced by the next iteration will replace the oldest information. In each iteration, a mini-batch of information uniformly sampled from memory will be used to update the parameters of the networks. This can significantly reduce the correlation between information and make the training process more stable.

It should be noted that, during the training, the calculation of the reward should be based on the real-time identification results of the system by the model networks. Therefore, the model networks should be added to the DDPG to assist in the training of action networks and critic networks. The specific structure is shown in Figure 8.14 and the procedure of the algorithm is listed in Table 8.7.

After the training, the agent can provide the optimal STATCOM-ADC parameter settings based on the system states. It should be noted that this chapter takes (8.34) as the reward for training. Therefore, the parameters provided by the agent have two characteristics:

1. The STATCOM-ADC parameter settings provided for one state of the system can ensure the stability of the system in this state.
2. When the state of the system changes, these STATCOM-ADC parameter settings can still ensure the stability of the system. It means that the parameters

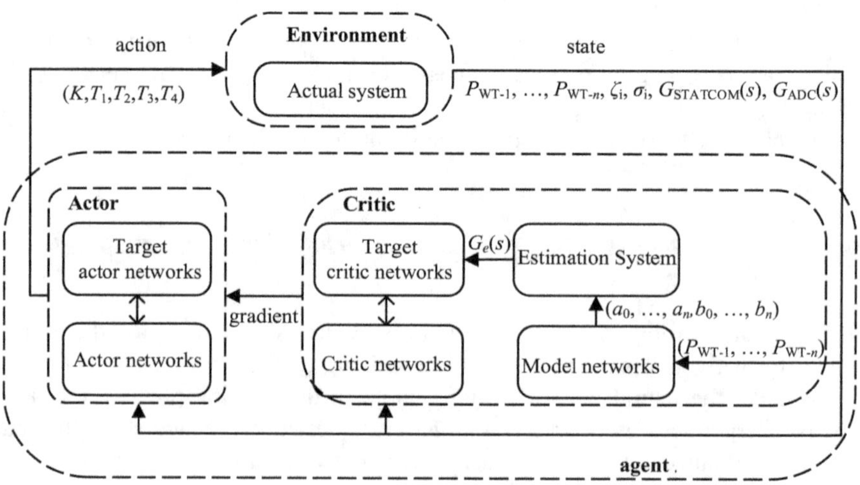

Figure 8.14 Adaptive robust agent training diagram [9]

Table 8.7 Policy learning based on model networks and DDPG

Input: The state of the power system: $(P_{\text{WT-1}}, \ldots, P_{\text{WT-}n}, G_{\text{STATCOM}}(s), G_{\text{ADC}}(s), \zeta_i, \sigma_i)$
Output: STATCOM-ADC parameter settings: (K, T_1, T_2, T_3, T_4)
1: **Initialize:**
Randomly initialize critic network and actor networks with weight w and μ, respectively
Randomly initialize critic network and actor networks with weight $w' \leftarrow w$ and $\mu' \leftarrow \mu$, respectively
Initialize the N-sized memory
2: **for** episode =1 **to** max-episode **do**
3: Initialize a random process ϑ for exploration
4: Initialize the state of the system
5: Observe the initial state of the system s_1
6: **for** time step = 1 **to** max-step size **do**
7: Select action $a_t = \pi(s_t) + \vartheta_t$
8: The action a_t is sent to the system and a simulation is carried out.
9: Obtain the reward a_t and new state s_{t+1}
10: Store the information (s_t, a_t, a_t, s_{t+1}) in the memory
11 If the memory is full:
12 Randomly sample mini-batch of information from memory
13 Update critic networks based on (8.44)
14 Update actor networks based on (8.46)
15 Update the target actor networks and target critic networks based on (8.47)
16: **end**
17: **end**

---

provided by the agent have robustness under different operating conditions, and they can be updated until their performance deteriorates.

Therefore, the policy learned by the agent has both adaptability and robustness. Compared with other adaptive control policies, the STATCOM-ADC parameter settings provided by the proposed agent need not be updated as soon as the system state changes, and they can be updated until they fail to ensure the stability of the system. The online application method listed below is based on the analysis mentioned above:

Step 1: Based on the dynamic trajectory of the system (e.g., the power angle of the generator), the eigenvalues of the system can be identified by the Prony algorithm.

Step 2: Analysis of the identified eigenvalues to determine whether the STATCOM-ADC parameters need to be updated or not. If not, the above steps are repeated, and if necessary, the strategy proceeds to the next step. In this paper, if all the eigenvalues satisfy the constraint in (8.34), it is considered that the system has satisfactory damping ratios and the STATCOM-ADC parameters need not be adjusted.

Step 3: Upload the power system status information. The agent provides the corresponding action based on the feedback information.

The above online application strategy, not only ensures the stability of the system but also avoids frequent adjustment of STATCOM-ADC parameters.

8.3.3 Numerical simulation results

In this section, to test the potential of the proposed approach in improving the stability of the power system, a practical WF located in Ningxia (a province in Northwest China) is adopted as a test system and the layout of the WF is shown in Figure 8.15. The test system consists of 66 direct-driven permanent magnet synchronous generators (DPMSGs) and the distribution of WTs in the wind farm is irregular due to the influence of the local geographical environment. More details about the WT can be found in Table 8.8. A 100 MVAr STATCOM is connected to the PCC. The STATCOM adopts the parameter settings in [13] and they are listed in Table 8.9.

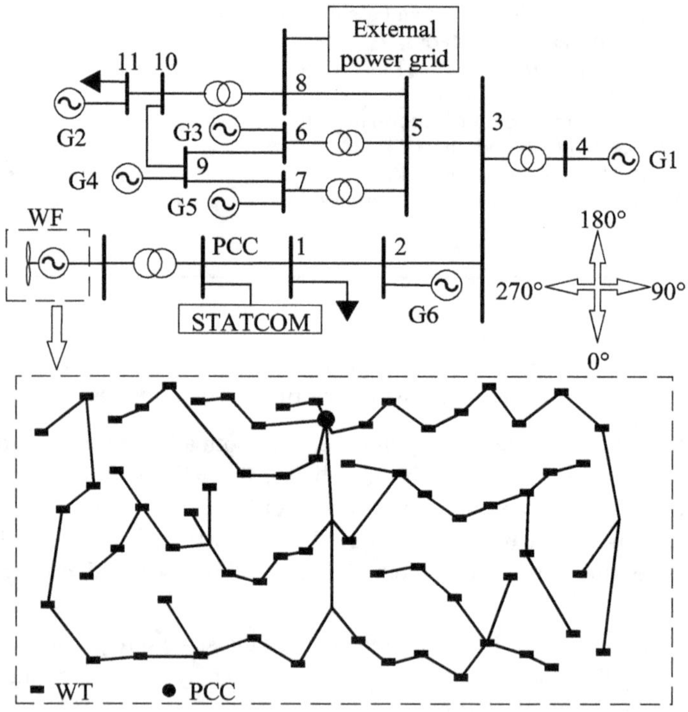

Figure 8.15 Transmission network for case study [9]

Table 8.8 Direct-driven permanent magnet synchronous generators WT specification in the wind farm

Parameter	Value	Parameter	Value	Parameter	Value
Cut-in wind speed	3 m/s	Rated wind speed	11.8 m/s	Cut-out wind speed	25 m/s
Rotor diameter	35.2 m	Rated power	1.5 MW	Rated rotor speed	15.4 rpm

Table 8.9 Parameters of STATCOM model

Parameter	Value	Parameter	Value	Parameter	Value
K_{pac}	55.36	K_{iac}	25.96	T_{ac}	10
K_{pdc}	18.09	K_{idc}	23.45	T_{dc}	10
V_{DC}	40 kV	C_{DC}	375 μF	—	—

The power grids outside the WF consist of two synchronous generators (G1 and G2), four equivalent wind farm models (G3, G4, G5, and G6), and two loads (The detailed information about these generators and loads can be found in [19]). The excess power generated by the generators will be sent to the external power grid through bus 8. For the convenience of calculation, an equivalent synchronous generator is used instead of an external power grid in the simulation process.

The success of the STATCOM-ADC tuning method depends on the accuracy of model networks in identifying the transfer function of the system. Therefore, the model networks should be carefully trained by trial and error to ensure identification accuracy. The model networks contain two hidden layers with 15 neurons and 20 neurons, respectively. The core parameters ρ, η, and α of the model networks are set to 0.5, 0.4, and 0.1, respectively. Moreover, a 1-year-long realistic wind speed data [20] is used to generate two datasets: a training set and a test set. In the case study, 90% of the data is allocated to the training set for the training of the model networks, and the other data is allocated to the test set for the validation of the model networks.

To excite the characteristics of the system, a pulse signal is used as the excitation signal. Taking this excitation signal as the input, the model networks are trained to minimize the difference between the output of the real system and the estimation system under different WT production. During the training, the order is set at 15.

Moreover, to test the ability of the proposed model networks against noise, a list of numbers is generated by random sampling within the range of -0.0065 and 0.0065, as shown in Figure 8.16. During the training, these data are attached to the output of the test system under the excitation signal to simulate noise.

To verify the effectiveness of the proposed identification method in suppressing noise, two methods are used to train the model networks:

1) Traditional identification method: The model networks are trained based on (8.37) and the noise is not processed.
2) Proposed identification method: The model networks are trained based on (8.38) and the effect of noise on identification is considered.

After the training, the model networks can accurately identify the transfer function of the system under different wind speeds. To test the effectiveness of the training model networks, four different wind speeds are selected from the test set and training set to form the cases, and details are shown as follows:

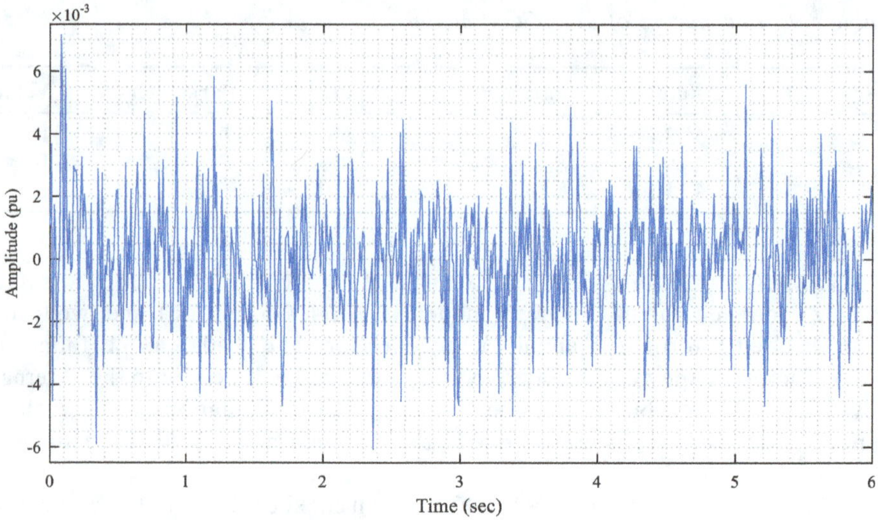

Figure 8.16 Noise data generated by random sampling [9]

Case 1: The wind speed is 8 m/s and the wind direction is 20°, which is selected from a training set.
Case 2: The wind speed is 5 m/s and the wind direction is 30°, which is selected from a training set.
Case 3: The wind speed is 6 m/s and the wind direction is 90°, which is selected from the test set.
Case 4: The wind speed is 12 m/s and the wind direction is 60°, which is selected from the test set.

Figure 8.17 shows the output of the estimation system trained by the above two methods. It can be seen that, compared with the traditional identification method, the proposed identification method makes the difference in output between the actual system and the estimation system much smaller under different wind speeds. To further compare the two identification methods, mean squared error (MSE) is introduced to quantify the difference between outputs and is listed in Table 8.10. Clearly, in the four cases, the MSE of the estimation system output trained by the proposed identification method is much less than the traditional identification method. Therefore, it means that the proposed identification method can effectively suppress the influence of noise, and it can successfully learn to identify the transfer function of the system under different wind speeds, which makes the estimation model have similar characteristics to the actual system.

The training of DDPG can be described as a process of getting the optimal control policy through trial and error. The purpose of DDPG training is to set up an agent that can give the optimal STATCOM-ADC parameters settings in real time under different wind speeds. As mentioned in the above section, the STATCOM-ADC parameters can be evaluated by (8.34). However, (8.34) cannot be applied

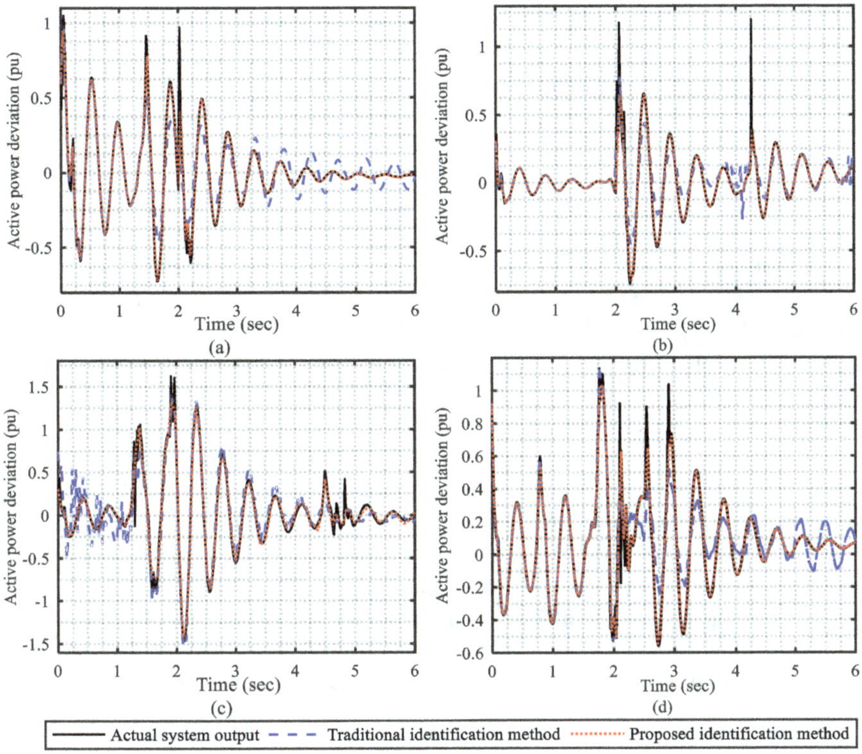

Figure 8.17 System response under different wind speeds: (a) Case 1; (b) Case 2; (c) Case 3; (d) Case 4 [9]

Table 8.10 MSE of the estimation system

Strategies	Case 1	Case 2	Case 3	Case 4
Traditional identification method	7.192×10^{-4}	5.618×10^{-4}	0.0012	7.498×10^{-4}
Proposed identification method	2.583×10^{-6}	8.874×10^{-6}	1.412×10^{-4}	3.841×10^{-6}

directly to the training of the agent, and it needs to be transformed as follows:

$$r(s_t, a_t) = -\left(J_{robust}(t) + J_{\text{punish}}(t)\right) \tag{8.48}$$

where $J_{\text{robust}}(t)$ is the robust function of STATCOM-ADC parameters settings.

$$J_{robust}(t) = \|G(s)/(1 + G(s)G_{\text{ADC}}(s)G_{\text{STATCOM}}(s))\|_{\infty} \tag{8.49}$$

where $J_{\text{punish}}(t)$ is the penalty function when the constraints of eigenvalues are not satisfied.

$$J_{\text{punish}}(t) = \sum_{i=1}^{n} f(\xi_i, \sigma_i)$$
$$f(\xi_i, \sigma_i) = \begin{cases} 0 if \xi_i \geq \xi_{spec}, \sigma_i \leq \sigma_{\text{spec}} \\ \delta else \end{cases} \tag{8.50}$$

To make the eigenvalues have a sufficient stability margin: $\zeta_{\text{spec}} = 0.05$, $\sigma_{\text{spec}} = -0.25$. The coefficient δ can be set to 5. Due to the limited range of the STATCOM-ADC parameters, the range for $K\omega$ is [5, 100] and for Tj is [0.05, 10]. During the training, they can be transferred to the upper and lower bounds of agent output. In this way, each parameter of STATCOM-ADC can meet the parameter bounds.

Before the training, the numbers of layers and neurons are carefully chosen. In this study, all four networks (actor networks, target actor networks, critic networks, and target-critic networks) in the DDPG adopt the same structure, which contains two hidden layers with 200 neurons and 30 neurons, respectively. Moreover, the core parameters of the DDPG algorithm are listed in Table 8.11.

The training of the agent follows the process shown in Table 8.7. Moreover, the data in the training set is used for the training of the agent. To ensure the effectiveness of the training, the agent was trained for 9,000 epochs to learn the control policy of the STATCOM-ADC and the training process of the agent is shown in Figure 8.18.

It can be observed from Figure 8.18 that the training process of the DDPG can be divided into three stages. The first phase is the data storage stage, which is from the 1st to the 800th epoch. During this stage, data is randomly selected and stored in memory. After the memory is filled up, the training process enters the second stage, which is from the 801st epoch to the 2,500th epoch. In this stage, based on the policy gradient, the parameters of the networks in DDPG are optimized to get the maximum cumulative reward. In this way, the agent learns to adjust STATCOM-ADC parameters to deal with the uncertainty of the wind power. After 2,500 epochs, it can be seen that the cumulative reward converges, which means that the agent has successfully learned the STATCOM-ADC parameter control policy to obtain higher rewards.

After the training, the agent can in real-time adjust the STATCOM-ADC parameters based on the present wind speed. For comparison, two other common strategies are used as examples to evaluate the superiority of the trained agent:

Table 8.11 Parameters used in DDPG algorithm

Parameter	Value	Parameter	Value
Discount factor (γ)	0.9	Learning rate for actor network (n_a)	0.001
Soft update coefficient (τ)	0.0001	Learning rate for critic network (n_c)	0.002
Experience replay memory capacity	8000	Mini-batch size	40
Step size of each episode	10	—	—

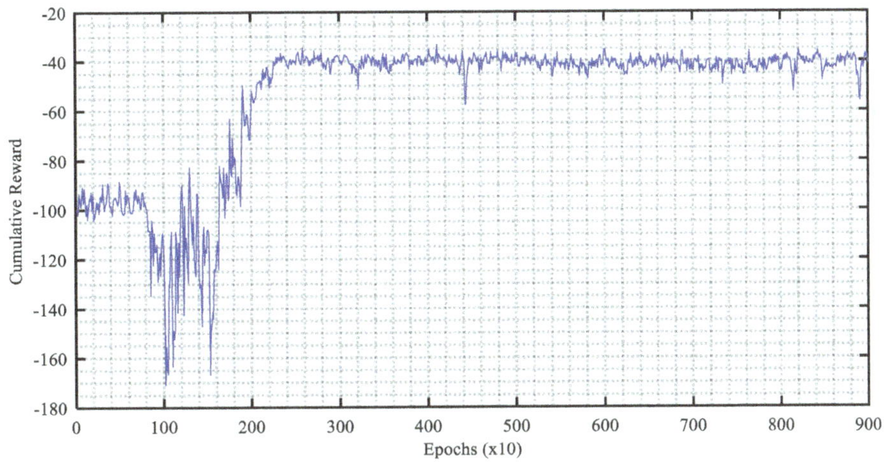

Figure 8.18 Cumulative reward change with an episode during the training process [9]

Conventional STATCOM-ADC: The parameters of STATCOM-ADC are adjusted by using the genetic algorithm proposed in [21].

Robust STATCOM-ADC: The parameters of STATCOM-ADC are adjusted by a robust parameter optimization strategy proposed in [22].

Subsequently, taking Case 1 as an example, the performances of the STATCOM-ADC parameters tuned by the proposed method and the other two methods are compared during the Fault 1 condition (Fault 1 is the three-phase ground fault occurring at the line between bus 2 and bus 3, starting at 2.0 s and ending at 2.2 s) as shown in Figure 8.19.

Figure 8.19 shows the dynamic response of the system with different STATCOM-ADC parameter settings. Obviously, without the STATCOM-ADC, both the relative power angle (power angle difference between G1, G2, and the equivalent generator) and the active power trajectory of the generators oscillate for a long time before they are restored to a steady state. After the STATCOM-ADC is applied to the system, the duration of the system oscillation is shortened. It means that the STATCOM-ADC can enhance the ability of the system to suppress oscillations. However, different parameter setting methods make the STATCOM-ADC show different performances. Clearly, compared with the Conventional STATCOM-ADC and the Robust STATCOM-ADC, the proposed method shows the best performance, and it can make the power angle and active power of generators reach a steady state in the shortest time. Therefore, it can be concluded that the proposed method is superior to other methods.

Moreover, the eigenvalue analysis method is also used to verify the control effect of the proposed agent. In this study, the Prony algorithm is applied to identify the electromechanical oscillation modes of the system and three dominant eigenvalues are obtained (Mode 1: $\xi = 3.45\%$, Mode 2: $\xi = 3.68\%$, Mode 3: $\xi = 7.53\%$). The

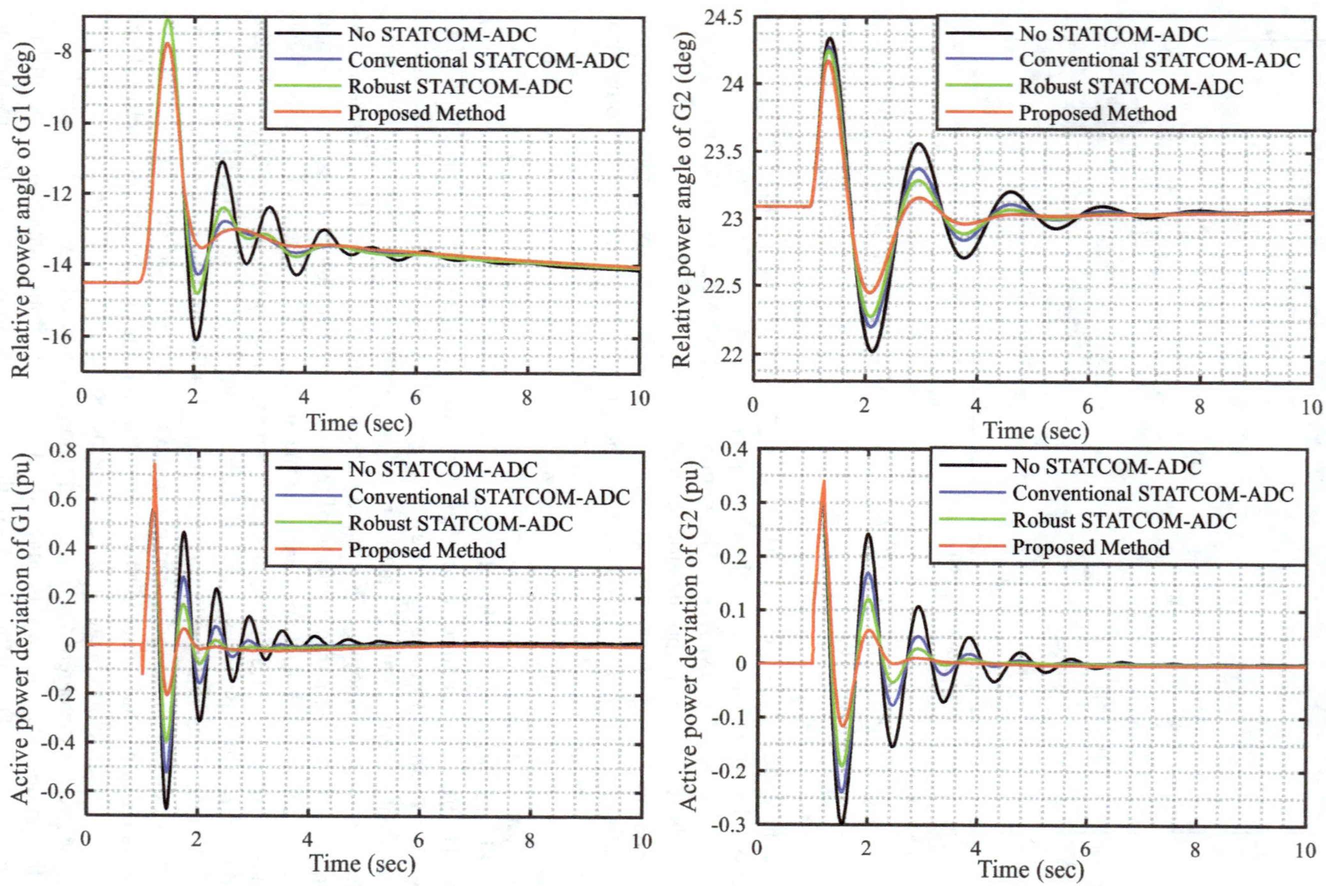

Figure 8.19 System dynamic response under Fault 1 [9]

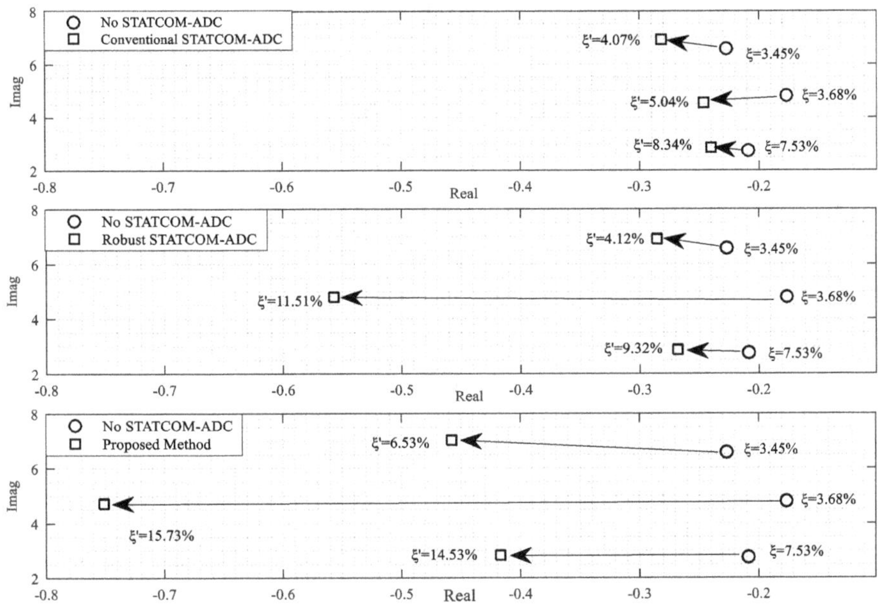

Figure 8.20 Dominant eigenvalues transfer by different strategies in Case 1 [9]

change of these three eigenvalues under different strategies is shown in Figure 8.20. It can be seen that, without the STATCOM-ADC, two dominant eigenvalues (Model 1 and Model 2) are poor damping modes ($\xi < 0.05$). After the STATCOM-ADC is configured, all the dominant eigenvalues move to the left and the damping increases, which makes Model 1 and Model 2 have a satisfactory damping ratio. However, compared with the other two methods, the proposed method can provide better damping for the dominant eigenvalues. It means that the proposed method can make the STATCOM-ADC show better control performance and make the system have larger stability margins. It is consistent with the conclusion of the time domain simulations in Figure 8.19.

One advantage of the proposed method is that the trained agent is adaptive, and it can provide appropriate parameter settings for the STATCOM-ADC based on the state of the system. To evaluate the adaptability of the proposed method under different wind speeds, the agent is tested in three cases (Case 2, Case 3, Case 4). For each case, different faults are used to excite the dynamic response of the system, and the simulation results are shown in Figures 8.21–8.23.

Fault 2: Three-phase to a ground fault occurring at the line between bus 3 and bus 5, starting at 2.0 s and ending at 2.2 s.

Fault 3: Double-phase to a ground fault occurring at the line between bus 7 and bus 9, starting at 2.0 s and ending at 2.2 s.

Fault 4: Single-phase to a ground fault occurring at the line between bus 10 and bus 11, starting at 2.0 s and ending at 2.2 s.

Figures 8.21–8.23 show the change in active power deviations of generators G1 and G2 under different cases. Clearly, as discussed above, without the STATCOM-ADC, time-domain simulation shows that the active power oscillations of the generator last for a long time, whereas the STATCOM-ADC with different

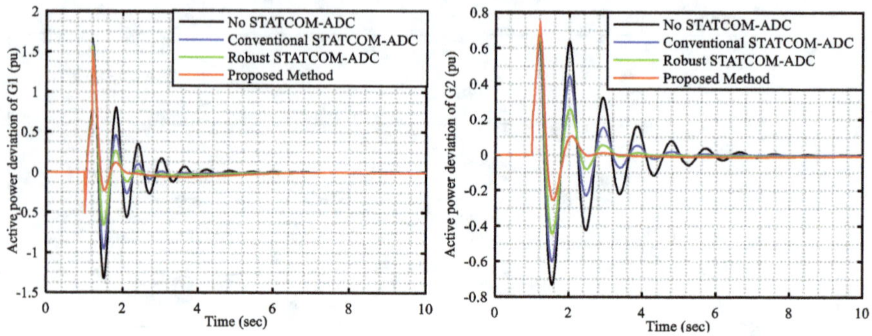

Figure 8.21 Power oscillations in Case 2 during Fault 2 [9]

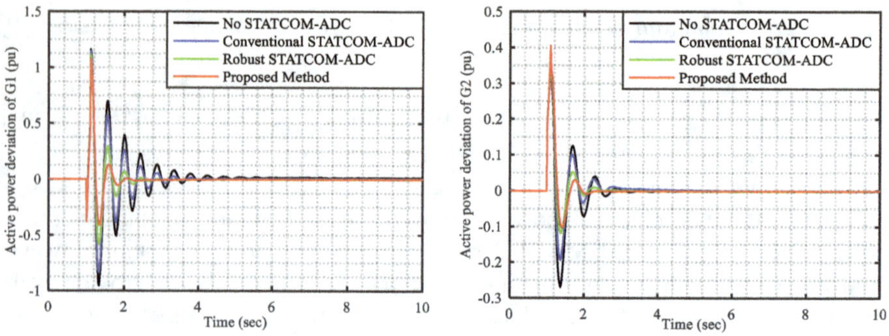

Figure 8.22 Power oscillations in Case 3 during Fault 3 [9]

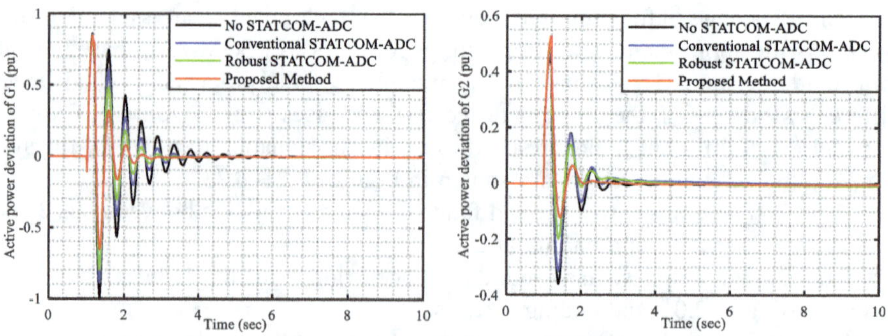

Figure 8.23 Power oscillations in Case 4 during Fault 4 [9]

sets of parameter settings can improve the ability of the system to suppress the active power oscillations. Among the three strategies, the parameter settings provided by the proposed method can show better performance than the Conventional STATCOM-ADC and the Robust STATCOM-ADC in all three cases. It means that the trained agent can make good decisions in an uncertain environment, and it can provide optimal parameter settings based on the state of the system.

Another advantage is that the control agent has acquired robustness. The parameter settings provided by this agent for one state of the system can still show good performance in the other states. To evaluate the adaptability of the proposed method, the parameter settings provided by the proposed agent for Case 1 are taken as an example to test its effectiveness in the other three cases (Case 2, Case 3, Case 4). Moreover, the other parameter settings provided by the traditional adaptive control strategy (TACS) [12] for Case 1 are used as an example for a case comparison.

Table 8.12 shows the distribution of the dominant eigenvalues. By comparing the damping changes of dominant modes under three methods, the following conclusions can be drawn:

1) For the parameter settings provided by TACS and the proposed agent, in Case 1, they can enhance the damping ratio of dominant eigenvalues and make poor modes (Mode 1 and Mode 2) have a satisfactory damping ratio. It means that both TACS and the proposed agent are adaptive, and they can provide appropriate parameter settings for the STATCOM-ADC based on the state of the system.
2) During Case 2, Case 3, and Case 4, the parameter settings provided by TACS have a limited effect on Mode 1 and they can only provide slight damping for Mode 1. Therefore, the damping ratio of Mode 1 is still poor ($\xi < 0.05$) in these three cases. It means that the parameter settings provided by the TACS according to one state of the power system have less control over the other states.
3) In Case 4, like the TACS, the proposed method has a limited effect on Mode 1. However, compared with the TACS, during Case 2 and Case 3, the parameter settings provided by the proposed agent can show good performance and ensure

Table 8.12 Comparison of dominant eigenvalues by two different strategies

Strategies	**Cases**	**Mode 1**		**Mode 2**		**Mode 3**	
		Real(σ)	**Damp(ζ)**	**Real(σ)**	**Damp(ζ)**	**Real(σ)**	**Damp(ζ)**
TACS	Case 1	0.538	7.84%	−0.747	15.66%	−0.541	18.58%
	Case 2	−0.285	4.12%	−0.609	13.21%	−0.225	8.03%
	Case 3	−0.301	4.75%	−0.572	12.64%	−0.264	9.20%
	Case 4	−0.261	3.85%	−0.681	14.11%	−0.208	7.53%
Proposed Method	Case 1	−0.458	6.53%	−0.751	15.73%	−0.417	14.53%
	Case 2	−0.418	6.27%	−0.529	11.66%	−0.341	12.04%
	Case 3	−0.335	5.31%	−0.569	12.58%	−0.237	8.23%
	Case 4	−0.256	3.76%	−0.594	13.52%	−0.259	9.12%

that all modes have satisfactory damping. It means that the proposed control agent has better robustness than the traditional TACS in different cases.

8.4 Deep reinforcement learning enabled bi-level optimization of hydro-governor parameters for ultra-low frequency oscillation control

Recent studies indicate that hydro-dominant power systems are easily subject to ultra-low frequency oscillation (ULFO), which has been observed in China as well as Nordic and Colombia. If ULFO is not properly controlled, there is a risk of system instability. Therefore, this chapter analyzes the theoretical mechanism of ULFO and designs a reinforcement learning-based optimization strategy for proportion integration differentiation (PID) parameters of the hydraulic turbine governor.

8.4.1 Modeling of hydro-governor system

This section first introduces a two-machine system. Then, based on it, the mechanism of ULFO is studied. Moreover, the mitigation effect of retuning PID parameter settings for ULFO is analyzed via the Routh criterion.

Figure 8.24 shows a two-machine system, which is introduced to act as the study system to investigate ULFO, including a hydropower unit, a thermal-power unit, and a load. The key parameter settings of this system are provided in the appendix. The hydropower unit mainly consists of a generator, a governor, and a turbine. Among them, the linearized dynamic equation of the generator can be described as follows:

$$T_J \frac{d\Delta\omega}{dt} = \Delta P_m - \Delta P_e - D_s\Delta\omega \tag{8.51}$$

Here, T_J represents the inertia time constant; $\Delta\omega$ is the rotor speed deviation; ΔP_m represents the mechanical power deviation; ΔP_e is the electromagnetic power deviation; D_s is the damping coefficient.

Assuming a frequency-dependent load as $\Delta P_e = K_L\Delta\omega$, where K_L represents the load frequency sensitivity and substituting $\Delta P_e = K_L\Delta\omega$ into (8.51), the transfer

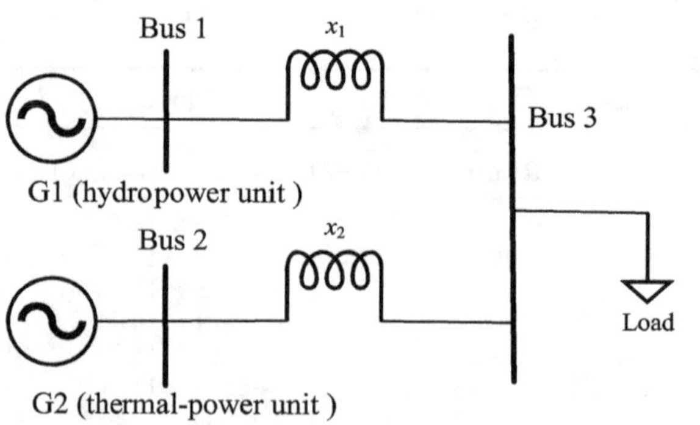

Figure 8.24 Schematic of the two-machine system [25]

function of the generator can be described as follows [24]:

$$G_{gen}(s) = \frac{\Delta\omega(s)}{\Delta P_m(s)} = \frac{1}{T_J s + D_s + K_L} = \frac{1}{T_J s + D} \tag{8.52}$$

where $D = D_s + K_L$; $G_{gen}(s)$ denotes the transfer function of the generator. The transfer function of the prime mover (governor and turbine) can be defined as [25]:

$$\begin{aligned} G_m(s) &= \frac{\Delta P_m(s)}{-\Delta\omega(s)} = G_{gov}(s)G_t(s) \\ G_{gov}(s) &= \frac{K_D s^2 + K_P s + K_I}{b_p K_I + s}\frac{1}{1 + T_G s} \\ G_t(s) &= \frac{1 - T_w s}{1 + 0.5 T_w s} \end{aligned} \tag{8.53}$$

where G_{gov} and G_t denote the transfer function of the governor and turbine, respectively; K_P, K_I, and K_D are the proportional, integral, and differential coefficients of the governor, respectively; b_p is the adjustment coefficient of the governor; T_G denotes the response time of the governor; T_w is the water hammer effect and it depends on the operating conditions of the hydro-turbine, that is given as [26],

$$T_w = \frac{\sum_{i=1}^{n} \frac{L_i}{A_i} Q_r}{g H_r} \tag{8.54}$$

Here, L_i and A_i denote the length and sectional area of the ith diversion pipeline, respectively; Q_r denotes the rated water flow; g denotes gravitational acceleration; H_r denotes the rated water head; n denotes the number of diversion pipelines.

For the thermal-power unit, this chapter ignores the boiler dynamic process and mainly considers the steam turbine and governor, that is [8]:

$$\Delta P_{mh}(s) = -G_h(s)\Delta\omega(s) = \frac{-K_a \Delta\omega(s)}{(1 + sT_g)(1 + sT_{ch})} \tag{8.55}$$

where K_a denotes the gain; T_g and T_{ch} denote the time-step of the hydraulic system and high-pressure cylinder, respectively.

A time domain simulation of the above study system under Fault 1, i.e., a two-phase short circuit fault at bus 3 from 2.0 s to 2.2 s is shown in Figure 8.25. It can be observed that the angle speed of two generators varies with the same phase and amplitude. No oscillations between two units are observed. It is different from traditional LFO. The Prony method [10] is utilized to identify the oscillation model, and the identification results show that an ULFO mode −0.0001 + j 0.403 exists in the system.

To investigate the cause of ULFO in the study system, the damping torque coefficient of two units' primary frequency regulation (PFR) is calculated. Among them, the detailed calculation process of the damping torque coefficient of

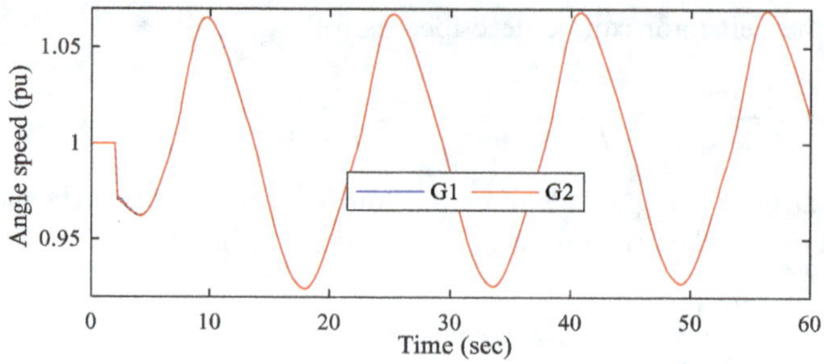

Figure 8.25 Dynamic response of G1 and G2 under Fault 1 [25]

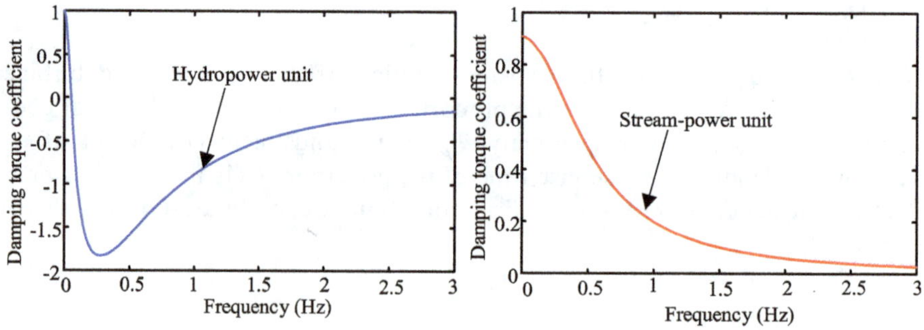

Figure 8.26 Damping torque curve of hydropower and stream power units [25]

hydropower is provided in [7] and it can be defined as:

$$D_H(s) = \frac{1 - 0.5T_G T_w \omega_d^2 - \omega_d^2 T_\omega (T_G + 0.5T_w)}{(1 - 0.5T_G T_w \omega_d^2) + \omega_d^2 (T_G + 0.5T_w)} \tag{8.56}$$

Bring the $s = j\omega_d$ into (8.53), the damping torque coefficient of the hydropower can be obtained [8]:

$$\Delta P_m = -D_T \Delta\omega - jD_{IM} \Delta\omega \tag{8.57}$$

Where

$$D_T = \frac{K_a (1 - T_g T_{ch} \omega_d^2)}{(1 - T_g T_{ch} \omega_d^2)^2 + (T_g + T_{ch})^2 \omega_d^2} \tag{8.58}$$

Submitting $\omega_d = 2\pi f$ into (8.56) and (8.58), the trajectory of damping torque coefficients D_H and D_T changed with frequency f can be obtained, and they are shown in Figure 8.26.

It can be observed that the hydropower unit will produce a negative damping below the ultra-low frequency band (below 0.1 Hz). In contrast, the thermal power unit would produce a positive damping in this band. Therefore, it can be concluded that the ULFO is strongly related to the hydropower unit.

To further investigate the relationship between ULFO and hydropower unit, the characteristic equation of the closed-loop transfer function of the hydropower unit is calculated based on (8.52) and (8.53) and it can be written as:

$$\begin{aligned}&(T_J s + D)(b_p K_I + s)(1 + T_G s)(1 + 0.5T_w s)+\\&(1 - T_w s)(K_D s^2 + K_P s + K_I) = 0\end{aligned} \tag{8.59}$$

After simplification, (8.59) can be rewritten as:

$$s^4 + a_1 s^3 + a_2 s^2 + a_3 s + a_4 = 0 \tag{8.60}$$

where a_1, a_2, a_3, and a_4 denote the Routh-Hurwitz criterion coefficient, and these coefficients are related to the water hammer effect coefficient T_w. Specifically, the trajectory of these coefficients and oscillation frequency ω with T_w change are shown in Figure 8.27. Note that, the detailed expressions of these coefficients and the calculation process of the oscillation frequency are provided.

It can be seen from Figure 8.27(a) that with the increase of T_w, all Routh-Hurwitz criterion coefficients decrease, and in particular, c_1 becomes negative. Based on the Routh-Hurwitz criterion [28], it can be concluded that the system is unstable and there is a negative damping oscillation mode in the system. In addition, Figure 8.27(b) shows that, with the increase in T_w, the oscillation frequency decreases below 0.1 Hz. It means that this dominant negative damping oscillation mode is ULFO mode.

In summary, the ULFO is strongly related to the hydropower units, and it is caused by the PFR of the hydropower units. More specifically, due to the water hammer effect, the hydraulic governor easily produces a negative damping torque (see Figure 8.26), thus making the system appear in negative ULFO mode.

On the other hand, Routh-Hurwitz coefficients are also related to the hydro-governor PID parameters. In fact, the PID parameter settings make these coefficients stay positive. Based on this consideration, we tested the trajectory of coefficient c_1

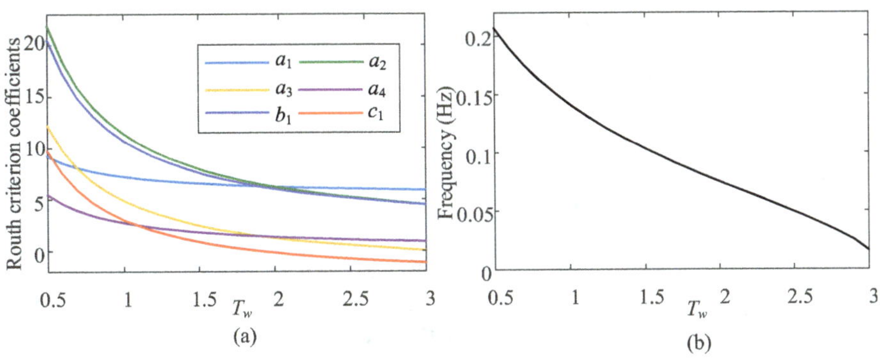

Figure 8.27 Dynamic performance indicators changed with the water hammer effect: (a) Routh criterion coefficients; (b) oscillation frequency [25]

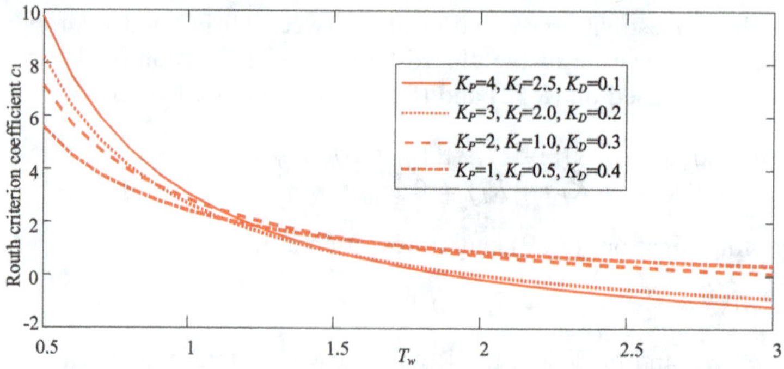

Figure 8.28 Routh criterion curve c_1 changed with different PID settings [25]

under different PID parameter setting conditions, and the results are shown in Figure 8.28. It can be seen that the tuning of PID parameters can make c_1 positive. In this way, the system becomes stable. This means that optimizing PID parameters contributes to preventing ULFO.

8.4.2 Formulation of governor parameter optimization

Based on the system linearization technology, the state matrix A can be obtained [29]. Then, (8.61) can be utilized to diagonalize matrix A to obtain eigenvalues of the system:

$$\begin{aligned} A &= V\Lambda \\ UV^T &= I \end{aligned} \tag{8.61}$$

where U and V denote the left and right feature vector, respectively; Λ denotes the diagonal matrix and it can be described as:

$$\begin{aligned} &\Lambda = \mathrm{diag}(\lambda_1, \lambda_2, \cdots \lambda_{\mathrm{ULFO}} \cdots, \lambda_m) \\ &\lambda_{\mathrm{ULFO}} = \sigma_{\mathrm{ULFO}} + j\beta_{\mathrm{ULFO}}; \xi_{\mathrm{ULFO}} = -\sigma_{\mathrm{ULFO}} / \sqrt{\sigma_{\mathrm{ULFO}}^2 + \beta_{\mathrm{ULFO}}^2} \end{aligned} \tag{8.62}$$

where λ_{ULFO} is the ULFO mode; n represents the number of eigenvalues; σ_{ULFO} and β_{ULFO} represent the real and imaginary parts of the λ_{ULFO}; ξ_{ULFO} is the damping ratio.

As mentioned above, optimizing the PID parameter settings contributes to the stability of ULFO mode λ_{ULFO}. However, some requirements should be paid attention to (1) less effect on the other oscillation modes: we should avoid weakening the damping of other modes when improving the damping of ULFO mode; (2) good dynamic performance for the PFR: previous studies indicate that if not properly designed, optimizing governor PID parameters may deteriorate the PFR dynamic performance, and thus weaken the frequency adjustment ability of the governor [30]; (3) robustness to the change of operating conditions: in the actual operation, the operating condition of the turbine is time-varying. The optimized PID parameters that show good performance under one operating condition may not work well for other conditions. Therefore, during the optimization process, extreme operating conditions should be considered. Based on the above considerations, the suppression of ULFO is modeled as a bi-level min-max optimization model with constraints, which is described as follows:

$$\begin{cases} \min J^{upper}\left(K_{P-i}, K_{I-i}, K_{D-i}, T^*_{w-i}\right) \\ K_{P-i}^{\min} \leq K_{P-i} \leq K_{P-i}^{\max} \\ K_{I-i}^{\min} \leq K_{I-i} \leq K_{I-i}^{\max} \\ K_{D-i}^{\min} \leq K_{D-i} \leq K_{D-i}^{\max} \\ T^*_{w,i} = \arg\max y_z^{lower}(K_{P-i}, K_{I-i}, K_{D-i}) \\ T_{w-i}^{\min} \leq T_{w-i} \leq T_{w-i}^{\max} \\ \xi_j'(K_{P-i}, K_{I-i}, K_{D-i}, T_{w-i}) > \xi_0 \\ i = 1, 2, ..., N \end{cases} \tag{8.63}$$

where $J\left(K_{P,i}, K_{I,i}, K_{D,i}, T_{w,i}\right)$ is the objective function of the optimization problem; $T_{w,i}$ is the time constant of the water hammer effect of the ith governor that is used to represent the operating conditions of the governor and limited between 0.5 and 3 [24]; ξ_j' is the damping of the jth oscillation mode except for the ULFO mode; ξ_0 denotes the desired damping ratio of the oscillation mode that is typically set as 0.05 [7]; N is the number of the governors; J^{upper} denotes the objective function of min-max optimization model and can be reformulated as [24]:

$$J^{upper}\left(K_{P,i}, K_{I,i}, K_{D,i}, T_{w,i}\right) = J_{STB} + J_{\mathrm{ITAE}}$$
$$\begin{cases} J_{STB} = \sum\limits_{\xi_{\mathrm{ULFO}} < \xi_{\mathrm{set}}} \left|\xi_{\mathrm{ULFO}} - \xi_{\mathrm{set}}\right| + \sum\limits_{\sigma_{\mathrm{ULFO}} > \sigma_{\mathrm{set}}} \left|\sigma_{\mathrm{ULFO}} - \sigma_{\mathrm{set}}\right| \\ J_{\mathrm{ITAE}} = \sum\limits_{i=1}^{N} \left(\int_0^T \left|t \cdot \left(\Delta P_m(t) - \Delta P_m{}^{\infty}\right)\right| dt\right) \end{cases} \tag{8.64}$$

where J_{STB} represents the stability of the ULFO. σ_{set} denotes the desired real part of the eigenvalues; ξ_{set} denotes the damping ratio of the eigenvalues; $\Delta P_m(t)$ represents the dynamic response of the governor under disturbance; $\Delta P_m{}^{\infty}$ represents the prime mover steady-state value under disturbances. In fact, J_{STB} can further be written as:

$$J_{STB} = \begin{cases} (\xi_{\mathrm{set}} - \xi_{\mathrm{ULFO}}) + (\sigma_{\mathrm{ULFO}} - \sigma_{\mathrm{set}}); \text{if } \xi_{\mathrm{ULFO}} < \xi_{\mathrm{set}}; \sigma_{\mathrm{ULFO}} > \sigma_{\mathrm{set}} \\ 0; \qquad \text{if } \xi_{\mathrm{ULFO}} \geq \xi_{\mathrm{set}}; \sigma_{\mathrm{ULFO}} \leq \sigma_{\mathrm{set}} \end{cases} \tag{8.65}$$

It can be seen from (8.65) that, when $\xi_{\mathrm{ULFO}} < \xi_{\mathrm{set}}; \sigma_{\mathrm{ULFO}} > \sigma_{\mathrm{set}}$, the J_{STB} can be defined as $(\xi_{\mathrm{set}} - \xi_{\mathrm{ULFO}}) + (\sigma_{\mathrm{ULFO}} - \sigma_{\mathrm{set}})$, by minimizing this target, the damping (ξ_{ULFO}) and real part (σ_{ULFO}) of ULFO mode would be pushed to be close to the predetermined damping and the real part. Note that the damping in ULFO mode is set to a higher value. By optimizing the PID parameter settings to make the damping of ULFO mode close to the predetermined damping, the damping of ULFO mode is improved. Once the damping and real part of ULFO mode reach the predetermined value, the J_{STB} becomes 0.

J_{ITAE} denotes the primary frequency control performance of the prime mover. By minimizing this target, the output deviation of the prime mover can be optimized to be close to the steady-state point. In this way, both the oscillation amplitude and oscillation time of the prime mover under disturbance can be optimized, which means that the dynamic performance of the primary frequency control can be improved.

8.4.3 Bi-level model solution method

For the bi-level optimization model, the lower-level model is reformulated as the MDP and solved by the DDPG algorithm. Subsequently, the trained agent is employed to assist a particle swarm optimization (PSO) to solve the upper-level model to obtain the optimal solution. The overall scheme of the optimization process is shown in Figure 8.29.

The core functions of the lower-level model shown in Figure 8.29 can be described as:

$$T^*_{w,i} = \arg\max y_z^{lower}(K_{P-i}, K_{I-i}, K_{D-i})$$
$$\begin{cases} T^{\min}_{w-i} \leq T_{w-i} \leq T^{\max}_{w-i} \\ \xi'_j(K_{P-i}, K_{I-i}, K_{D-i}, T_{w-i}) > \xi_0 \\ i = 1, 2, \ldots, N \end{cases} \tag{8.66}$$

In each episode, the upper-level model would provide a set of PID parameter settings $(K_{P-i}, K_{I-i}, K_{D-i})$ to the lower-level model. Based on this, the lower-level model searches for the extreme operating condition $T^*_{w,i}$ with the worst performance under this PID setting condition. In this way, both the upper-level variables and lower-level variables are determined, and the objective function (8.64) can be calculated and sent to the upper-level model.

In fact, the function of the lower-level model is to find the corresponding extreme operating condition under every PID setting condition. Therefore, the key to solving this lower-level model is to find the policy function $T^*_{w-i} = \pi(K_{P-i}, K_{I-i}, K_{D-i})$ that maximizes the objective function. It is a decision-making problem in an uncertain environment. Here it is reformulated as an MDP, and the key elements related to MDP are defined as a tuple, <*S*, *A*, *P*, *R*>, where *S* denotes the state. In this study, it is composed of the PID parameter settings of each hydro-governor; *A* denotes action, and it is represented as the operation conditions of the governor; *P* denotes the transition probability; *R* denotes the reward, and it is used to evaluate the action taken by the agent at each time step. In this study, the reward is defined as:

$$r(s_k, a_k) = J\left(K^k_{P,i}, K^k_{I,i}, K^k_{D,i}, T^k_{w,i}\right) + \sum_{j=1}^{m} \sum_{\xi_0 < \xi'_j} \left(\xi_0 - \xi'_j\left(K^k_{P,i}, K^k_{I,i}, K^k_{D,i}, T^k_{w,i}\right)\right) \tag{8.67}$$

The reinforcement learning algorithm is a common method to solve this MDP. The background of reinforcement learning is described as follows:

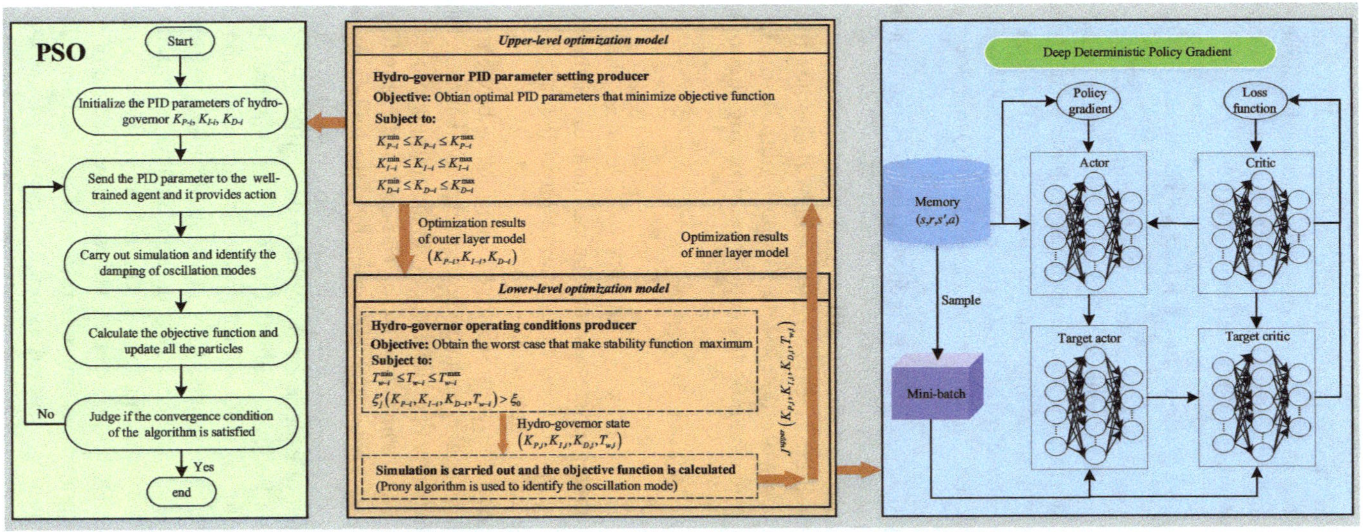

Figure 8.29 The overall scheme of the optimization process [25]

1) Q-Learning: It is a widely used reinforcement learning method, where an agent is trained to learn the optimal control policy to maximize the cumulative reward. In Q-learning, each state-action pair (s_k, a_k) is assigned a Q-value, stored in the Q-table, and denoted as $Q(s_k, a_k)$. For a given state, the higher Q-value for an action denotes a higher potential cumulative reward, and it can be utilized to update the Q-value via a Bellman equation [14]:

$$Q(s_k, a_k) \leftarrow Q(s_k, a_k) + \alpha[r(s_k, a_k) + \gamma \max Q(s_{k+1}, a_{k+1}) - Q(s_k, a_k)] \quad (8.68)$$

where γ is the discount rate; α denotes the learning rate. The ε-greedy policy is adopted in Q-learning, where the agent will choose the action with the highest Q-value during the training process. Note that Q-learning is only suitable in cases where the state-action space is small. The increase in state-action space size would make the Q-table too big, resulting in each Q-value being rarely updated. In this context, a deep Q-network is proposed to address it.

2) Deep Q-Network (DQN): DQN is proposed for solving high-dimension state-action problems via combining deep neural networks (DNN) with the Q-learning method. Specifically, a DNN is utilized to approximate the Q-table, named Q-network, represented as $Q(s_k, a_k, \theta)$, where θ denotes the parameters of the DNN. The Q-network takes state as input, outputs the Q-value for each state-action pair, and it can be trained by minimizing the loss function:

$$L(\theta) = \mathbb{E}\left[y - Q(s_k, a_k, \theta)^2\right] \quad (8.69)$$

where $y = r(s_k, a_k) + \gamma \max Q(s_{k+1}, a_{k+1}, \theta')$ denotes the target Q-value. To stabilize the training process, it can be calculated via the target Q-network $Q(s_{k+1}, a_{k+1}, \theta')$, where θ' denotes the parameter of the target Q network and it is updated via the soft-update method [31].

Moreover, a replay buffer is employed in DQN to break the identically distributed state samples and reduce the correlation between them, leading to improved data efficiency. Specifically, during the training process, all information is saved as an experience $e = (s_t, a_t, r_t, s_{t+1})$ and stored in the memory $D = \{e_1, e_2, \cdots, e_M\}$. After the buffer is full, the oldest experience will be replaced by the newly obtained one. Subsequently, at each iteration, the agent will sample a mini-batch of experience from the replay buffer.

3) Deep Deterministic Policy Gradient (DDPG): The standard DQN method only works effectively for solving control problems with continuous states and low-dimension discrete action sets. It is not suitable to solve the PID parameter settings optimization problem. To this end, DDPG is introduced and it can achieve better performance in solving control problems with continuous action space compared to DQN. Figure 8.29 shows the procedure of the DDPG and it consists of two eponymous ingredients: an actor network is utilized to fit the state s_k to action a_k, denoted as policy function $\mu(s_k|\theta^\mu)$ via adjusting the network parameter $\theta^\mu = \{W_1^a, b_1^a, \ldots, W_n^a, b_n^a\}$; a critic network is utilized to

fit the action-value function $Q(s_k, a_k|\theta^Q)$ via adjusting network parameter $\theta^Q = \{W_1^c, b_1^c, \ldots, W_m^c, b_m^c\}$.

During the training process, these two networks are trained against each other. Among them, the critic network can be updated by the loss function [32]:

$$L(\theta^Q) = \frac{1}{N}\sum_{i=1}^{n}\left[\left(Q(s_k, a_k|\theta^Q) - y_k\right)^2\right] y_k = r(s_k, a_k)$$
$$+ \gamma Q\left(s_{k+1}, \mu\left(s_k|\theta^\mu\right)|\theta^Q\right)\theta_{k+1}^Q = \theta_k^Q + \alpha^Q \cdot \nabla_{\theta^\mu} L\left(\theta_k^Q\right) \tag{8.70}$$

where N denotes the number of mini-batch; a^Q is the learning rate of the critic networks. Then, the parameters of the actor network can be updated by the gradient descent [9]:

$$\begin{aligned} \nabla_{\theta^\mu} J(\theta^\mu) &= \frac{1}{N}\sum_{i=1}^{N} \nabla_\alpha Q(s, a|\theta^Q)|_{s=s_k, a=\mu(s_k)+\mathcal{N}} \nabla_{\theta^\mu}\mu(s|\theta^\mu)|_{s=s_k} \\ \theta_{k+1}^\mu &= \theta_k^\mu + \alpha^\mu \cdot \nabla_{\theta^\mu} J\left(\theta_k^\mu\right) \end{aligned} \tag{8.71}$$

where α^μ is the learning rate of the actor network; $\mathcal{N}$ denotes the Gaussian noise.

Moreover, to stabilize the training process, both target actor-critic pair networks are added in DDPG, which can be parameterized by $\theta^{\mu'}$ and $\theta^{Q'}$. At each iteration, soft updating is utilized to synchronize its parameters to the target actor-critic networks [32]:

$$soft\ update \begin{cases} \theta^{Q'} \leftarrow \tau\theta^Q + (1-\tau)\theta^{Q'} \\ \theta^{\mu'} \leftarrow \tau\theta^\mu + (1-\tau)\theta^{\mu'} \end{cases} \tag{8.72}$$

where τ denotes the soft-update rate.

After offline training, the well-trained agent can learn the optimal control policy and provide extreme operating conditions ($T_{w,i}$) for each PID parameter setting of the system.

$$T_{w-i}^* = \mu(K_{P-i}, K_{I-i}, K_{D-i}|\theta^\mu)\ i = 1, 2, \ldots, N \tag{8.73}$$

In this way, the proposed min-max optimization model in (8.73) can be transformed into a single-level mathematical programming problem with constraints, i.e.,

$$\begin{cases} \min J\left(K_{P-i}, K_{I-i}, K_{D-i}, T_{w-i}^*\right) \\ K_{P-i}^{\min} \le K_{P-i} \le K_{P-i}^{\max} \\ K_{I-i}^{\min} \le K_{I-i} \le K_{I-i}^{\max} \\ K_{D-i}^{\min} \le K_{D-i} \le K_{D-i}^{\max} \\ T_{w-i}^* = \mu(K_{P-i}, K_{I-i}, K_{D-i}|\theta^\mu) \qquad i = 1, 2, \ldots, N \end{cases} \tag{8.74}$$

It can be seen from (8.73) that the bi-level model is converted into a nonlinear optimization problem. A heuristic algorithm is a good choice to solve that and here a PSO algorithm [33] is used. The detailed solution processes of the optimization model are as follows:

Step 1: Define the solution space and fitness function: This study used the PSO to find the optimal PID parameter settings for the governor. In this context, the particle position is designed as the PID parameter settings ($K_{P\text{-}i}$, $K_{I\text{-}i}$, $K_{D\text{-}i}$). The fitness function is applied to evaluate the training error and the goodness of a given solution, which is defined as the objective function and is shown in (8.74).

Step 2: Initialize the random swarm location and velocities: Before beginning to search for the optimal position, each particle is initialized with the random PID parameter setting within the allowable ranges. Moreover, the direction and length of movement of the particle at each episode is named velocity, which is also initialized.

Step 3: Calculate the fitness of each particle: The PID parameter setting carried by each particle is transmitted to a well-trained DDPG agent. Then, the actor network in DDPG can provide the extreme operating condition (T^*_{w-i}) with the worst performance for each particle under the corresponding PID parameter setting conditions. Next, both PID parameter settings and operating condition data of each particle can be updated to the study system to calculate the fitness via (8.74).

Step 4: Update the particle position and velocity: the velocity and position of each particle can be updated via:

$$\begin{cases} v_{i+1}^d = \omega v_i^d + c_1 r_1 \left(p_i^d - x_i^d\right) + c_2 r_2 \left(p_g - x_i^d\right) \\ x_{i+1}^d = x_i^d + v_i^d \end{cases} \tag{8.75}$$

where v_i^d is the velocity of the dth particle in ith iteration; x_i^d is the position of the dth particle in ith iteration; P_g is the best position all particles in the population obtained up to the nth iteration; p_i^d is the optimal position of the dth particle got up to nth iteration, ω is the inertia factor; c_1 and c_2 are the acceleration coefficients; r_1 and r_2 are random numbers in [0, 1]. In this chapter, $c_1 = c_2 = 2$; $\omega = 0.6$ [24].

Step 5: Find the optimal solution: Repeat Steps 2–3 until the minimum error is met, or the maximum number of iterations is reached. Output the final result as the solution to the above optimization problem.

8.4.4 Numerical simulation results

In this section, the performance of the proposed method is tested on the IEEE 10-machine 39-bus system. All generators adopt a fifth-order model and a simplified excitation system [30]. Eigenvalue analysis results show that a ULFO mode 0.00045 + j 0.55 exists in the system. The generators strongly related to this mode are G1, G2, G5, and G8. To dampen that oscillation, the ULFO suppression problem is a bi-level PID parameter settings optimization model, as shown in Figure 8.29. Then, the DDPG and PSO algorithms are combined to solve this model. Simulations are carried out via Python and power system analysis software package (PSASP). Among them, the study system is constructed in PSASP. PSO and DDPG are modeled in Python.

8.4.4.1 Performance evaluation of the trained agents

The numbers of layers and neurons for the networks in the DDPG algorithm are set as follows: both actor and critic networks adopt the same structure, which contains three hidden layers; the numbers of neurons for the hidden layers are 128, 64, and 64, respectively. The hyperparameters of the DDPG algorithm can be found in Table 8.13.

To obtain the corresponding extreme operating conditions ($T_{w,i}$) under different PID parameter settings that maximize the objective function, numerous scenarios are constructed for training. Specifically, we generate scenarios by randomly sampling from the upper and lower limits of the PID parameter settings of each hydro-turbine ($0 \leq K_{P-i} \leq 25$, $0 \leq K_{I-i} \leq 0.5$, $0 \leq K_{D-i} \leq 3$). Note that 95% of these PID parameter settings are used for training, while the remaining 5% are applied for testing. The training process of the DDPG algorithm is shown in Figure 8.30, where the blue solid lines and the shaded areas represent the

Table 8.13 Hyperparameters of the DDPG algorithm

Parameters	Values
Learning rate for actor network	0.001
Learning rate for critic network	0.002
Experience replay memory capacity	8,000
Step size of each episode	12
Mini-batch size	40

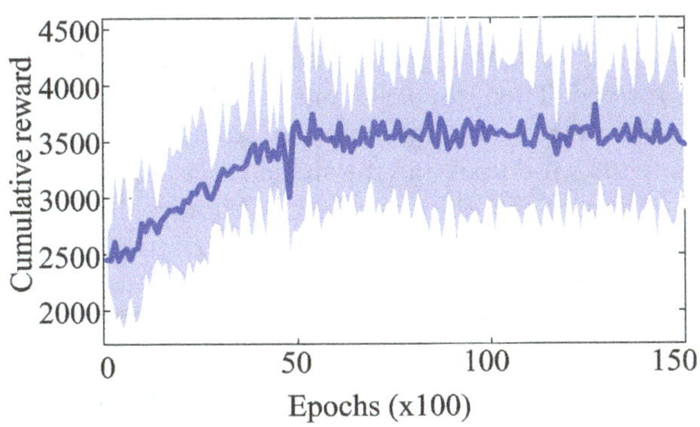

Figure 8.30 Cumulative reward changes with an episode during the training process [25]

Table 8.14 Computational efficiency comparison of different methods

Methods	Average solution time of lower-level model
Optimization method in [16]	13.4 s
Optimization method in [19]	3.7 s
Trained agent	0.05 s

average accumulated reward and the standard deviation of the reward, respectively. At the beginning of the training process, as the parameters of DDPG are randomly initialized, the agent does not learn the optimal policy that satisfies all constraints, leading to a low accumulated reward. As the training process goes further, the agent interacts with the system to obtain experiences to optimize the parameters of DDPG, and the cumulative reward increases gradually. After about 5,000 episodes, the algorithm converges.

After training, the well-trained agent can provide each PID parameter set to the corresponding extreme operating condition. After that, this agent can be applied to assist the PSO algorithm in solving the upper-layer model. To show the computational efficiency of this agent, we compared the solution methods of the lower-level optimization model in [6] and [34] with our trained agent. The comparison results are shown in Table 8.14. It can be observed that the proposed method achieves the highest computational efficiency.

Moreover, we further investigate the performance of the PSO and differential evolution (DE) to test the performance of these two methods. To balance the training time and optimization effect, after careful trial and error, the population size of these two methods is set to 30. Except for the population size, the hyperparameters of the algorithm have a great impact on the optimization performance. To deal with it, each hyperparameter in the algorithm will first be discretized into several points in its feasible region. Then, these discrete points form a test table using the orthogonal method.

In this way, nine sets of hyperparameter settings for the DE and PSO methods can be formed. Then, the PSO and DE algorithms can be trained with each set of hyperparameter settings, and the simulation results are shown in Figure 8.31, where the solid lines and the shaded areas represent the average and the standard deviation of the objective function under different trials, respectively. Moreover, some quantitative indicators are introduced to evaluate the optimization performance of PSO and DE methods, as shown in Table 8.15. It can be observed that the performances of both PSO and DE are almost similar in terms of the best fitness value obtained in the 18 runs. Specifically, the DE method is slightly better than the PSO method. However, the PSO method performs better compared to DE in terms of the average, the worst, and the standard deviation of the objective function.

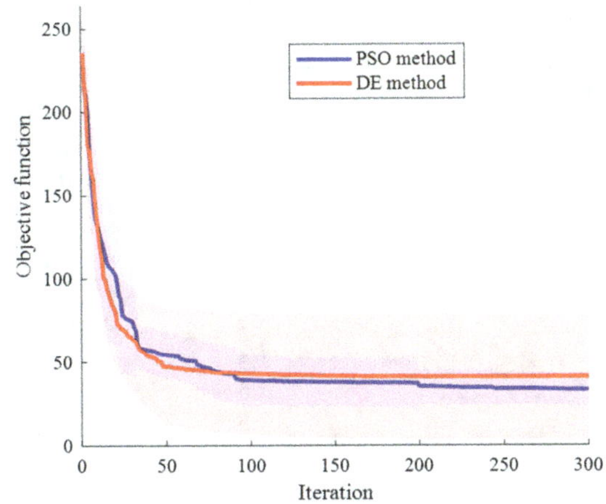

Figure 8.31 *Convergence process of DE and PSO algorithms with different hyperparameter settings [25]*

Table 8.15 *Quantitative indicators of optimization solutions of PSO and DE methods with different hyperparameter settings*

Values	**PSO method**	**DE method**
Best	26.42	25.01
Average	33.18	41.07
Worst	53.30	140.27
Standard deviation	9.71	37.68

8.4.4.2 Control performance comparisons

To test the effectiveness of the proposed method, five cases are tested by setting different time constants for the hydraulic turbine T_w, see Table 8.16. Other PID parameter setting tuning methods are compared as well, including the optimization method shown in [30], called Robust Method-I, and the one in [24], called Robust Method-II. The optimized parameters are used in the time-domain simulations to validate the damping control performances. Taking G1 and G2 as examples, their dynamic responses under Fault 1, i.e., a double-phase short circuit fault at Bus 3 from 2.0 s to 2.2 s, are shown in Figure 8.32. The Prony method is employed to identify the damping of ULFO mode under different PID parameter settings, and the identification results are presented in Table 8.17. It can be observed from Table 8.17 and Figure 8.32 that ULFO exists in the system before PID parameter tuning. After PID parameter optimization, the damping of ULFO mode is improved and the oscillation is suppressed. However, the PID parameter settings optimized by different methods show different performances. Compared with the other two

Table 8.16 Time constant (T_w) of hydro-turbines in different cases

Gen	Case 1	Case 2	Case 3	Case 4	Case 5
G1	0.5	2.5	1.5	1.5	2.5
G2	1.5	1.5	2.5	0.5	0.5
G3	0.5	1.5	1.5	1.5	0.5
G4	2.5	2.5	2.5	1.5	1.5
G5	1.5	2.5	0.5	1.5	0.5
G6	1.5	0.5	0.5	1z.5	2.5
G7	0.5	1.5	0.5	0.5	2.5
G8	0.5	0.5	1.5	1.5	1.5
G9	1.5	2.5	1.5	2.5	1.5
G10	2.5	0.5	1.5	0.5	2.5

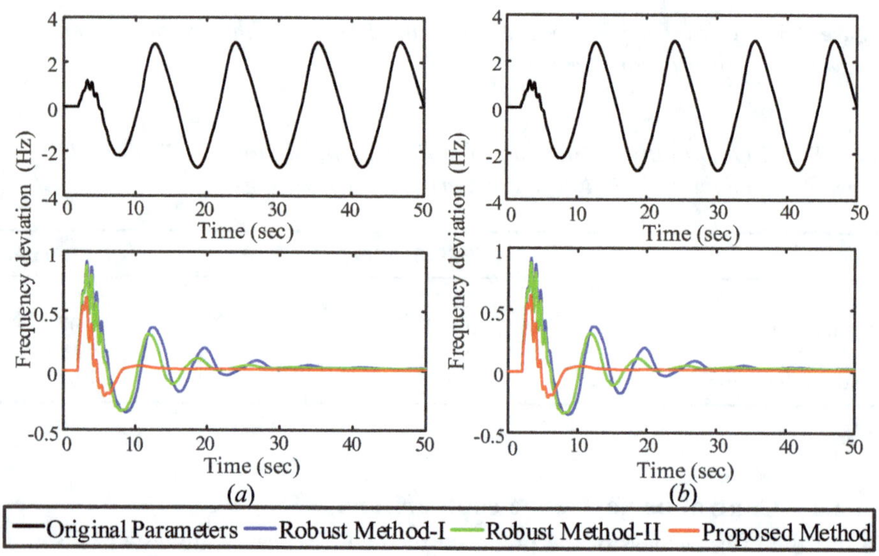

Figure 8.32 Frequency deviations of units under Fault 1: (a) G1; (b) G2 [25]

robust optimization methods, the proposed method achieves the best damping in ULFO mode and allows the generators to reach a steady state in the shortest time.

To further elaborate on the results, the damping characteristic curves and step responses of G1 and G2 are shown in Figures 8.33 and 8.34. It can be observed that the original parameters make the units achieve the best dynamic response, but the worst damping performance of ULFO. The parameter settings re-tuned by the three optimization methods can improve the damping characteristics of the governor at the cost of reduced dynamic characteristics. Among them, the proposed method not only improves the damping characteristics of the governors but also retains a better dynamic response of PFR than the other two methods.

Table 8.17 Damping of ULFO modes using different optimization methods

Methods	Real part	Imag part	Damping
Original parameter	0.00045	0.55	−0.082%
Robust Method-I	−0.038	0.57	6.72%
Robust Method-II	−0.056	058	9.63%
Proposed Method	−0.090	0.53	16.93%

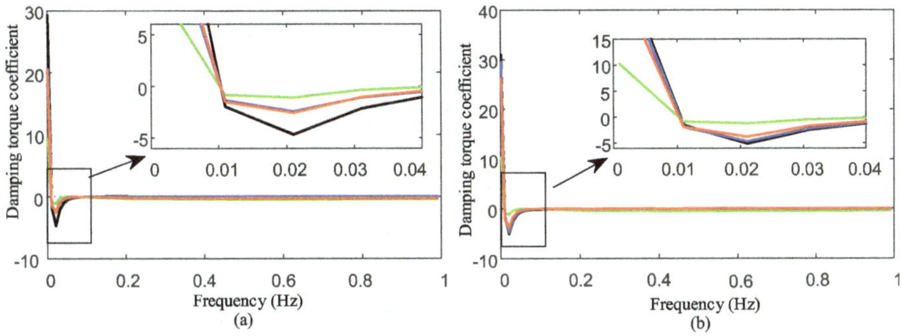

Figure 8.33 Dynamic characteristic of G1: (a) damping curve; (b) step response [25]

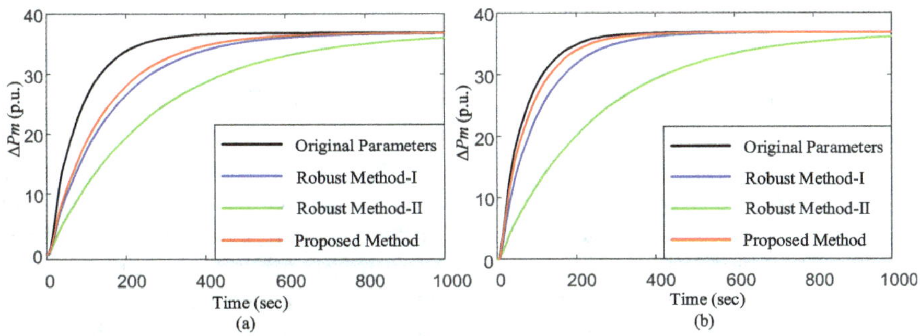

Figure 8.34 Step response of primary frequency regulation: (a) G1; (b) G2 [25]

8.4.4.3 Robustness to various operating conditions

To test the robustness of all three methods under different disturbances, four different two-phase short-circuit faults are applied with a 100 ms fault duration, where the faulted buses are 8, 16, 23, and 25. They are named Fault 2, Fault 3, Fault 4, and Fault 5, respectively. The simulation results are shown in Figure 8.35. It can be seen that all methods can dampen out the oscillations. Among them, the proposed method takes the least time to dampen the oscillations as compared to the other two robust methods. The proposed method also has the smallest oscillation magnitude, demonstrating the best performance of damping out the UFLO under different faults.

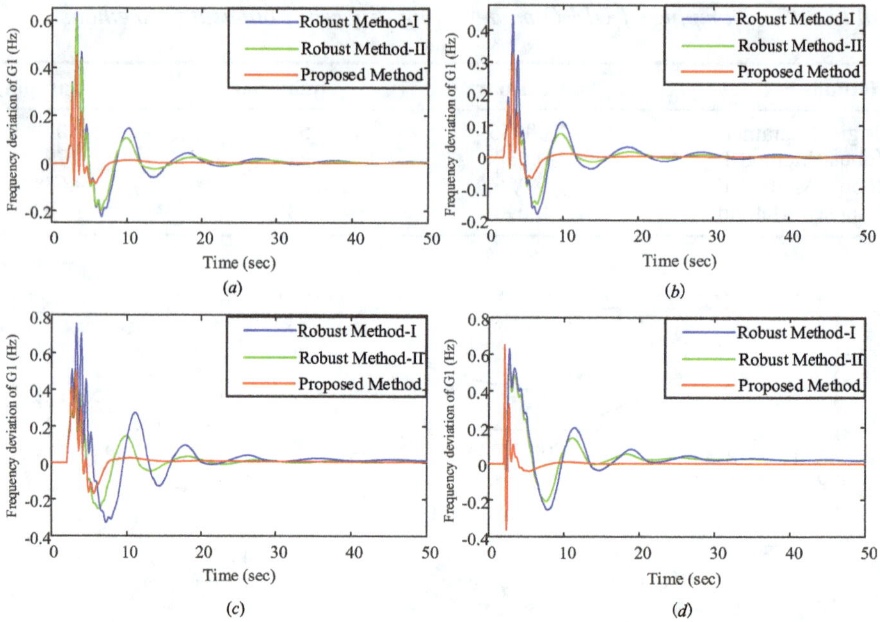

Figure 8.35 Dynamic response of G1 under different disturbances: (a) Fault 2; (b) Fault 3; (c) Fault 4; (d) Fault 5 [25]

Furthermore, different T_w settings for Cases 2–5 in Table 8.15 are taken as comparative cases to test the robustness of the proposed method under different cases. For each case, time-domain simulations are carried out using different PID parameter settings. The results are shown in Figure 8.36. It can be concluded that the PID parameters optimized by the proposed method can achieve better performance in comparison with two other optimization methods. A Monte Carlo approach is used to sample the time constant T_w of turbines between 0.5 and 3 to produce 300 examples to further explore the performance of the suggested method. For each case, the ULFO mode is calculated and plotted in the complex plane. Moreover, the probability density function (PDF) of ULFO nodes is also calculated, see Figure 8.37. It can be seen that before PID parameter optimization, the damping of the ULFO mode is less than 0 for some cases and not located in a secure region ($\xi > 5\%$). This means that there is a risk of ULFO-induced system instability during the operation of the system. Figure 8.37(b)–(d) shows the distributions of ULFO modes under different PID parameter settings. Some statistical indexes are also listed in Table 8.18. Clearly, compared to two other robust optimization methods, the proposed method can make the ULFO mode achieve a better mean value and a smaller standard deviation of the damping ratio, yielding a larger stability margin.

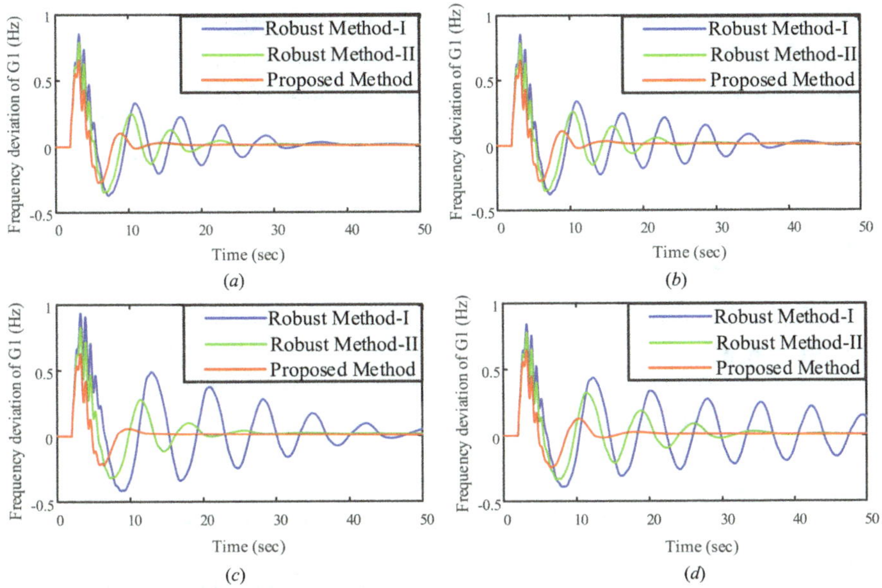

Figure 8.36 Dynamic response of G1 under different cases: (a) Case 2; (b) Case 3; (c) Case 4; (d) Case 5 [25]

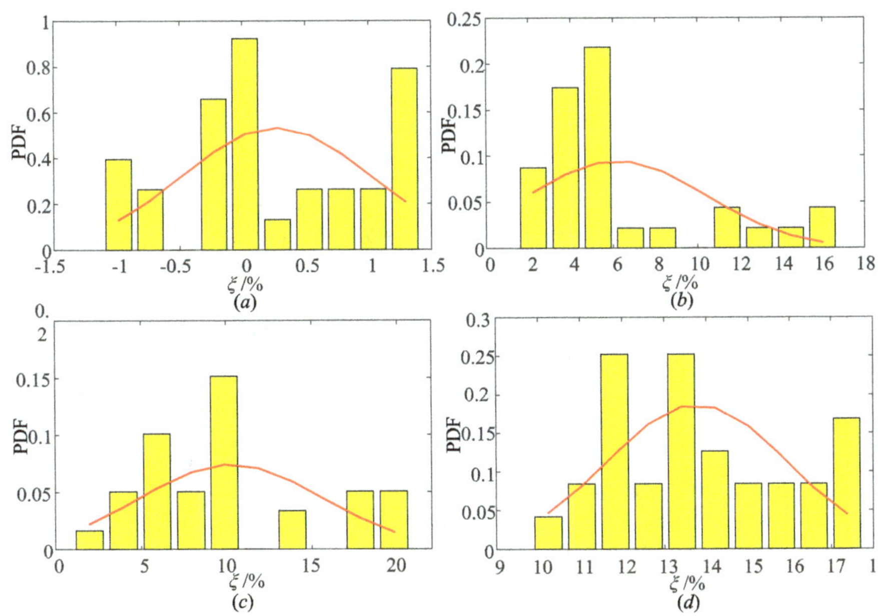

Figure 8.37 Damping probability density function of ULFO mode under different cases: (a) Original parameters; (b) Robust Method-I; (c) Robust Method-II; (d) Proposed Method [25]

Table 8.18 Statistical hypothesis for damping ratio of ULFO mode under 300 operating conditions

Methods	Mean value	Standard deviation
Original parameter	0.0027	0.0075
Robust Method-I	0.0618	0.0420
Robust Method-II	0.1025	0.0536
Proposed Method	0.1378	0.0213

8.5 Summary

This chapter focuses on designing stability control methods based on artificial intelligence techniques for renewable energy integration systems. Among them, for the wind power grid-connected system, Sections 8.2 and 8.3 present the effects of different controllers (CPSS, STATCOM-ADC) on the low-frequency oscillation performance of the system and train the agents by reinforcement learning methods to realize adaptive control under different wind speed conditions, respectively. For the hydropower dominant system, Section 8.4 describes the mechanism of governor PID influence on ultra-low-frequency oscillations and the use of a Bi-level optimization strategy (the inner optimization is based on a reinforcement learning approach) to re-tune the PID parameters in order to ensure that the system still has a good stable performance under extreme operating conditions.

References

[1] M. J. Morshed and A. Fekih, "A probabilistic robust coordinated approach to stabilize power oscillations in DFIG-based power systems," *IEEE Transactions on Industrial Informatics*, vol. 15, no. 10, pp. 5599–5612, 2019.

[2] T. Surinkaew and I. Ngamroo, "Adaptive signal selection of wide-area damping controllers under various operating conditions," *IEEE Transactions on Industrial Informatics*, vol. 14, no. 2, pp. 639–651, 2018.

[3] M. Derafshian and N. Amjady, "Optimal design of power system stabilizer for power systems including doubly fed induction generator wind turbines," *Energy*, vol. 84, pp. 1–14, 2015.

[4] D. Cao, W. Hu, J. Zhao, G. Zhang, Z. Liu, Z. Chen, and F. Blaabjerg, "Reinforcement learning and its applications in modern power and energy systems: a review, " *Journal of Modern Power Systems and Clean Energy*, vol 8, no. 6, pp. 1029–1042, 2020.

[5] G. Zhang, W. Hu, J. Zhao, D. Cao, Z. Chen and F. Blaabjerg, "A novel deep reinforcement learning enabled multi-band PSS for multi-mode oscillation control," *IEEE Transactions on Power Systems*, vol. 36, no. 4, pp. 3794–3797, 2021.

[6] G. Zhang, W. Hu, D. Cao, Q. Huang, Z. Chen and F. Blaabjerg, "A novel deep reinforcement learning enabled sparsity promoting adaptive control method to improve the stability of power systems with wind energy penetration," *Renewable Energy*, vol. 178, pp. 363–376, 2021.

[7] Y. F. Du, L. Jiang, Y. Li and Q. Wu, "A robust optimization approach for demand side scheduling considering uncertainty of manually operated appliances," *IEEE Transactions on Smart Grid*, vol. 9, no. 2, pp. 743–755, 2018.

[8] G. Zhang, W. Hu, D. Cao, Q. Huang, J. Yi, Z. Chen and F. Blaabjerg, "Deep reinforcement learning-based approach for proportional resonance power system stabilizer to prevent ultra-low-frequency oscillations," *IEEE Transactions on Smart Grid*, vol. 11, no. 6, pp. 5260–5272, 2020.

[9] G. Zhang, W. Hu, D. Cao, J. Yi, Q. Huang, Z. Liu, Z. Chen and F. Blaabjerg, "A data-driven approach for designing STATCOM additional damping controller for wind farms," *International Journal of Electrical Power & Energy Systems*, vol. 117, pp. 105620, 2020.

[10] Y. Chen, Y. Liu, Z. Tang, J. Hou, Y. Zhang, W. Mo and H. Chen, "Analysis of ultra-low frequency oscillation in Yunnan asynchronous sending system," in *IEEE Power & Energy Society General Meeting*, Chicago, IL, pp. 1–5, 2017.

[11] M. Netto and L. Mili, "Robust data filtering for estimating electromechanical modes of oscillation via the multichannel Prony method," *IEEE Transactions on Power Systems*, vol. 33, no. 4, pp. 4134–4143, 2018.

[12] G. Zhang, J. Yi, Y. Teng, P. Wang and Q. Huang, "Coordinated optimization of multi-machine power system stabilizers under multiple operating conditions," *Power System Technology*, vol. 42, no. 9, pp. 2797–2805, 2018.

[13] T. K. Chau, S. S. Yu, T. Fernando, H. H.-C. Iu and M. Small, "A load-forecasting-based adaptive parameter optimization strategy of STATCOM using ANNs for enhancement of LFOD in power systems," *IEEE Transactions on Industrial Informatics*, vol. 14, no. 6, pp. 2463–2472, 2018.

[14] S. M. Elazim and E. S. Ali. "Optimal location of STATCOM in multi-machine power system for increasing load ability by Cuckoo Search algorithm," *International Journal of Electrical Power & Energy Systems*, vol. 80, pp.240–251, 2016.

[15] Z. Yan and Y. Xu, "Data-driven load frequency control for stochastic power systems: a deep reinforcement learning method with continuous action search," *IEEE Transactions on Power Systems*, vol. 34, no. 2, pp. 1653–1656, 2019.

[16] A. N. C. Supriyadi, H. Takano, H. Murata and T. Goda, "Adaptive robust PSS to enhance stabilization of interconnected power systems with high renewable energy penetration," *Renewable Energy*, vol. 63, pp. 764–774, 2014.

[17] N. Wang, J Li, W Hu, B Zhang, Q Huang and Z Chen, "Optimal reactive power dispatch of a full-scale converter based wind farm considering loss minimization," *Renewable Energy*, vol. 139, pp. 292–301, 2019.

[18] D. K. Chaturvedi and O. P. Malik, "Generalized neuron-based adaptive PSS for multi-machine environment," *IEEE Transactions Power Systems*, vol. 20, no. 1, pp. 358–366, 2005.

[19] T. P. Lillicrap, J. J. Hunt, A. Pritzel, N. M. O. Heess, T. Erez and Y. Tassa, "Continuous control with deep reinforcement learning," in *Proceedings of the International Conference on Learning Representations*, pp. 1–14, 2016.

[20] A. F. Dizche, A. Chakrabortty and A. Duel-Hallen, "Sparse wide-area control of power systems using data-driven reinforcement learning," in *American Control Conference*, Philadelphia, PA, USA, pp. 2867–2872, 2019.

[21] P. Wang, Z. Zhang, Q. Huang, N. Wang, X. Zhang and W.-J. Lee, "Improved Wind farm aggregated modeling method for large-scale power system stability studies," *IEEE Transactions on Power Systems*, vol. 33, no. 6, pp. 6332–6342, 2018.

[22] J. Li, N. Wang, D. Zhou, W. Hu, Q. Huang, Z. Chen and F. Blaabjerg. "Optimal reactive power dispatch of permanent magnet synchronous generator-based wind farm considering levelised production cost minimisation," *Renewable Energy*, vol. 145, pp. 1–12, 2020.

[23] A. Movahedia, A. H. Niasara and G. B. Gharehpetianb. "Designing SSSC, TCSC, and STATCOM controllers using AVURPSO, GSA, and GA for transient stability improvement of a multi-machine power system with PV and wind farms," *International Journal of Electrical Power & Energy Systems*, vol. 106, pp. 455–466, 2019.

[24] Y. H. Nie, Y. D. Zhang, Y. Zhao, B. B. Fang and L. L. Zhang, "Wide-area optimal damping control for power systems based on the ITAE criterion," *International Journal of Electrical Power & Energy Systems*, vol. 106, pp. 192–200, 2019.

[25] G. Zhang, J. Zhao, W. Hu, *et al.*, "Deep reinforcement learning enabled bi-level robust parameter optimization of hydropower-dominated systems for damping ultra-low frequency oscillation," *Journal of Modern Power Systems and Clean Energy*, vol. 11, pp. 1770–1783, 2023.

[26] L. Chen, X. Lu, Y. Min, Y. Zhang, Q. Chen, Y. Zhao and C. Ben, "Optimization of governor parameters to prevent frequency oscillations in power systems," *IEEE Transactions on Power Systems*, vol. 33, no. 4, pp. 4466–4474, 2018

[27] Z. Liu, W. Yao, J. Wen, and S. Jie, "Effect analysis of generator governor system and its frequency mode on inter-area oscillations in power systems," *International Journal of Electrical Power & Energy Systems*, vol. 96, pp. 1–10, 2018.

[28] C. Jiang, J. Zhou, P. Shi, W. Huang and D. Gan, "Ultra-low frequency oscillation analysis and robust fixed order control design," *International Journal of Electrical Power & Energy Systems*, vol. 104, pp. 269–278, 2019.

[29] W. Yang, J. Yang, W. Zeng, R. Tang, L. Hou, A. Ma and Z. Zhao, "Experimental investigation of theoretical stability regions for ultra-low frequency oscillations of hydropower generating systems," *Energy*, vol. 186, pp. 1–11, 2019.

[30] G. Chen, F. Tang, H. Shi, R. Yu, G. Wang, L. Ding, B. Liu and X. Lu, "Optimization strategy of hydro-governors for eliminating ultralow-frequency oscillations in hydro-dominant power systems," *IEEE Journal of Emerging and Selected Topics in Power Electronics*, vol. 6, no. 3, pp. 1086–1094, 2018.
[31] J. Duan, D. Shi, R. Diao, H. Li, Z. Wang, B. Zhang, D. Bian and Z. Yi, "Deep-reinforcement-learning-based autonomous voltage control for power grid operations," *IEEE Transactions on Power Systems*, vol. 35, no. 1, pp. 814–817, 2020
[32] Y. Hashmy, Z. Yu, D. Shi and Y. Weng, "Wide-area measurement system-based low frequency oscillation damping control through reinforcement learning," *IEEE Transactions on Smart Grid*, vol. 11, no. 6, pp. 5072–5083, 2020.
[33] J. Park, K. Lee, J. Shin and K. Y. Lee, "A particle swarm optimization for economic dispatch with nonsmooth cost functions," *IEEE Transactions on Power Systems*, vol. 20, no. 1, pp. 34–42, 2005.
[34] S. Mei, "A game model for robust optimization of power systems and its application," *Proceedings of the CSEE*, vol. 33, no. 19, pp. 47–56, 2013.

Chapter 9

Conclusions and outlook using AI in power systems

Weihao Hu[1], *Guozhou Zhang*[2], *Zhenyuan Zhang*[1], *Sayed Abulanwar*[3,4] *and Frede Blaabjerg*[5]

9.1 Summary and conclusion

In summary, this book has focused on applying AI technology to deal with a series of challenges caused by the integration of renewable energy. Among them, we first overview the development trend of renewable energy in the power system and introduce several important AI technologies in Chapter 1. After that, Chapters 2–8 provide some examples of utilizing AI methods to address some issues like fault diagnosis, power electronic reliability, dual active bridge modulation, voltage control of an active distributed network, energy management, and stability-oriented control. Specifically, Chapter 2 analyzes the characteristics of electric machine faults, and applies the few-shot learning method and Gaussian-process-enabled method to do fault diagnosis. In Chapter 3, deep learning methods are utilized for power electronic monitoring, optimization, and control. Moreover, to improve energy conversion efficiency, Chapter 4 explores reinforcement learning-enabled modulation strategies for dual active bridges. In Chapter 5, the voltage fluctuation in an active distributed network caused by the integration of photovoltaic (PV) and wind turbine (WT) is considered, and a reinforcement learning multi-agent deep reinforcement learning (DRL) algorithm is used to solve it. Chapters 6 and 7 apply the DRL method for the energy management of hybrid energy systems and microgrids, respectively. Chapter 8 pays attention to the small signal stability of renewable energy systems, and the DRL-enabled controllers are designed to enhance the small signal stability of the systems.

Overall, the integration of renewable energy will contribute to reducing carbon emissions to alleviate the environmental energy crisis, but it increases the

[1]School of Mechanical and Electrical Engineering, University of Electronic Science and Technology of China, China
[2]College of Engineering and Technology, Southwest University, China
[3]Electrical Engineering Department, Faculty of Engineering, Mansoura University, Egypt
[4]Faculty of Engineering, Horus University-Egypt, Egypt
[5]Department of Energy Technology, Aalborg University, Denmark

nonlinearity and uncertainty in the power system, which makes the operation, management, and control of such a system more difficult. Especially, most conventional approaches use physical model-based methods, which highly depend on accurate system parameters and topology. However, due to the large-scale, high-order, and nonlinear time-varying characteristics of the actual power grid, it is difficult to establish a full-scale, detailed model. With the use of phasor measurement units (PMU), a large amount of electricity data with high volumes, mutual correlations, and complex structures can be observed and saved, which can be applied for the extraction of experience and further used for the operation and planning of the power system with a renewable energy system. In this way, the disadvantages of the physical model-based method can be overcome. To this end, data-driven methods are proposed. Among them, the AI method can learn from historical data to construct an adaptive agent to deal with the newly encountered conditions. In recent years, AI has achieved rapid development in optimization and control problems, around power electronics, active distribution networks, and the main power network. In this context, this book has demonstrated some state-of-the-art applications of AI technology for renewable energy systems, and power electronics. The results of these studies indicate that AI can better deal with the uncertainties and intermittency of complex systems, which makes it demonstrate better control and optimization performance using renewable energy systems.

9.2 Challenges and future outlook using AI in the power system

Although AI technology has been successfully applied in renewable power systems and some positive results have been obtained in related research, there are still some interesting problems that need to be further studied, including but not limited to:

1. Since renewable energy systems have a high requirement for safety, the physical constraints should be better handled when building the AI-based model instead of directly adding soft constraints to the reward function. A safe AI-based agent is a suitable way to deal with optimization and control problems by solving a constrained Markov decision process (MDP). Other ways to embed physical knowledge in the AI-based agent may also improve its reliability and drive for real-world implementation.
2. The offline training of an AI-based agent relies on accurate physical models, while the online training may affect the operation of renewable energy systems. The meta-learning method and surrogate model are two ways to reduce dependence on a physical model without impacting the operation of the system. However, they require a certain amount of historical data. Transfer learning may be another promising alternative by training the AI-based agent offline and transferring the learned control strategy to real-world environments with few recorded samples and iterations.
3. With the integration of different electrical equipment, renewable energy systems are becoming more complex and larger, with more operation conditions

and control options. A single-agent AI algorithm adopts a centralized framework that relies heavily on complete communication links and thus is not able to deal with communication delays and also scale up to large systems. Multi-agent AI can partially mitigate this issue by adopting a centralized training, decentralized execution framework. However, existing multi-agent AI algorithms face large challenges when dealing with very large systems that require large populations. Further research may be done toward advanced multi-agent AI algorithms with novel population scaling mechanisms to enable the AI method to scale up to very large power systems.

4. A lot of control and optimization problems in renewable energy systems have typical hierarchical structures, as well as the decision-making process of human beings. The hierarchical framework can reduce the deployment cost of complete communication devices under centralized control and avoid the isolation issue of local control, making it another promising way to control large systems. Owing to the complexity of hierarchical structure and the lack of a general hierarchical framework, applications of AI technology for hierarchical control are rare in renewable energy systems. Future research may be applied to an AI-based hierarchical control framework for large systems.
5. With the increased integration of power electronic devices, distributed energy resources (DERs), and flexible loads, the complexities and uncertainties in the modern power system are growing. Classical offline training and online execution of AI have difficulties dealing with the continuously generated dynamics. Meta-learning and continuous learning can be integrated with the AI algorithm to achieve lifelong learning ability. This helps the continuous transformation of online data into powerful knowledge, which can successively enhance the control behavior of an AI agent. Therefore, the robustness and adaptability to system dynamics can be enhanced, and the training time can be shortened in complex scenarios.
6. Most of the AI methods belong to the data-driven category. When AI methods are utilized for the long-term prediction of electrical parameters, load demand, and electricity price in the power system, which are highly dependent on the quality of the historical data. However, there are some unavoidable issues with the collected data, e.g., packet dropout, misalignment, and so on, which will affect the prediction performance. In addition, the power system is time-varying, and some new system dynamic characteristics are not included in the historical data, which makes the predicted model trained using historical data hardly achieve high-precision prediction during the online application process. To this end, incremental learning may be utilized as an alternative to address it. In future work, some studies may focus on embedding incremental learning into the prediction method. This contributes to solving the stability-plasticity dilemma, avoiding catastrophic forgetting, and achieving lifetime learning.
7. Although AI-controlled power systems can enhance the stability and robustness of the system, on the other hand, the application of AI in power systems would also increase security risks. For example, to solve the target tasks in a complex system, the mathematical model is established in a black box manner with an AI algorithm. This will reduce the interpretability of the system and

increase the difficulty of system mechanism analysis. Additionally, the AI-based control agent needs to receive the system states to provide the corresponding action. To achieve this goal, additional communication networks need to be constructed and exposed to open cyberspace, which is vulnerable to malicious attacks because of the distributed characteristics of the renewable energy system. Once the system states cannot be correctly sent to the AI-based control agent, it hardly works. In this context, there is a risk of the system losing control.

8. With the development of high-voltage direct current (HVDC) and high-voltage alternating current (HVAC) transmission technologies, research on modular multilevel converters (MMCs) is becoming increasingly popular. However, because of the circuit complexity of such converters, the cost of fault diagnosis and control is high. Therefore, artificial intelligence-based methods can be studied to reduce the cost of sensors required for fault diagnosis and the amount of computation required to control them. For example, deep learning can be used to improve the recognition of fault characteristics, thereby determining the type and even location of the fault. Deep reinforcement learning can also be used to enhance the perception of the operating state and action decision-making of such converters. In addition, with the increasing complexity of power electronic systems, similar ideas can be used to solve various problems under multi-machine operation.

Index

www.ingramcontent.com/pod-product-compliance
Lightning Source LLC
LaVergne TN
LVHW011843200926
848484LV00002B/396
* 9 7 8 1 8 3 9 5 3 7 7 4 5 *